LABORATORY MA

CAT

ROBERT B. CHIASSON
University of Arizona

WILLIAM J. RADKE
University of Central Oklahoma

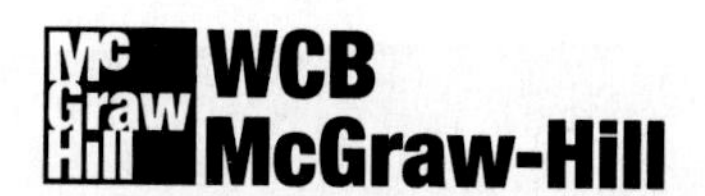

Boston, Massachusetts   Burr Ridge, Illinios   Dubuque, Iowa
Madison, Wisconsin   New York, New York   San Francisco, California   St. Louis, Missouri

Ashely Hughes

WCB/McGraw-Hill

*A Division of The **McGraw-Hill** Companies*

Cover/interior design by Rokusek Design

Cover illustration by Phil Sims

Library of Congress Catalog Card Number: 94–79883

ISBN 0–697–24926–3

Printed in the United States of America

10 9 8 7 6 5 4 3 2

Contents

List of Figures v
Preface vii
Introduction ix

1 Histology 1

2 External Anatomy and Skin 10

3 Skeletal System 13

4 Muscular System 48

5 Digestive System 90

6 Respiratory System 105

7 Circulatory System 110

8 Excretory System 132

9 Reproductive System 135

10 Nervous System 138

11 Receptors 151

12 Endocrine Glands 157

Appendix 1 Sheep Heart 161
Appendix 2 Sheep Brain 166
Appendix 3 Ox Eye 172
Glossary 179
Index 184

List of Figures

I.1 Diagram of anatomical planes and directions for the cat and human. xii
1.1 The compound microscope. 2
1.2 Epithelia. 5
1.3 Connective tissues. 6
1.4 Muscle tissues. 7
1.5 Nervous tissues. 8
2.1 Diagrammatic composite section of mammalian skin. 11
3.1 Lateral view of the cat skeleton. 14
3.2 Skull and mandible of the cat and human in lateral views. 15
3.3 Skull of the cat in lateral view with the mandible removed. 16
3.4 Cat and human skulls in dorsal and superior views. 17
3.5 Cat and human skulls in frontal and anterior views. 18
3.6 Cat and human crania in ventral and inferior views. 19
3.7 Cat and human skulls in sagittal views. 21
3.8 Coronal sections of the skulls of the cat and human allowing views of the cranial floors. 22
3.9 Lateral views of the cat and human hyoid apparatuses. 25
3.10 Dental patterns of the cat and human. 26
3.11 Lateral views of the cat skeleton and vertebral column. 29
3.12 Lateral views of the human skeleton and vertebral column. 30
3.13 Typical cervical vertebrae of the cat and human. 31
3.14 Typical thoracic and lumbar vertebrae of the cat and human. 32
3.15 Lateral view of the cat and human sacral vertebrae. 33
3.16 Pectoral girdle and limbs of the cat, frontal view. 35
3.17 Comparison of human and cat girdles and limbs, frontal and anterior views. 36
3.18 Lateral view of the cat pectoral girdle and limb, and frontal view of the carpal bones. 37
3.19 Pectoral girdle and limb of the human, lateral view. 38
3.20 Lateral view of the cat pelvic girdle and limb and frontal view of the tarsal bones. 41
3.21 Lateral view of the human pelvic girdle and limb. 42
3.22 Anterior view of the human skeleton. 43
3.23 Posterior view of the human skeleton. 44
3.24 Dorsal view of the cat pelvic vertebrae, girdle, and femur. 45
3.25 Lateral views of the pelvic limb and girdle of human and cat. 45
3.26 Caudal view of the cat knee joint. 46
3.27 Medial and lateral views of the cat knee joint. 47
4.1 Dissection incisions for removing the skin of the cat. 50
4.2 Superficial lateral muscles of the cat. 51
4.3 Lateral view of the cat jaw muscles. 52
4.4 Diagrammatic section through the head of the cat. 53
4.5 Ventral view of the throat and neck muscles of the cat. 54
4.6 Lateral muscles of the cat shoulder, brachium, and forearm. 55
4.7 Pectoral limb muscles of the cat shoulder, superficial and deep. 56
4.8 Ventral view of the superficial and deep musculature of the cat. 60
4.9 Muscles of the medial surface of the cat pectoral limb. 61
4.10 Diagrammatic frontal view of the cat extrinsic shoulder ("sling") muscles. 62
4.11 Deeper lateral muscles of the cat shoulder, brachium, and forearm. 63
4.12 The deepest lateral muscles of the cat shoulder, brachium, and forearm. 65
4.13 Deep medial muscles of the cat pectoral limb. 69
4.14 Anterior view of the superficial and deep musculature of the human. 70
4.15 Ventral view of the deep neck muscles of the cat. 72

4.16 Lateral views of the deep neck muscles of the cat. 74
4.17 Lateral views of the cat deep trunk and vertebral muscles. 75
4.18 Lateral view of the cat superficial trunk and limb musculature. 79
4.19 Lateral view of the cat deep trunk and limb musculature. 80
4.20 Superficial muscles of the cat hip and thigh. 82
4.21 Middle layer of muscles of the cat hip, thigh, and shank. 84
4.22 Deep muscles of the cat hip, thigh, and shank. 85
4.23 Medial view of the muscles of the cat hip and thigh. 87
5.1 Dissection of the visceral cavities of the cat. 91
5.2 Cat serous membranes. 92
5.3 Stereodiagrams illustrating stages in the development of some of the cat abdominal membranes. 93
5.4 Salivary glands of the cat. 94
5.5 Diagrammatic illustration of the acinar salivary glands of the cat and their ducts. 95
5.6 *M. lingualis proprius* (intrinsic tongue musculature). 97
5.7 Dorsal view of the cat tongue and pharynx. 97
5.8 Ventral view of the abdominal viscera with the greater omentum removed and with the liver and most of the jejunum and ileum removed. 98
5.9 Internal view of the stomach, esophagus, and duodenum. 99
5.10 Photomicrograph of a cross section of the cat fundic stomach. 100
5.11 Photomicrograph of a cross section of the cat duodenum. 100
5.12 Microstructure of the mammalian liver. 102
5.13 Ventral view illustrating the relationships of the pancreas to the digestive tract. 103
6.1 Ventral view of the thoracic viscera of the cat. 106
6.2 Diagram of the distribution of bronchi in the cat lungs. 107
6.3 Microstructure of the mammalian alveoli and their ducts. 108
7.1 Microstructure of a median sized artery and vein. 110
7.2 Venous valves. 111
7.3 The heart of the cat. 112
7.4 Major arteries of the neck and shoulder. 114
7.5 Major arteries and veins of the trunk of the cat. 117
7.6 The celiac artery, its branches, and some other visceral blood vessels. 118
7.7 Details of the arteries, veins, and nerves of the iliac and inguinal regions of the cat. 119
7.8 Major veins of the cat neck and shoulder. 122
7.9 Superficial veins of the cat head and neck. 123
7.10 Medial view of the arteries, veins, and nerves of the cat axilla and brachium. 124
7.11 Major venous drainage of the trunk of the cat. 124
7.12 The hepatic portal system of the cat. 128
7.13 Lymphatic vessels of the cat. 130
8.1 Microstructure of the mammalian kidney and a diagram of a kidney unit. 133
9.1 Male and female reproductive systems of the cat. 136
10.1 Dorsal view of the cat brain. 140
10.2 The cat brain in lateral view. 140
10.3 Ventral view of the cat brain. 143
10.4 Sagittal section of the cat brain. 143
10.5 Sagittal section of the human brain. 144
10.6 Sagittal section of the cat pituitary gland. 144
10.7 The brachial plexus and associated autonomic nerves in ventral view. 146
10.8 Lumbosacral plexus and abdominal autonomic nerves. 147
10.9 Celiac and cranial mesenteric plexi. 149
11.1 Muscles of the right eye, dorsal view. 152
11.2 Muscles of the right eye, ventral view. 152
11.3 Sagittal section of the eye. 153
11.4 Sagittal section and detail of the mammalian eye. 154
11.5 The middle ear of the cat. 155
12.1 Endocrine glands of the cat. 157
H.1 Ventral views of the sheep heart. 162
H.2 Dorsal views of the sheep heart. 163
B.1 Dorsal view of the sheep brain. 166
B.2 Lateral view of the sheep brain. 167
B.3 Ventral view of the sheep brain. 169
B.4 Midsagittal view of the sheep brain. 169
B.5 Cross section of the sheep midbrain. 170
E.1 Rostral view of the undissected ox eye. 172
E.2 Directions and planes of the mammalian eye. 173
E.3 Features of the ox eye. 174
E.4 Instructions for sectioning the ox eye. 176
E.5 Features seen in a gross dissection of the eye. 177
E.6 Diagrams illustrating accommodation by the eye. 178
E.7 Diagrams illustrating myopia and hyperopia. 178

Preface

Purpose of this Manual

This manual is designed for undergraduate mammalian or human anatomy courses for which human cadavers are unavailable. The differences and similarities in anatomical structure among mammals can help explain why we humans are limited in some of our activities, and how we might compensate for some of these limitations. For example, a "four-legged" runner starts a race from a normal standing position, while a human "bipedal" runner bends forward from the hips. A human standing in a normal erect position has twisted the ligaments of the hip joint so the limb cannot be further extended to move the body forward. Bending forward from the hip untwists the ligament, thus allowing the extension that propels the body forward. Other examples will be discussed in special boxes found throughout this manual.

It is assumed that most classrooms will have access to human skeletal material and microscope slides of human tissues. Nevertheless, the dissection emphasis is on the cat, and many aspects of the human are omitted.

The cat is easily the most popular specimen for mammalian dissection in the United States and is widely used as a substitute for human dissection where human cadavers are unavailable. The cat's size makes it a convenient dissection specimen. Your dissection of the cat should be performed with respect and care in consideration of the cat as a representative of your own body.

Important Features of this Manual

1. Important terms that the student should memorize are presented in **BOLD CAPITALS** when they first appear in the text.
2. The dissection instructions for the cat are presented in *italics*.
3. The text descriptions of most human and cat systems have been integrated, rather than placing the human description in an appendix.
4. Special comparisons between the cat and human that are of general or philosophical interest are presented in text boxes.
5. A pronunciation guide is presented in brackets after each significant term when it first appears in the text. A key to the pronunciation guide is presented in the introduction.
6. A glossary of terms with their Latin and Greek roots is presented after the appendices.
7. This manual is spiral bound so that it will lie flat on the lab table next to the specimen, thus freeing the student's hands for dissection.
8. Instructions for the use of the microscope and a description of the types of tissues making up the mammalian body are presented in chapter 1.

Introduction

Introduction to the Manual

Pronunciation Key

A pronunciation guide is provided for the terminology used in this manual. Each term is spelled phonetically immediately after it first appears in the text. The following general rules were used in creating this guide.

1. The syllable with the strongest accent appears in capital letters; for example: a-NAT-*o*-m*e*.
2. Vowels to be pronounced in the long form are in italics, as in: bl*a*de and b*i*te.
3. Unmarked vowels are pronounced in the short form, as in: mitt and drum.
4. Other indicators of sounds are: oo as in blue, yoo as in cute, oy as in foil.

Learning the correct spellings of terms can be easier if just a few moments are spent studying the proper pronunciations. Additional information regarding pronunciations, word origins, and meanings can be found in a variety of sources, including (1) Borror, D. J. 1960, *Dictionary of Word Roots and Combining Forms,* Palo Alto, Calif., N-P Publications; (2) a medical dictionary such as *Dorland's Illustrated Medical Dictionary* (any recent edition), Philadelphia, W. B. Saunders; and (3) anatomical nomenclatures such as *Nomina Anatomica Veterinaria,* 3rd ed., 1983, revised by the International Committee on Veterinary Gross Anatomical Nomenclature, Ithaca, N.Y., or *Nomina Anatomica* 1956, Sixth International Congress of Anatomists held in Paris, Baltimore, Williams & Wilkins Co.

Terminology

Consequential terminology (those terms you should learn) are presented in **BOLD CAPITALS.** Other terminology is in **bold;** your instructor may require you to know these as well. Much of the manual is arranged in outline form to show the relationships between terms. A term that appears as a subheading under another term is a component of that other term. Generally, the relationships are anatomical and the subterm is, therefore, a part of the primary term.

Dissection instructions not found at the beginning of a chapter are presented in *italics.* Cautionary statements, those important to your health or to prevent equipment damage, are in SMALL CAPITALS. Always read them carefully and consult your instructor if there is a question.

Directions and Planes (fig. I.1)

Several principles and rules for anatomical terminology were adopted by the Editorial Committee of Nomina Anatomica Veterinaria (N.A.V.) for domestic mammals including the cat. In particular, the terms **cranialis** (kr*a*-n*e*-A-lis) and **caudalis** (kaw-DA-lis) are to be used when referring to directions on the neck or trunk.

The terms anterior, inferior, posterior, and superior differ in meaning when used in human anatomy and quadruped anatomy. The term **anterior** (an-T*E*-r*e*-*o*r) as applied to the human refers to that part of the body that usually "goes through the door first." Thus, your face, chest, abdomen, and the front of your thighs and legs are all anterior. The back of your head, trunk, buttock, thighs, and legs are **posterior.** Most of what is anterior of the human is **ventral** (VEN-tral) in other mammals and much of that which is posterior of the human is **dorsal** (D*O*R-sal) in other mammals. Consequently, dorsal and ventral are infrequently used in human anatomical terminology. Human terminology also uses **superior** (syoo-P*E*-r*e*-*o*r) and **inferior** (in-F*E*-r*e*-*o*r) as anatomical directions to designate structures nearer to the head or nearer to the feet. The quadruped cranialis is superior in the human and the quadruped caudalis is inferior in the human. Consequently, the N.A.V. has rejected these human anatomy terms for quadrupeds to avoid confusion, except when referring to some structures of the eye and ear.

For quadrupeds, cranialis pertains to the head and **craniad** (KR*A*-n*e*-ad) is forward, toward the head. Caudalis pertains to the tail and **caudad** (KAW-dad) is backward, toward the tail. Craniad literally means "toward the head," but some parts of the head may be

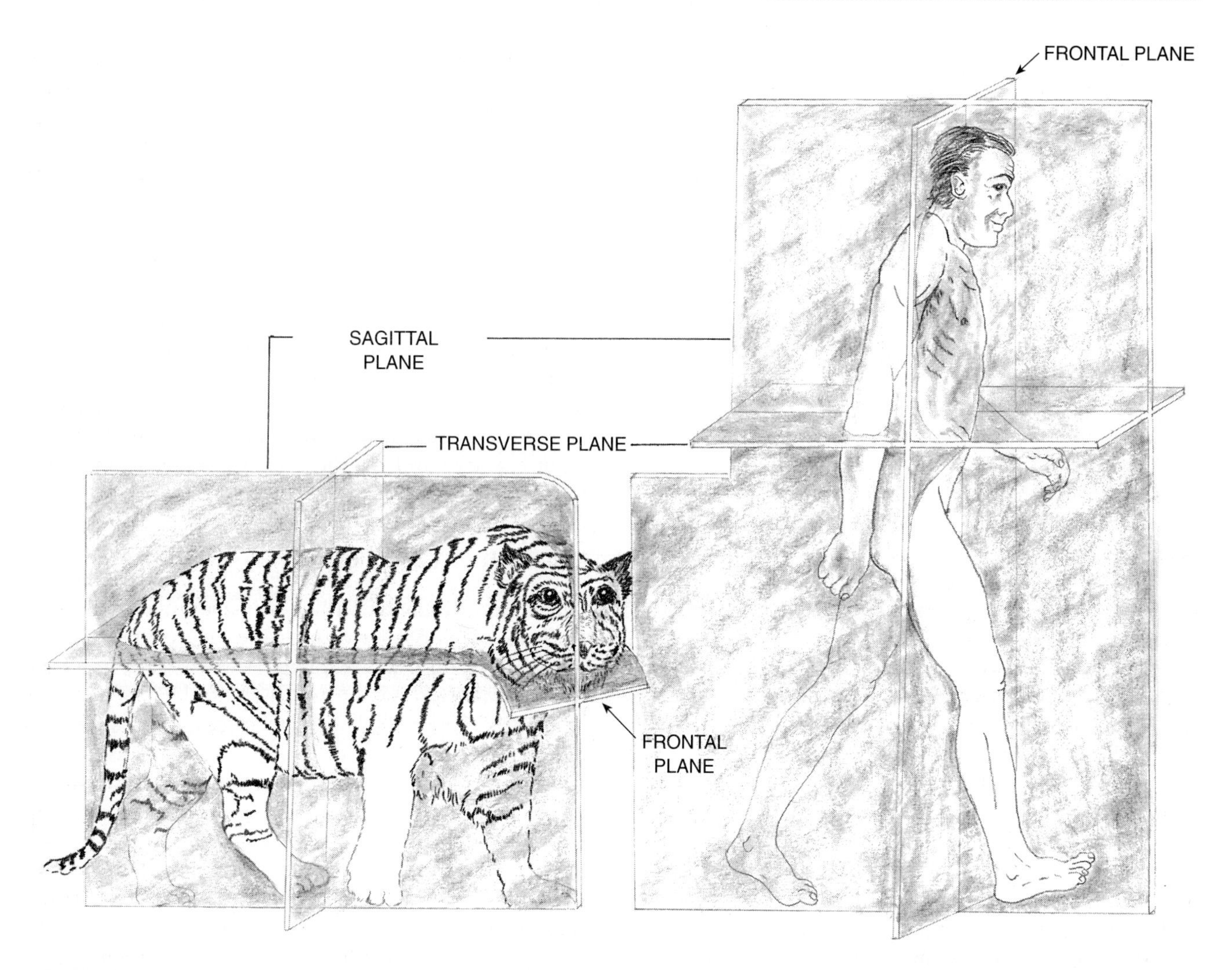

FIGURE I.1. Diagrammatic comparison of the anatomical planes and directions for the cat and human.

rostral (ROS-tral) to other parts. The human arrangement has the face directed anteriorly (that would have been ventral in the cat) but the base of the cranium (the brain stem and occipital region) are superior (cranial in the cat). This arrangement may be analogized to bending the cat's head 90° to obtain the human head. This requires differences between the two forms in the skull, brain and many other features.

Distal (DIS-tl) and **proximal** (PROX-si-mal) are relative terms pertaining to a point of reference. (In this case the point of reference is the midline of the body.) The wrist is distal to the elbow and the elbow is distal to the shoulder. In a similar fashion, the shoulder is proximal to the elbow and the elbow is proximal to the wrist.

Dissection Techniques and Specimen Selection

Before beginning the dissection of the cat you should become familiar with a few general principles of dissection. In the dissection of any animal, take care so that delicate structures will not be mutilated during the first stages of the work. Proper care should be used in each step of the process.

1. Read directions carefully before dissecting any part of the animal. The dissecting probe is one of the most important tools you have; use it freely and USE YOUR SCALPEL SPARINGLY. Blunt dissection preserves the specimen and produces superior results. If you are unable to find a structure after reading the instructions, call on the laboratory instructor for help.

2. There is variation in the arrangement of any animal's organs; the illustrations represent the usual conditions found in the cat, but you are sure to find some exceptions. Point out unusual arrangements of structures to your instructor.
3. Each student should have the following dissecting instruments:
 Scalpel, good quality (preferably with removable blade)
 Hemostat, inexpensive as possible
 Scissors, one blunt tip, good quality
 Probe, blunt tip
 Dissecting needles, six or more
 Disposable latex examination gloves, box of 100

Additional information regarding dissection techniques will be provided in the introduction to the muscular system. As you continue your study of the systems, specific dissection instructions will follow as needed in the chapter. This more specific guidance is presented in italics.

If all of the dissections and investigations described in this manual are to be covered, some special materials will be necessary. Microscope slides of the tissues, skin, blood, vessels, etc., are available from biological supply firms. The study of a few of these slides will require a compound microscope with total magnification to 400×. Study of some dissections (brains, for example) may require a dissecting microscope.

Prepared skeletal materials are available from most biological supply firms. Currently it is difficult to obtain human bone but some high quality plastic models are adequate. Preserved cats should be at least doubly injected (arterial and venous), but triple injection (for hepatic portal vessels) is best for complete study.

Introduction to the Mammal

Mammals are members of a large and diverse class of vertebrates (VER-te-br*a*ts), or animals with backbones. All mammals belong to the class Mammalia (mah-M*A*-l*e*-ah) and share two common characteristics, growing hair and feeding the young with milk. The root of the word mammal (*mamma*) is from Latin and means "teat." Like all vertebrates, mammals have dorsal nerve cords, embryonic notochords, and incomplete pharyngeal gill slits as embryos. There are about 3500 extant species in 1000 genera (more than 2000 genera are extinct). The living mammals are separated into the Prototheria (Pr*o*-t*o*-TH*E*-r*e*-ah) or egg-layers and the Theria, the live-bearers. The Theria form two groups, the Metatheria (pouched mammals) and the Eutheria, or mammals with a placenta (such as cats and humans) of which there are 18 extant orders.

Further separation of mammals is based on the type of reproductive system, dentition, and other anatomical features. The taxonomy of mammals is too complex to present here in its entirety. The mammalian orders are presented below. Taxonomic categories below order are presented only for the cat and man.

Mammals arose from reptiles via a well-documented transition. The reptiles that produced mammals were an ancient line. Modern mammals are, thus, far removed from modern reptiles. The origin of the early mammal-like reptiles occurred during the Triassic period (230 million years ago) and was overshadowed by the dominant dinosaurs. Near the end of the Triassic, intense competition with the dinosaurs left only a few small mammal-like reptiles (the therapsids) to become the first true mammals. The mammals went on to prosper and become one of the Cenozoic's (65 million years ago) dominant vertebrate forms, perhaps because of their intelligence, efficient locomotion, homeothermy, live-bearing behavior, care of the young, or because of the extinction of the dinosaurs.

The cat belongs to the order Carnivora, as do dogs, bears, and raccoons. These agile animals stalk their prey rather than running them down. Teeth are specialized for puncturing and shearing, as the animals are flesh eaters. It is supposed that the cat was domesticated as long ago as 2400 BC, probably in Egypt. However, a variety of wild cats in diverse regions throughout the world were kept in captivity and the origin of the modern cat may be from several interbreeding populations.

Primates is the order of man and of lemurs, monkeys, and apes. Man takes pride in supposing himself to be the most "advanced" or "highest" of all vertebrates, but there really are few differences between monkeys, apes, and man. All of these higher primates have large cerebrums, good eyesight, small noses, and hands modified for grasping. Man is a "great ape," as is the gibbon, orangutan, chimpanzee, and gorilla. Like the other great apes, humans have a large body size and lack a tail. Chimpanzees are our closest relatives (although obviously not our ancestors) and share more than 97 percent of our DNA.

The following is based on Romer and Parsons, 1986.

CLASS **Mammalia** (mam-M*A*-l*e*-a). Mammals.
 Endothermic, hair, mammary glands, single lower jawbone.
 SUBCLASS Prototheria.
  INFRACLASS Allotheria.
   ORDER Monotremata. Egg-laying mammals.
 SUBCLASS **Theria** (TH*E*R-*e*-a).
  INFRACLASS Metatheria.
   ORDER Polyprotodonta. American opossum.

ORDER Peramelida. Bandicoots.
ORDER Diprotodonta. Wombats, kangaroos.
INFRACLASS **Eutheria** (*u*-TH*E*R-*e*-a).
ORDER Insectivora. Shrews, moles.
ORDER Dermoptera. "Flying lemur."
ORDER Chiroptera. Bats.
ORDER **Primates** (pr*i*-M*A*-t*e*z). Monkeys, apes, humans.
SUBORDER Lemuroidea. Lemurs.
SUBORDER Tarsioidea. *Tarsius.*
SUBORDER Platyrrhini. The nostrils open to the side of the rostrum. New world monkeys, "organ-grinders monkey," marmosets.
SUBORDER **Catarrhini.** The nostrils open ventrally on the rostrum. Old world monkeys and apes.
FAMILY Pongidae. Anthropoid apes. Gibbons, orangutans, gorillas, chimpanzees.
FAMILY **Homonidae** (h*o*-MON-i-d*a*). Man [*Homo sapiens* (H*O*-mo S*A*-p*e*-enz)].
ORDER **Carnivora** (kar-NIV-*o*r-a). Cats, dogs, bears, raccoons.
SUBORDER Fissipeda. Terrestrial carnivores.
FAMILY Canidae. Dogs.
FAMILY Ursidae. Bears.
FAMILY Procyonidae. Raccoons, pandas.
FAMILY Mustelidae. Weasels, minks, sables, skunks.
FAMILY Viverridae. Civets, mongooses.
FAMILY Hyaenidae. Hyena.
FAMILY Felidae. Cats.
SUBFAMILY **Felinae.** GENUS *Felis.* Sixteen species including the domestic cat *Felis catus* (or *Felis domesticus* [F*E*-lis d*o*-MES-ti-kus]), panthers (*Panthera*), leopards, lions, tigers, cheetahs (*Acinonyx*).
SUBORDER Pinnipedia. Marine carnivores. Seals, sea lions, walruses.
ORDER Cetacea. Whales, dolphins.
ORDER Edentata. Sloths, armadillos.
ORDER Tubulidentata. Aardvarks.
ORDER Pholidota. Pangolins, scaly anteaters.
ORDER Rodentia. Rats, mice, voles, capybaras, gophers, squirrels.
ORDER Lagomorpha. Rabbits, hares, pikas.
ORDER Hyracoidea. Hyraxes.
ORDER Perissodactyla. Odd-toed ungulates. Horses, zebras, tapirs, rhinoceros.
ORDER Artiodactyla. Even-toed ungulates. Pigs, cattle, antelopes, hippopotamuses, sheep, goats.
ORDER Proboscidea. Elephants.
ORDER Sirenia. Dugongs, manatees.

Suggested Readings

Blakiston's New Gould Medical Dictionary. (any recent edition). New York: McGraw-Hill Book Company.

Dorland's Illustrated Medical Dictionary. 25th ed. 1980. Philadelphia and London: W. B. Saunders Company.

Internat. Comm. Vet. Gross Anat. Nomen. 1983. *Nomina Anatomica Veterinaria.* 3d ed. International Committee on Veterinary Gross Anatomical Nomenclature, World Assoc. Veterinary Anat., Ithaca, N.Y.

Martin, L. D. 1980. Functional morphology and the evolution of cats. *Trans. Nebr. Acad. Sci.* 8:141–54.

Simpson, G. G. 1945. Principles of classification and a classification of mammals. *Bull. Am. Museum. Nat. Hist.* 85; xvi:1–350.

Walker, E. P., ed. 1968. *Mammals of the World.* 2d ed. Baltimore, Md.: Johns Hopkins University Press.

Warfield, M. S., and W. I. Gay. 1984. The cat as a research subject. *The Physiologist.* 27(4):177–89.

CHAPTER 1

Histology

Many important functional aspects of anatomical systems cannot be appreciated by a gross dissection alone. Structures that may be observed by the use of a microscope and prepared histological slides are described under the appropriate gross anatomical heading. If slides are not available for demonstration or concentrated study, the student should consult a good histology textbook. The observation of histological structures will require the use of a compound microscope. Before starting your study of cells and tissues for this chapter, read the following rules regarding the use of the microscope.

Using the Compound Microscope

(figure 1.1)

Because of the fragility and expense of microscopes, it is important to develop good habits for their use.

A. Always transport your microscope by PLACING ONE HAND UNDER THE BASE WHILE GRASPING THE ARM OF THE SCOPE WITH THE OTHER HAND.

B. Place the microscope on your laboratory table and connect the power plug to the outlet. Then, with the help of figure 1.1, locate the following:
   1. Ocular lenses
   2. Objective lenses
   3. Illuminator
   4. Illuminator switch
   5. Stage
   6. Stage clips/mechanical stage
   7. Coarse adjustment
   8. Fine adjustment
   9. Diaphragm
   10. Mechanical stage controls

C. To view any microscope slide, follow these instructions:
   1. Using the coarse adjustment, increase the distance between the stage and the objective lenses to the maximum.
   2. Rotate the objective lenses until the scanning objective (4×) snaps into place over the opening in the stage.
   3. Place the slide on the stage and secure the slide with clips or the spring-loaded brackets of the mechanical stage. If you have a microscope with a mechanical stage, use the mechanical stage controls to move the slide. If not, move the slide directly with your hand.
   4. Using the coarse adjustment, bring the objective as close to the slide as possible without touching it.
   5. Then, while looking through the ocular lens, slowly rotate the coarse adjustment until an image is observed.
   6. Sharpen the image using the fine adjustment. If there seems to be an improper amount of light, adjust the diaphragm to produce the best image. On scanning power, the diaphragm should be nearly closed. Using the most common lenses, the image you see is magnified 40 times. That is 4× objective × 10× ocular = 40×. Not all lenses are of these powers. Your scanning objective may be 2× or 3× and the ocular may be 5× or 15×. HINT: If you have a monocular microscope, practice viewing with both eyes open to avoid eyestrain from squinting and also to avoid seeing your own eyelashes in the field of view. One way to do this is to leave both eyes open but cover the unused eye with your hand.
   7. To increase the magnification, turn to the 10× objective (low dry power). If your microscope is **parfocal** (par-FO-kal) you will only need to use the fine adjustment to sharpen the image (10× objective × 10× ocular = 100×). DO NOT USE THE COARSE ADJUSTMENT. Again, center the object you wish to view.

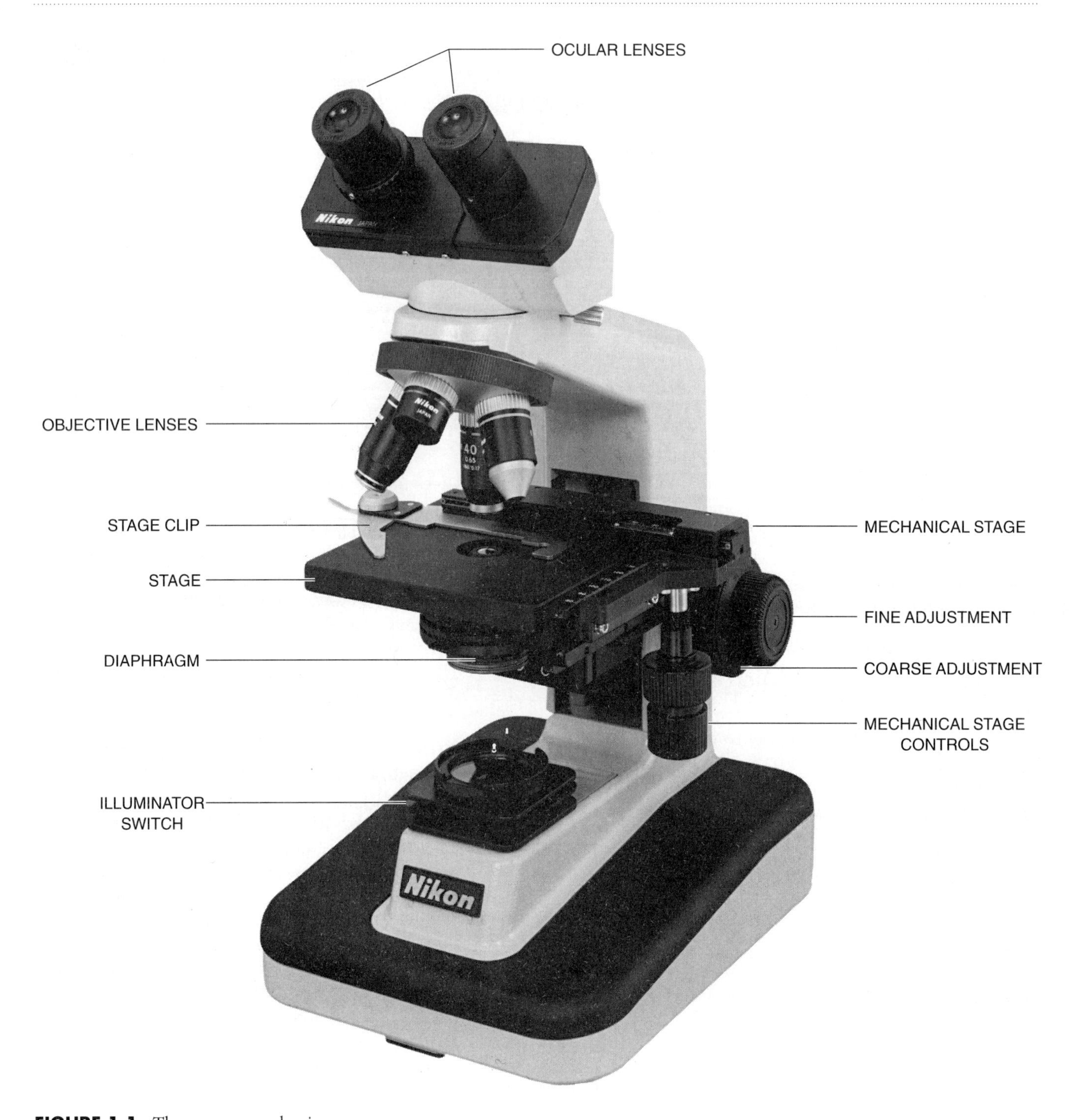

FIGURE 1.1. The compound microscope.
Courtesy of Nikon Corporation.

8. Finally, rotate the high dry objective (40×) into position and then focus with the fine adjustment only. (USE OF THE COARSE ADJUSTMENT MAY DRIVE THE LENS THROUGH THE COVERSLIP OF THE SLIDE AND DAMAGE BOTH THE LENS AND THE SLIDE.) The magnification here is 40× objective × 10× ocular = 400×. You will most certainly have to increase the amount of light by opening the diaphragm when using this magnification. Finally, center the object you wish to view.

D. What to do if you cannot see an image:
   1. Check the light source. Is it on?
   2. Check the diaphragm. Is it open?
   3. Make sure the objective lens is properly snapped into position. Is it aligned with the opening in the stage?
   4. Clean the lenses (WITH LENS TISSUE ONLY).
   5. Consult your instructor.

E. When you finish:
   1. Remove the slide from the stage.
   2. Clean the lenses (WITH LENS TISSUE ONLY).
   3. Leave the low power or scanning objective in place.
   4. Wrap the power cord around the microscope BUT NOT OVER FRAGILE PARTS (or follow your instructor's directions).
   5. Return the microscope to storage.

Cells and Tissues

Cell Morphology

An electron microscope is required to view most of the structures mentioned here; it is not expected that this instrument will be available to the student. Consequently, the features listed are intended only to serve as partial explanation for those structures that are seen (at least under special circumstances) by light microscopy.

A. **PROTOPLASM** (PR*O*-to-plazm). A general term for the material making up the cell.
   1. **CYTOPLASM** (CI-to-plazm). The material of the cell between the nucleus and cell membrane (plasmalemma). The material may be basophilic or neutrophilic, but is usually acidophilic (stains red with eosin). Ribonucleic acid (RNA) is basophilic (stains blue to purple with hematoxylin).
   2. **NUCLEOPLASM** (NYOO-kl*e*-*o*-plazm). The material surrounded by the double-layered nuclear membrane (inner and outer). Chromatin stains dark with hematoxylin (strongly basophilic) due to the high DNA content. If the protein content increases (at times in nucleolus), the material becomes acidophilic.
   3. **CELL MEMBRANE** (0.008 μm thick). Highly elastic but regulates osmosis. All living animal cells are permeable, nonliving cells with keratin or fat are *less* permeable. Cell membranes have a very high fat content and therefore a high energy storage (and cost).
   4. **NUCLEAR MEMBRANE.** A double-walled structure (inner and outer membranes) and therefore thicker than the cell membrane. The outer membrane is continuous with the endoplasmic reticulum.

B. **CELL ORGANELLES**
   1. **ENDOPLASMIC RETICULUM [ER]** (en-d*o*-PLAZ-mik re-TIK-*u*-lum). Rough (RER) or smooth (SER), a system of interconnected membranes from the outer nuclear membrane to the cell membrane. Some interconnections may be transitory. Not seen with light microscopy.
   2. **GOLGI** (G*O*L-j*e*) **APPARATUS.** Similar to ER but with circular arrangement. Stains with osmic acid or silver. Involved in cell secretion and protein formation.
   3. **MITOCHONDRIA** (m*i*-to-KON-dr*e*-a). Energy transformation occurs on the cristae lining the inner walls of these organelles. ATP is formed here. Not seen by ordinary light microscope preparation methods but special methods are available to stain the enzyme forming ATP (ATPase), thus identifying concentrations of mitochondria.
   4. **CENTROSOMES** (SEN-tr*o*-s*o*m). Best seen during mitosis. Contains paired centrioles that separate to form mitotic apparatus. Nine rodlets and basal corpuscles arranged in a cylinder similar to cilia (= centriole).
   5. **LYSOSOMES** (L*I*-s*o*-s*o*m). Contain enzymes that transform ingested substances and degrade waste material.

6. **VESICLES** (VES-ih-kl). Small packets pinched off from RER or Golgi, or pinocytotic packets from cell membrane. Similar to lysosomes but less dense.
7. **FIBRILS** (FI-bril): Three types. They have different appearances and roles in cells as follows:
   a. **Tonofibrils** (TON-*o*-*i*-bril). In epithelial cells, disappear during mitosis. Clumped submicroscopic protein filaments.
   b. **Neurofibrils.** In nerve cells, function unknown. Stained by methylene blue and silver stains.
   c. **Myofibrils** (m*i*-*o*-F*I*-bril). Actin and myosin in muscle cells. Function in muscle contraction.
8. **CILIA** (SIL-*e*-ah). Appendages on the "free" surface of cells. Each cilium has nine peripheral double filaments and a single central double filament. The central filament is absent in nonmotile cilia. Each cilium also has a hollow basal body and striated rootlets at the base of each filament.
9. **MICROVILLI** (m*i*-kr*o*-VIL-*i*). Nonmotile and very short.
10. Other structures found in cells:
    Microtubules
    Desmosomes (DES-mo-s*o*m)—macula adhaerens (MAK-*u*-lah ad-H*E*R-enz) or **adhesion plate,** localized.
    Zonula adhaerens (Z*O*N-*u*-lah)—terminal bar, completely around cell.
    Tight junction.
    Gap junctions—allow ion communication between epithelial cells.

Fundamental Tissues

These tissues make up the various organs of the vertebrate body. Two or more of them may be combined to form an organ.

1. **EPITHELIA.** See Cell Arrangement for epithelial types (figure 1.2).
2. **CONNECTIVE** and **SUPPORTING TISSUES** (figure 1.3). A vast array of different tissues ranging from blood to bone, but their common entity is the presence of cells suspended in a matrix. The other fundamental tissues are primarily cells.
3. **MUSCLE** (figure 1.4). Contractile cells that form striated, smooth, or cardiac muscles.
4. **NERVOUS TISSUE** (figure 1.5). Glial cells and neurons of the nervous system.

Cell Shapes and Arrangements for Epithelial Tissues

Cell Shapes (figure 1.2)

1. **SQUAMOUS** (SKWA-mus) **CELLS.** Flat with their thickest region near the nucleus. These cells resemble a miniature fried egg.
2. **CUBOIDAL** (ku-BOYD-al) **CELLS.** Approximately equal in height and width.
3. **COLUMNAR** (ko-LUM-nar) **CELLS.** Have much greater height than width.

Cell Arrangements and Examples (figure 1.2)

1. **Simple squamous.** Mesothelium, endothelium of blood vessels.
2. **Simple cuboidal.** Kidney tubules.
3. **Simple columnar.** Stomach, small intestines, gallbladder.
4. **Simple columnar ciliated.** Oviduct.
5. **Pseudostratified ciliated columnar.** Trachea.
6. **Stratified columnar.** Sublingual duct, urethra near vestibule.
7. **Stratified squamous.** Epidermis, mouth, esophagus.
8. **Pseudostratified columnar.** Urethra, parotid duct.
9. **Transitional.*** Urinary bladder.

Organ Systems

Each of the ten organ systems is composed of two or more organs. Both the gross structure and the tissues will be illustrated with the description for each organ system. The following list names the organ systems and the fundamental tissues that form them, in their relative degree of importance.

1. **SKIN** (or integument). Epithelia (epidermis), connective and supporting, and muscle (dermis).
2. **SKELETON.** Connective and supporting tissues.
3. **MUSCULATURE.** Muscle and connective tissues.
4. **CIRCULATORY** (including blood). Connective, muscle, and epithelial tissues.

* Actually intermediate between columnar and squamous.

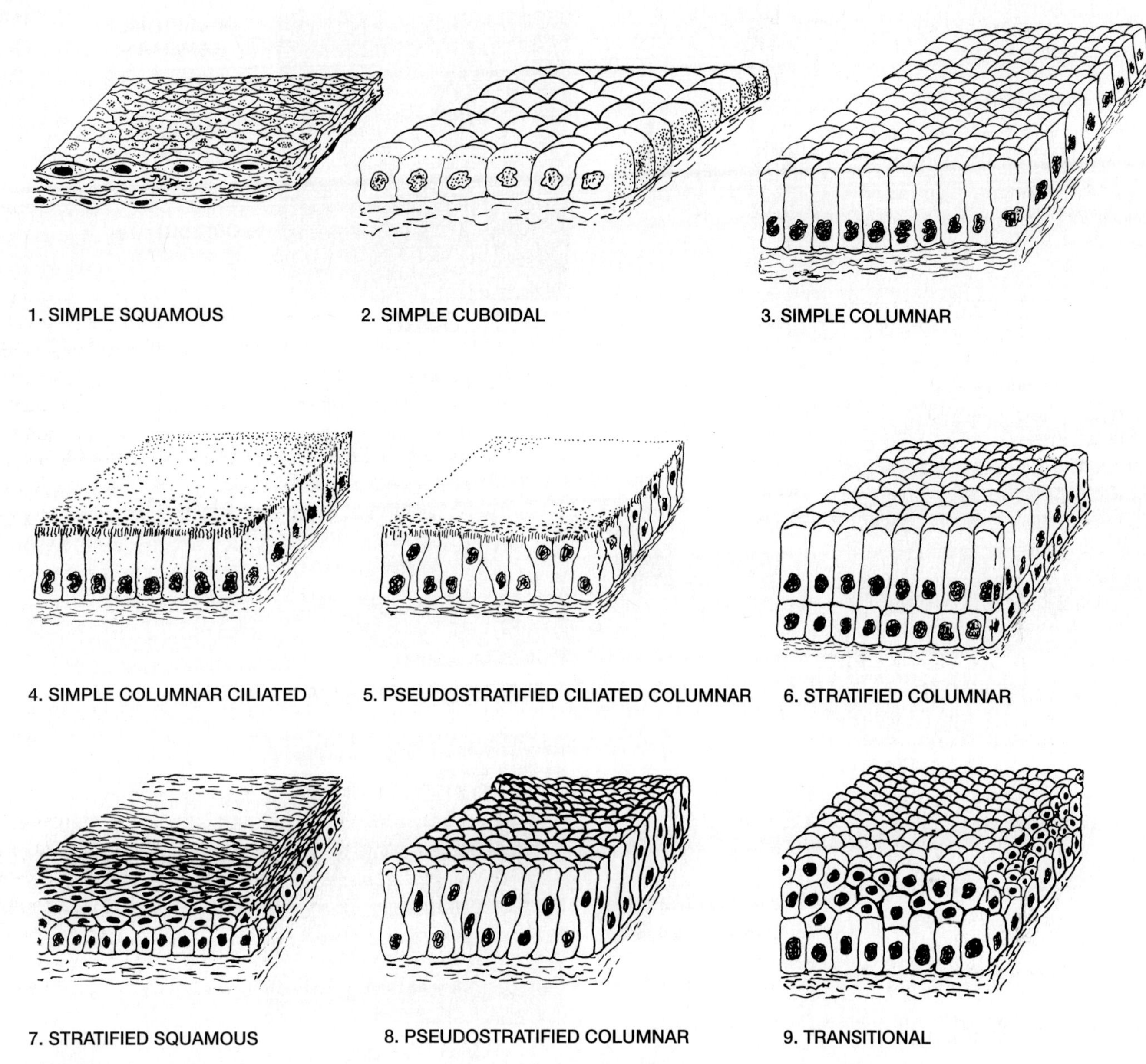

FIGURE 1.2. Epithelia. (1) Simple Squamous. (2) Simple Cuboidal. (3) Simple Columnar. (4) Simple Columnar Ciliated. (5) Pseudostratified Columnar Ciliated. (6) Stratified Columnar. (7) Stratified Squamous. (8) Pseudostratified Columnar. (9) Transitional.

From Chiasson and Odlaug, *Laboratory Anatomy of the Fetal Pig*.

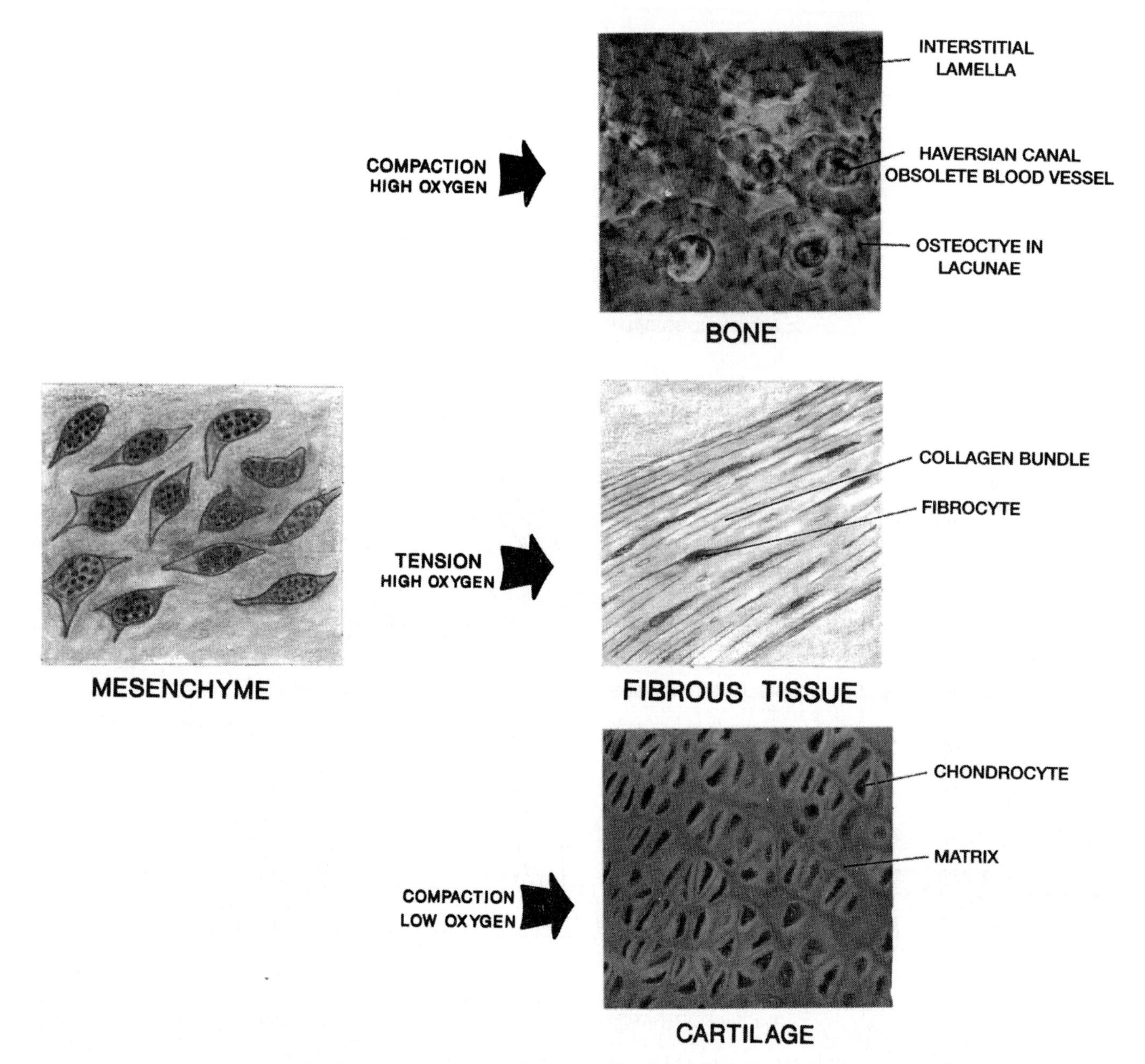

FIGURE 1.3. Connective tissues. Diagrams illustrating the origin and appearance of the supportive connective tissues. The common fibroblast [or mesenchyme cell] is able to form either bone, cartilage, or fibrous tissue depending upon the environment it is subjected to. Pressure and a good oxygen supply (vasculature) produces bone; pressure and a poor oxygen supply produces cartilage; and tension plus a good oxygen supply produces fibrous tissue.

From Chiasson and Odlaug, *Laboratory Anatomy of the Fetal Pig.*

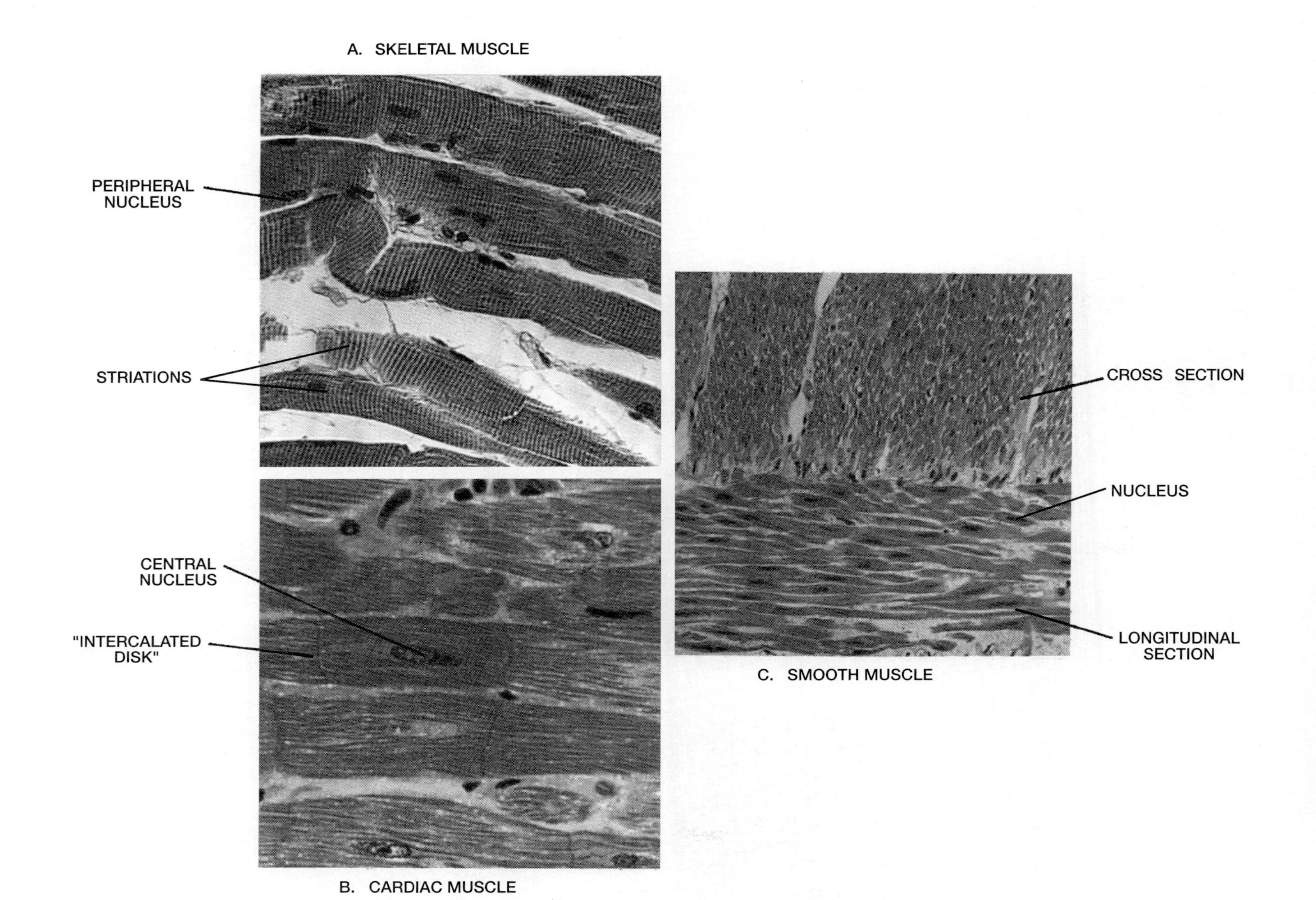

FIGURE 1.4. Muscle Tissues. (A) Striated (skeletal) muscle. (B) Cardiac (striated) muscle. (C) Visceral (smooth or non-striated) muscle.
From Chiasson and Odlaug, *Laboratory Anatomy of the Fetal Pig*.

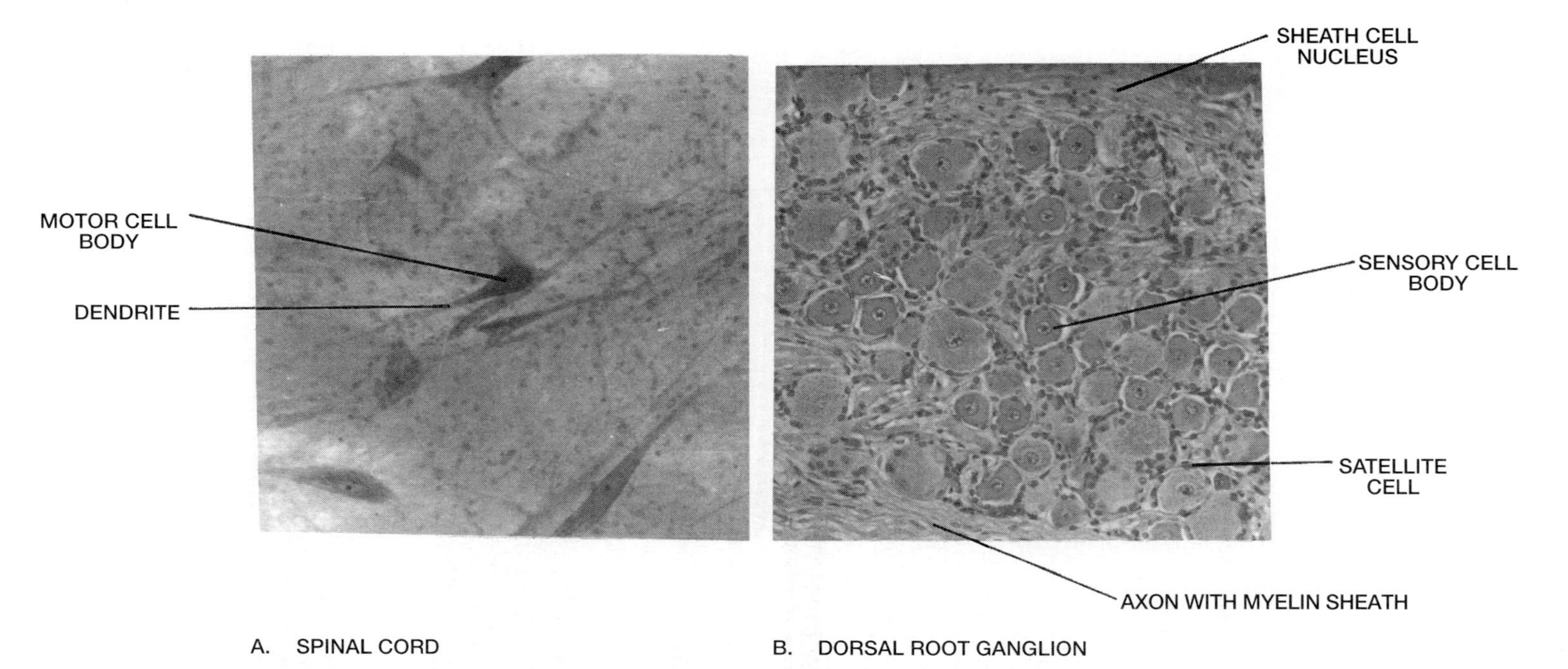

FIGURE 1.5. Nervous Tissues. (A) Motor neurons, (B) sensory neurons and glial satellite cells.
From Chiasson and Odlaug, *Laboratory Anatomy of the Fetal Pig*.

5. **DIGESTIVE.** Epithelial, muscle, and connective tissues.
6. **RESPIRATORY.** Epithelial and connective tissues.
7. **EXCRETORY.** Epithelial and connective tissues.
8. **REPRODUCTIVE.** Mostly epithelial and connective tissues.
9. **NERVOUS.** Nervous and connective tissues.
10. **ENDOCRINE.** Modified epithelial, nervous, and connective tissues.

Suggested Readings

Telford, I. R., and Bridgman, C. F. 1995. *Introduction to Functional Histology*. Harper Collins College Publishers.

CHAPTER 2

External Anatomy and Skin

External Anatomy

Mammals have two unique external characteristics that distinguish them from all other vertebrates: (1) all mammals have hair at some time during their development and (2) all female mammals possess mammary glands with external openings for nourishing the young.

Orders of mammals can be recognized rather easily because of differences in external appearances; the external anatomy of a cat is quite different from that of a rat. The external features that separate mammals into orders include the number of digits on the feet, the method of walking or other locomotion, characteristics of the teeth, and others.

Posture and Locomotion

The position of the hand and foot in locomotion varies considerably, but is constant within a single order of mammals.

To increase stride length and therefore speed of locomotion, the distance covered by each step is increased. This growth is accomplished by increasing the number of limb segments involved in each step. The cat walks on the digits with the heel elevated above the ground, a posture known as **digitigrade** (DIJ-i-ti-gr*a*d). When the heel rests on the ground, as it does with humans, the posture is called **plantigrade** (PLANT-i-gr*a*d). If the mammal walks on the very tip of modified digits (as in horses and cattle), the posture is called **unguligrade** (ung-GU-li-gr*a*d).

Examine your cat and locate the following features:

A. **APPENDAGES.** The terms applied to the appendages of the cat, a rather typical terrestrial quadruped, are also applied to the human and are defined in the same way.
   1. **FORELIMB.** The pectoral appendage consists of:
      a. **BRACHIUM.** The proximal segment of the pectoral appendage, supported by the humerus.
      b. **ANTEBRACHIUM.** The middle segment of the pectoral appendage, supported by the radius and ulna.
      c. **MANUS.** The distal segment of the pectoral appendage, supported by the carpals and metacarpals. The cat has five **digits** on each hand (manus) supported by phalanges. Each digit is supplied with a pad, a **torus digitalis** (T*O*-rus dij-i-TAL-is) (not illustrated). Both the manus and pes are equipped with **claws** formed from a dermal core and having an epidermal sheath.
   2. **HINDLIMB.** The pelvic appendage consists of the:
      a. **THIGH.** The proximal segment of the pelvic appendage, supported by the femur.
      b. **LEG.** Also called the shank, the middle segment of the pelvic appendage, supported by the tibia and fibula.
      c. **PES.** The distal segment of the pelvic appendage, supported by the tarsals and metatarsals. Supported by phalanges, there are four **digits** on each pes, or foot. Each has a pad called a **torus tarsalis** (tar-SAL-*is*) (not illustrated).

B. **HEAD.**
   1. **MOUTH.** Surrounded by the lips, the upper of which is cleft by a groove called the **philtrum** (FIL-trum).
   2. **EXTERNAL NARES** (NA-r*ez*). These open to the **NASAL CAVITIES.**
   3. **VIBRISSAE** (v*i*-BRIS-*e*). Long hairs located on various regions on the face. They are tactile (sensory) and provide spatial information as the animal moves in the dark.
   4. **EYES.** They have **upper** and **lower eyelids** and a **nictitating** (NIK-ti-t*a*-ting) **membrane** that moves laterally from the medial corner of the eye.
   5. **PINNA** (PIN-a). The large, flaplike portion of the external ear.

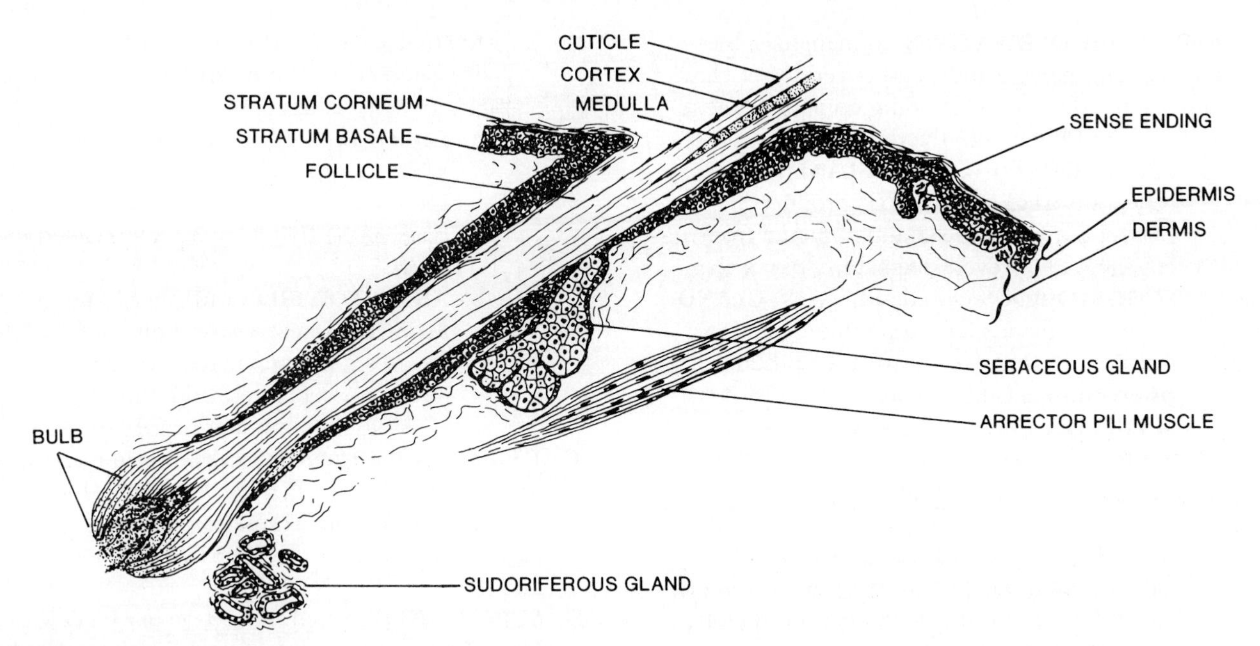

FIGURE 2.1. Diagrammatic composite section of mammalian skin. The cat lacks sudoriferous (sweat) glands except on the surface of the feet.
From Chiasson and Radke, *Laboratory Anatomy of the Vertebrates.*

C. **TRUNK.** Divided into a cranial **THORAX** (TH*O*-raks) and a caudal **ABDOMEN** (AB-d*o*-men) at approximately the point where the most caudal rib can be felt.
   1. **NIPPLES** (NIP-elz). Usually four pairs in the cat, running in lines along the ventral surface, from the axilla to the groin. These are the external openings of the mammary glands. A female that has recently given birth will have large nipples; males and juvenile females have small nipples.
   2. **UROGENITAL** (*u*-r*o*-JEN-i-tal) **OPENINGS.** Mammals have separate urogenital and anal openings (see chapters 8 and 9).
      a. **VESTIBULE** (VES-ti-b*ul*). In the female, the urinary and genital passages open to this common area.
      b. **PENIS** (P*E*-nis). The male urogenital organ contains a common urinary and genital passage, the **urethra** (*u*-R*E*-thra). The penis is generally retracted into a midventral **prepuce** (PR*E*-p*us*).
      c. **SCROTUM** (SKR*O*-tum). A sac that contains the **testes** of the male, lying just cranial to the anal opening.
      d. **ANAL** (A-nal) **OPENING.** This opening is located at the base of the tail, dorsal to the urogenital opening.

The Skin (figure 2.1)

Examine the slide section of either cat or human skin under low power (10× objective) of the compound microscope. Locate the following features:

A. **EPIDERMIS.** The outermost layer of the skin. The epidermis is formed from ectoderm and is subdivided into as many as five layers.
   1. **STRATUM BASALE** (STRA-tum b*a*-SAL). The innermost layer of the epidermis, where cells are produced continually and migrate toward the surface. This layer also gives rise to skin glands and hair, as well as to the epidermal components of nails and claws.
   2. **STRATUM SPINOSUM** (sp*i*-N*O*-sum) (not illustrated). Immediately superficial to the stratum basale.
   3. **STRATUM GRANULOSUM** (gran-*u*-L*O*-sum) (not illustrated). The cytoplasm of these cells is granular.
   4. **STRATUM LUCIDUM** (LOO-si-dum) (not illustrated). Occurs in thick skin, immediately deep to the stratum corneum. The cytoplasm of this layer's cells is clear and enucleate, hence the name.
   5. **STRATUM CORNEUM** (K*O*R-n*e*-um). The outer layers of cells with cornified cytoplasm. This layer provides many of the protective functions associated with the skin.

B. **EPIDERMAL DERIVATIVES.** In mammals, these include hair, nails, glands, and coverings of claws and horns. Although part of the epidermis, most are large and bulge into the dermis.

1. **SEBACEOUS** (*se*-B*A*-shus) **GLAND.** Also called an **oil gland.** The oils produced protect the hair and stratum corneum from cracking and prevent premature loss of hair.
2. **SUDORIFEROUS** (soo-d*o*-RIF-er-us) **GLAND.** The sweat glands have a thermoregulatory function. In many mammals they produce **pheromones** (FER-*o*-mons).
3. **MAMMARY** (MAM-er-*e*) **GLAND** (not illustrated). This "gland" in mammals (the terms **mammal** and **mamma** are derived from this word) is actually an accumulation of small ducted glands called **alveolar** (al-V*E*-*o*-lar) **glands**, each with a duct leading to the surface and forming an elevated region called the **nipple.**
4. **HAIR.** The principal covering of the cat, and most mammals. Hair is a feature of mammals only, and occurs in all mammals at some time during their lives. *If a light-colored human or cat hair is placed on a dry slide and covered with a cover glass, it may be possible to observe some features.* The outer layer of the hair is the **cuticle;** it consists of scalelike cells covering a middle **cortex** (K*O*R-teks), which is made of a layer of elongated cells. The **medulla** (me-DUL-a), a core of cells, is deepest.
   a. **FOLLICLE** (FOL-lih-kl). Hair grows within a capsule of epidermal cells called the follicle. Actual mitosis occurs at the base of the follicle in a region called the **bulb.**
   b. **ARRECTOR PILI** (a-REK-t*o*r P*I*-l*e*) **MUSCLE.** These visceral muscles and the blood supply are actually dermal structures separated from the hair follicle by a basal membrane.

C. **DERMIS.** A layer of skin derived from mesoderm that is much thicker than the epidermis and composed of fibrous connective tissues. This region also contains the blood vessels, muscles, and nerves of the skin.

D. **SUBCUTANEOUS** (sub-k*u*-T*A*-n*e*-us) **LAYER** or **hypodermis** (not illustrated). Lying between the dermis and the skeletal muscles, this layer contains loose connective tissue and adipose tissue for lipid storage.

Suggested Readings

Creed, R. F. S. 1958. The histology of mammalian skin, with special reference to the dog and cat. *Vet. Rec.* 70:171–75.

Kenshalo, D. R., D. B. Duncan, and C. Weymark. 1967. Thresholds for thermal stimulation of the inner thigh, footpad, and face of cats. *J. Comp. Physiol. Psychol.* 63:133–38.

Strickland, J. H. 1963. The integumentary system of the cat. *Amer. Jour. Vet. Res.* 24(102):1018–29.

Winkelmann, R. K. 1958. The sensory endings in the skin of the cat. *Jour. Comp. Neurology* 109(2):221–32.

CHAPTER 3

Skeletal System

The **SKELETON** (SKEL-eh-ton) of a mammal consists of fibrous tissue, bone, and cartilage. The framework of the body is made up primarily of bones with some cartilage. The skeleton of the external ear, the end of the nose, and the respiratory tract are mainly cartilage.

CARTILAGE (KAR-ti-lij) has firmness and elasticity that makes it an excellent supportive tissue for the respiratory tract and a cushion for joints. A large portion of the bony skeletal system is originally formed as cartilage in the embryo and is replaced by bone through the process of cartilage resorption and bone deposition.

BONE (bon) is formed by impregnating other types of connective tissues with calcium salts, which not only provides a hard, strong support for the body but also serves as a reservoir for calcium and phosphorus, which are essential for muscle contraction and nerve conduction.

Individual bones may be classified as either membrane or cartilage bone. **Membrane bone** (dermal bone) is formed by the deposit of calcium salts directly into the embryonic membranous connective tissue; **cartilage bone (endochondral** (en-d*o*-KON-dral) **bone)** is formed in a similar manner but is preceded by a cartilage stage so that the bone replaces the recently absorbed cartilage. **Periosteal** (per-*e*-OS-t*e*-al) **bone** is cartilage bone that is formed as a collar around long cartilage bones. Bones such as the femur or humerus are often referred to as compound bones since they consist of both periosteal and cartilage bone.

Study both mounted and disarticulated skeletons of the cat (figure 3.1) and, if available, the human (see figures 3.12, 3.22, 3.23). The mounted skeleton will show the relationship of each bone to the other bones. While examining the disarticulated skeleton you should periodically refer to the articulated skeleton for placement and relationships of individual bones.

The skeleton is arranged in two parts: an **AXIAL** (AK-s*e*-al) portion, including bones of the skull, spinal column, and thorax, and an **APPENDICULAR** (ap-en-DIK-*u*-lar) portion consisting of bones of the pectoral and pelvic girdles and limbs.

Axial Skeleton

Both humans and cats face forward, but humans face forward from an erect or bipedal position, which places the face at a 90 degree angle to the vertebral column. The cat's face is a forward continuation of the horizontal vertebral column. The bipedal stance together with the greater development of the brain is the basis for most differences between the cat (or any other mammal) and human skulls.

Both the cat and human skulls have a **NEUROCRANIUM** (n*u*-ro-KR*A*-ne-um) enclosing the brain and continuing from the vertebral column (**superior** for the human and **rostral** for the cat).

The major sense organs (nose, eye, and ear) are also protected by the neurocranium, but the eye and nose are in differing positions in the two mammals.

The jaws, hyoid, and larynx are parts of the **SPLANCHNOCRANIUM** (SPLANK-n*o*-kr*a*-n*e*-um).

A. **NEUROCRANIUM** (figures 3.2–3.6). Composed of the skull bones enclosing the brain. The roof and upper side walls of the neurocranium are composed of dermal bones, and the floor and lower walls are of cartilage bone.

> The number of bones in the cat and human adult skull is not the same. The embryological origins and numbers are similar, but fusions between bones occur in the adult human.

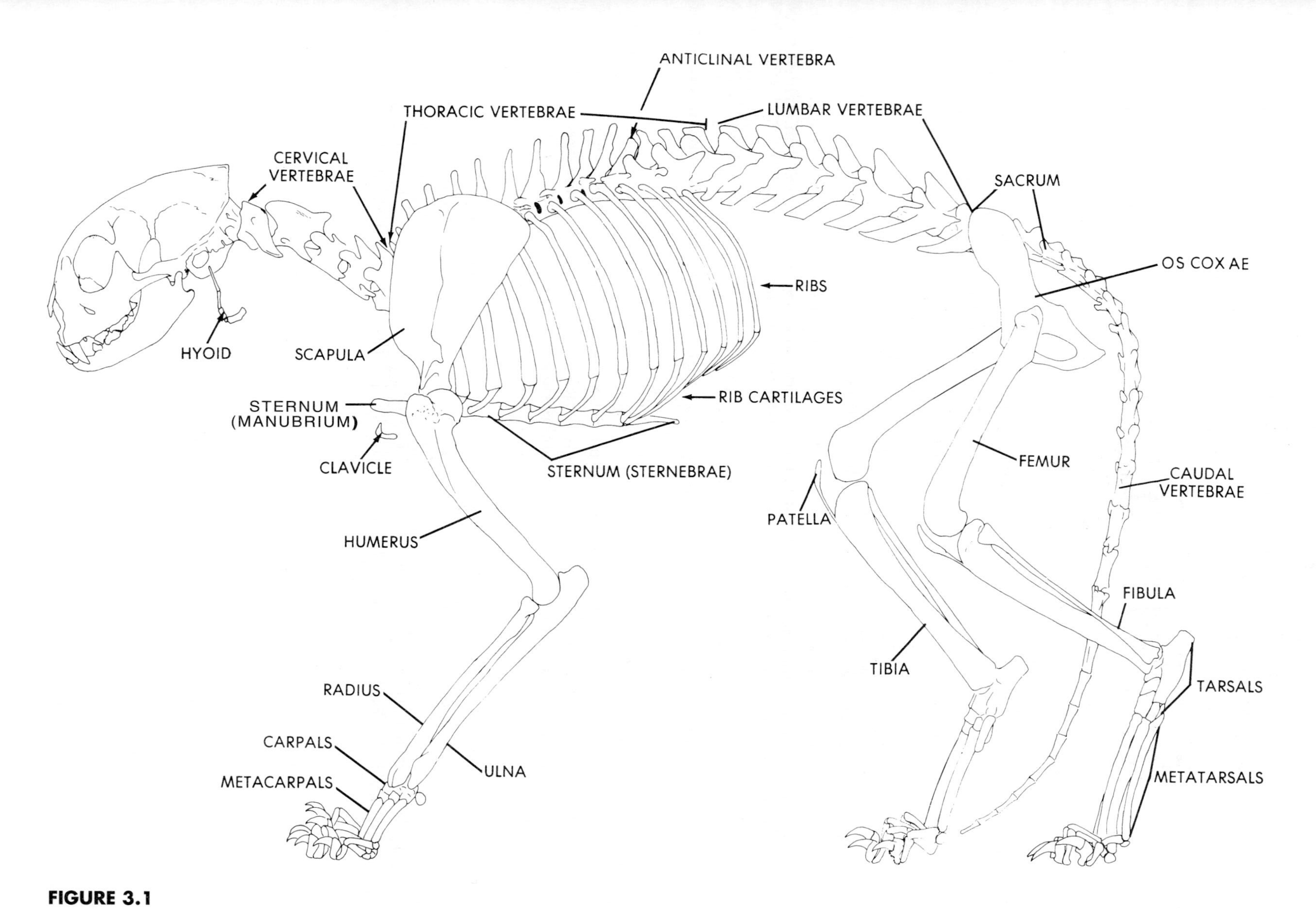

FIGURE 3.1
Lateral view of the cat skeleton. See figure 3.12 for a lateral view of the human skeleton.

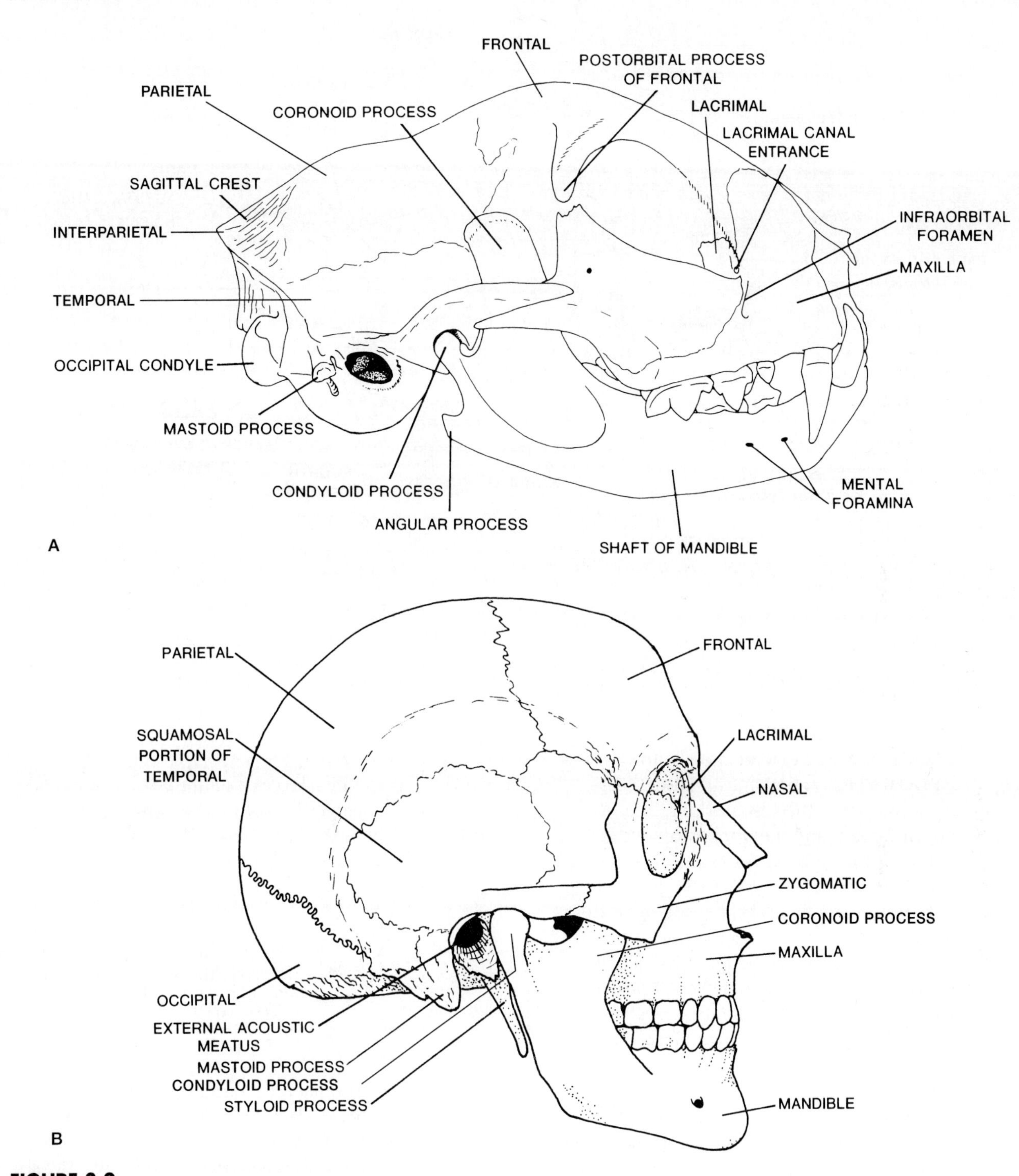

FIGURE 3.2
Skull and mandible of the cat (A) and human (B), lateral views.

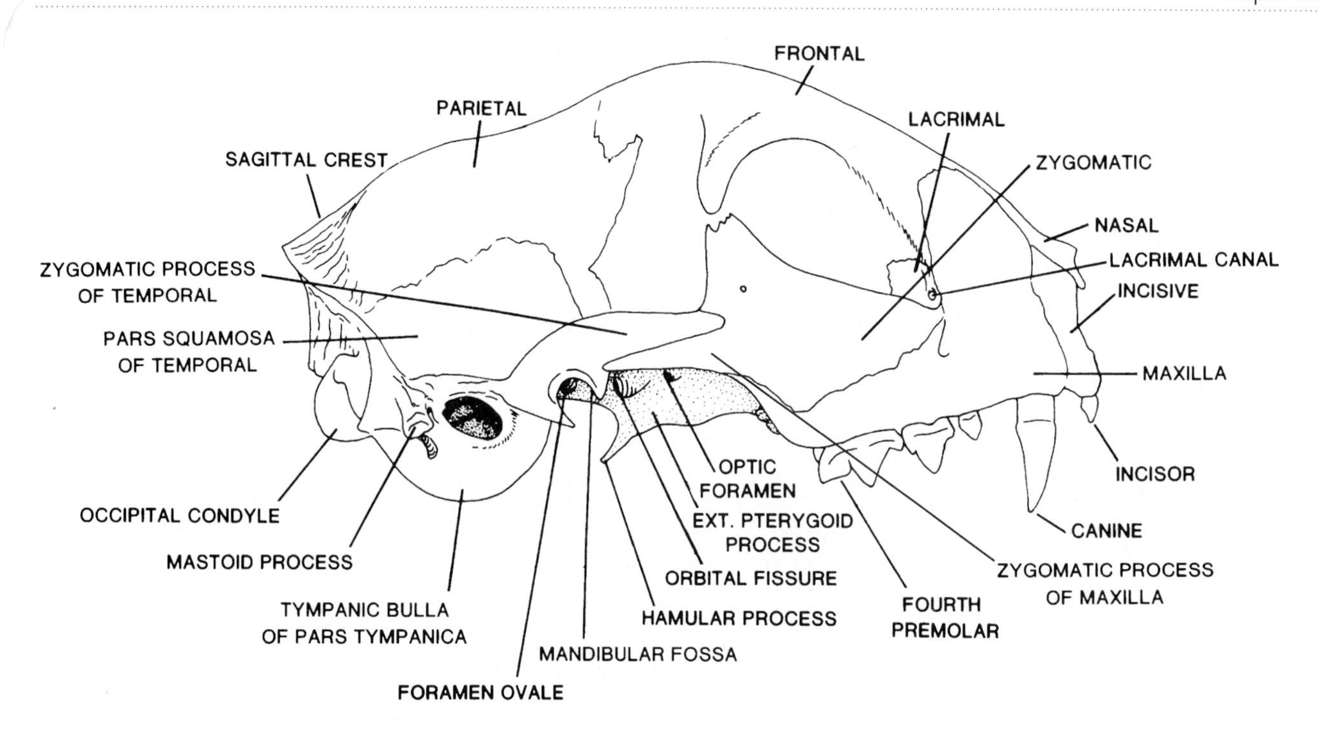

FIGURE 3.3
Skull of the cat in lateral view with the mandible removed.

1. The **FRONTAL** (FRUN-tal) is a single bone in adult humans, but cats have paired frontal bones with **POSTORBITAL** (p*o*st-*O*R-bi-tal) **PROCESSES** that extend laterally toward the **ZYGOMATIC** (z*i*-g*o*-MAT-ik) **ARCH.** The frontal(s) forms the rostral roof of the neurocranium and the upper wall of the **orbit** (or-bit) (a bony socket for the eyeball).
2. **PARIETAL** (pa-R*I*-e-tal) **BONES.** Paired units in both the human and cat, just behind the frontals. The caudal border of each cat parietal has a shelf of bone on its medial surface called the tentorium (ten-T*O*-r*e*-um), which separates the cerebral hemispheres from the cerebellum. In the human, the tentorium is a part of the nonossified, folded **dura mater.**
3. An **INTERPARIETAL** (in-ter-pah-R*I*-e-tal) **BONE** occurs in the cat but not in the human. It is an unpaired triangular-shaped bone in the midline of the skull between the two parietals and the occipital bone.

   The midline surface of this bone forms a **sagittal crest** in the cat. The caudal border, together with the supraoccipital bone, forms a transverse lambdoidal (lam-DOYD-al) ridge or crest. These crests serve for the cranial attachment of the **nuchal** (NYOO-kal) **ligament** that supports the head of quadruped mammals much like cables support a suspension span of a bridge. The vertebral attachment of the nuchal ligament is to the elongated neural spines of the last cervical and first thoracic vertebrae (see vertebral column, page 28).

> The human skull is superior to the vertebral column, rather than "slung" forward from the vertebrae and supported by the nuchal ligament. Consequently, humans lack both the interparietal bone and the strong sagittal and lambdoidal crests.

4. **TEMPORAL** (TEM-p*o*-ral) **BONES.** Complex paired units forming the lateral walls of the braincase. Each temporal bone is composed of three parts.
   a. **PARS SQUAMOSAL** (parz skwa-M*O*-sl). Borders the frontal, parietal, occipital, and sphenoid bones in addition to the pars tympanica of the temporal.

      A forward projecting arm of the squamosal, the **ZYGOMATIC PROCESS,** articulates with the **ZYGOMATIC** (cheek) **BONE.** The ventral base of the zygomatic process

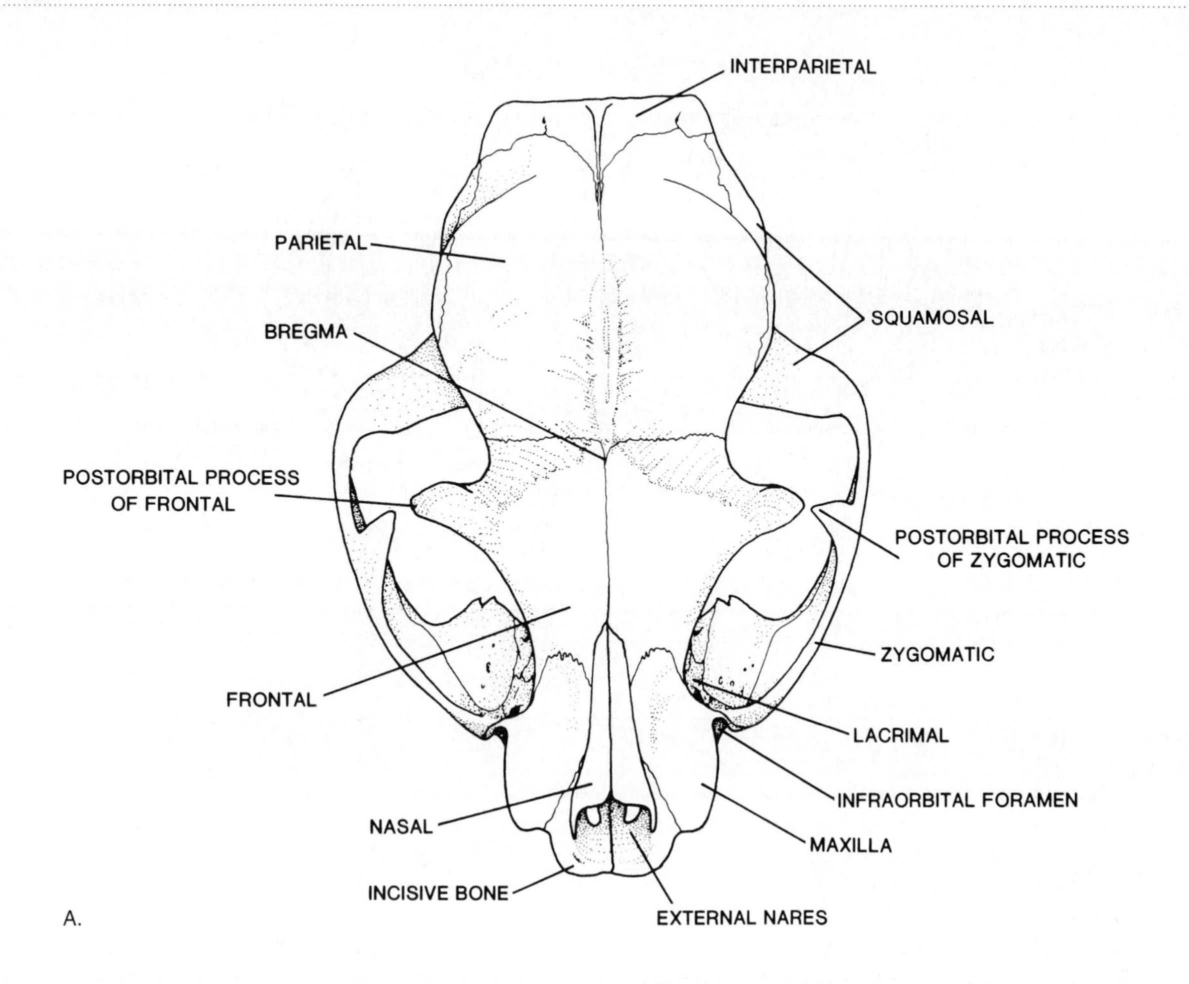

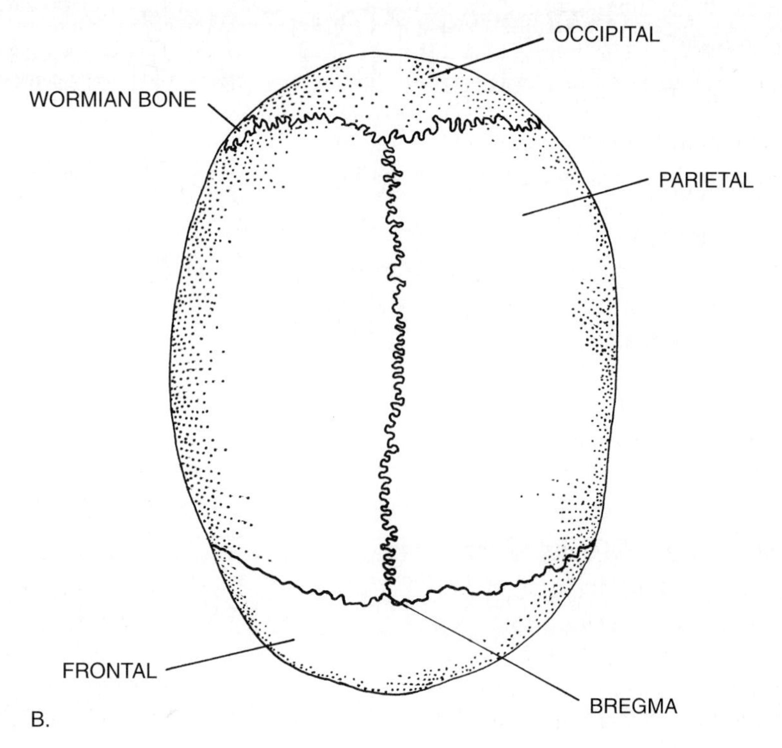

FIGURE 3.4
Cat (A) and human (B) skulls, dorsal/superior views.

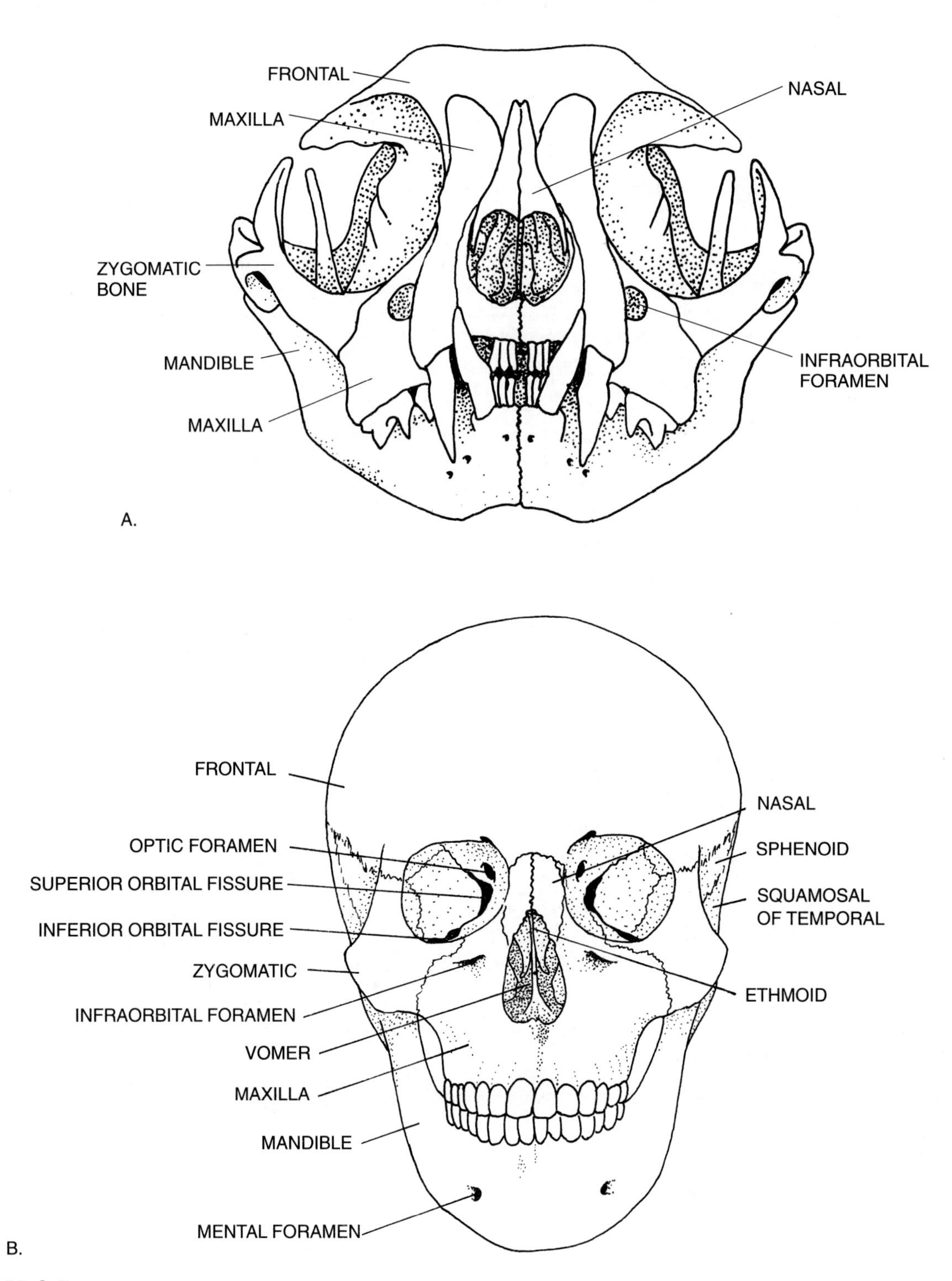

FIGURE 3.5
Cat (A) and human (B) skulls, frontal/anterior views.

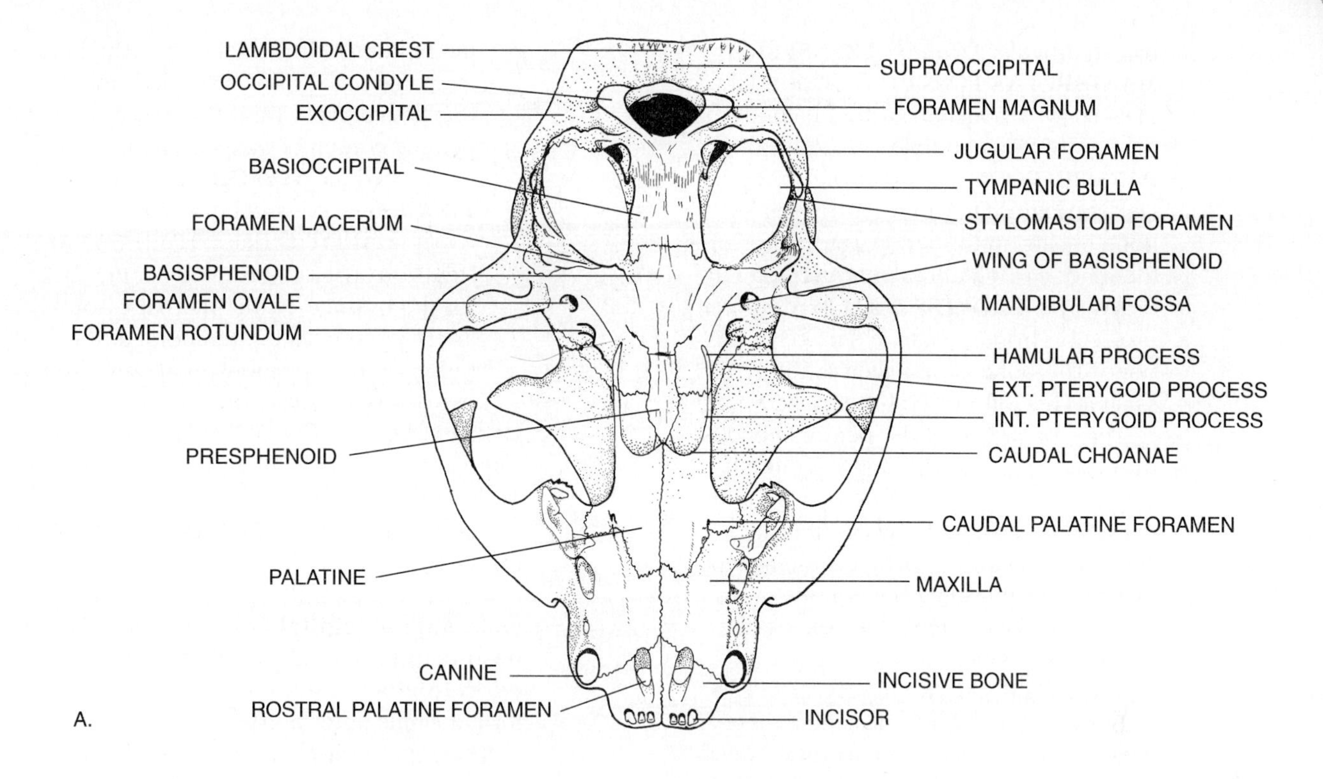

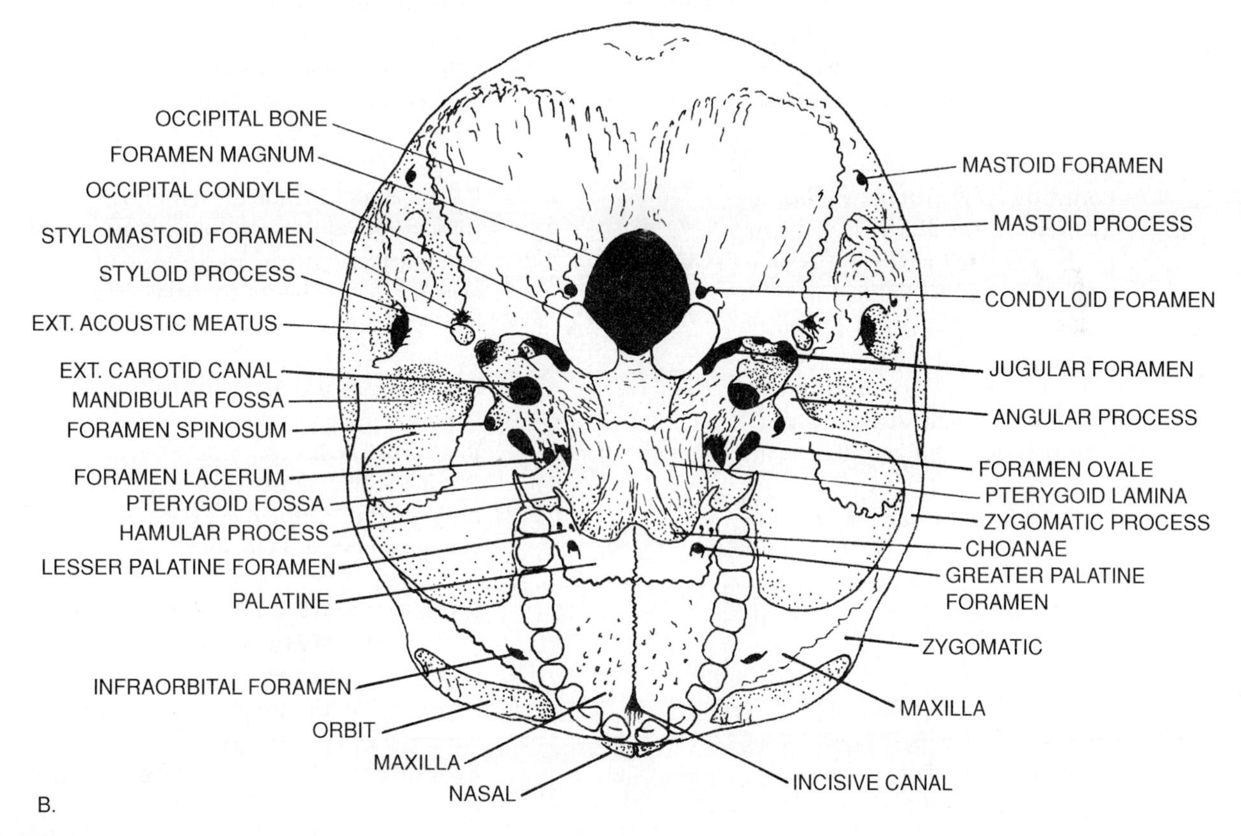

FIGURE 3.6
Cat (A) and human (B) crania, ventral/inferior views.

has an articular face, the **TRANSVERSE MANDIBULAR FOSSA** (man-DIB-*u*-lar FOS-a), for articulation with the condylar process of the mandible.

b. **PARS PETROSA** (pe-TR*O*-sa) **(petromastoid).** An irregularly shaped bone enclosing the sensory structures of the inner ear. The human **MASTOID** (MAS-toyd) **PROCESS** is a stubby ventral projection from the lateral surface of this bone. The bone is best seen in the sagittal view of the skull (figure 3.7).

The caudal end of the petrous bone is fused with the mastoid bone. Although some textbooks refer to the petrous and mastoid as separate bones in the cat, *Nomina Anatomica* considers them to be two parts of the same bone in the human, so the mastoid is a process of the petrosal bone.

The **INTERNAL ACOUSTIC MEATUS** (a-KOOS-tik m*e*-*A*-tus), providing the passage for the facial and acoustic nerves and the internal auditory artery, is a large foramen near the center of the medial face of the pars petrosa. Laterally, the pars petrosa is covered by the pars tympanica.

c. **PARS TYMPANICA** (tim-PAN-i-ka). Forms the ventrolateral wall of the temporal bone and houses the middle ear bones (**MALLEUS** [MAL-*e*-us], **INCUS** [ING-kus], and **STAPES** [ST*A*-p*e*z] see figure 11.5). The external acoustic meatus is the large lateral channel to the tympanic cavity. The **auditory** (AW-dih-tor-*e*) (eustachian) **tube** passes ventromedially through the rostral border of the temporal bone.

The tympanic bone of the cat is "expanded" into an **AUDITORY BULLA** (BUL-ah) that is actually in two parts. A lateral ectotympanic ring surrounds the external acoustic meatus and is fused with a smoothly rounded inner portion, the endotympanic ring. Caudal to the acoustic meatus of the cat is a socket for the tympanohyal portion of the hyoid apparatus and a stylomastoid foramen for the passage of the facial nerve. The human tympanic bulla is not expanded, and the human tympanohyal is fused to the petrous portion of the petromastoid where it is termed the **STYLOID** (ST*I*-loyd) **PROCESS** (see hyoid bone, page 24).

The **CAROTID** (ka-ROT-id) **CANAL** (figure 3.6B, not illustrated on the cat) for the passage of the internal carotid artery penetrates the temporal at the union of the squamous, tympanic, and petrous portions.

5. **OCCIPITAL** (ok-SIP-i-tal). A cartilage bone forming the caudal wall and part of the floor of the neurocranium. The center of this bone is a large foramen, the **FORAMEN MAGNUM** (MAG-num), for the passage of the spinal cord. The occipital develops as four separate units, **supraoccipital** (syoo-pra-ok-SIP-i-tal), **basioccipital** (b*a*-*se*-ok-SIP-i-tal), and two **exoccipitals** (ek-sok-SIP-i-tal), that fuse to form a single adult bone.

The supraoccipital is above the foramen magnum and serves as a site of insertion for musculature moving the head. The cranial (superior) border of the supraoccipital is sutured to the parietal (and interparietal in the cat) and forms the lambdoidal ridge or crest.

The **exoccipitals** are on either side of the foramen magnum and have large occipital condyles that articulate with the cranial articular processes of the atlas. The basioccipital is the caudal floor of the neurocranium. Laterally, the junction of the basioccipital and exoccipitals is extended ventrally as long fingers of bone, the **PARAMASTOID** (p*a*ra-MAS-toyd) **PROCESSES.** Between each paramastoid process and occipital condyle is a **HYPOGLOSSAL FORAMEN** for the passage of the hypoglossal nerve. Rostrally, the **BASISPHENOID** (b*a*-se-SF*E*-noyd) articulates with the **PRESPHENOID** (pr*e*-SF*E*-noyd); this juncture is often fused. The gap between the basioccipital and the pars tympanica of the temporal bone is the **JUGULAR** (JUG-*u*-lar) **FORAMEN.** Although cerebral veins pass through the jugular foramen to join the

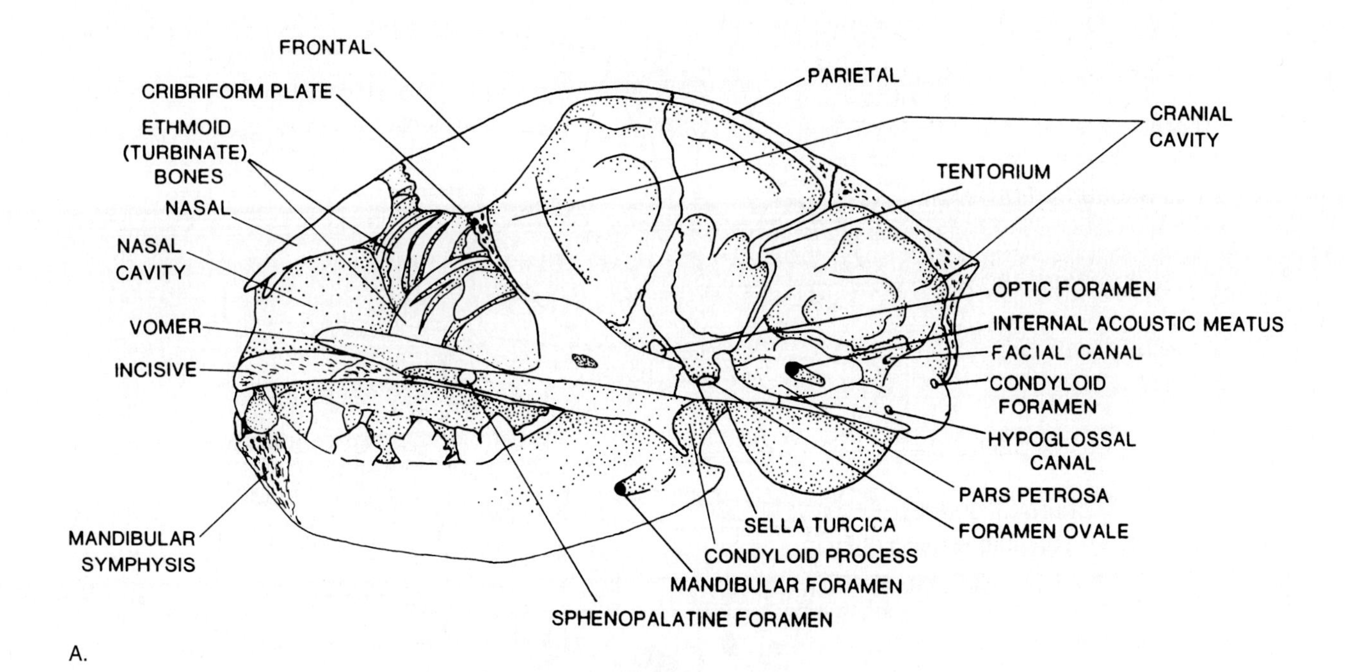

A.

FRONTAL
PARIETAL
SPHENOID
NASAL
SELLA TURCICA
PERPENDICULAR PLATE OF ETHMOID
INTERNAL ACOUSTIC MEATUS
JUGULAR FORAMEN
MAXILLA
STYLOID PROCESS
OCCIPITAL
MANDIBULAR FORAMEN
PALATINE
MANDIBLE

B.

FIGURE 3.7
Cat (A) and human (B) skulls, sagittal views.

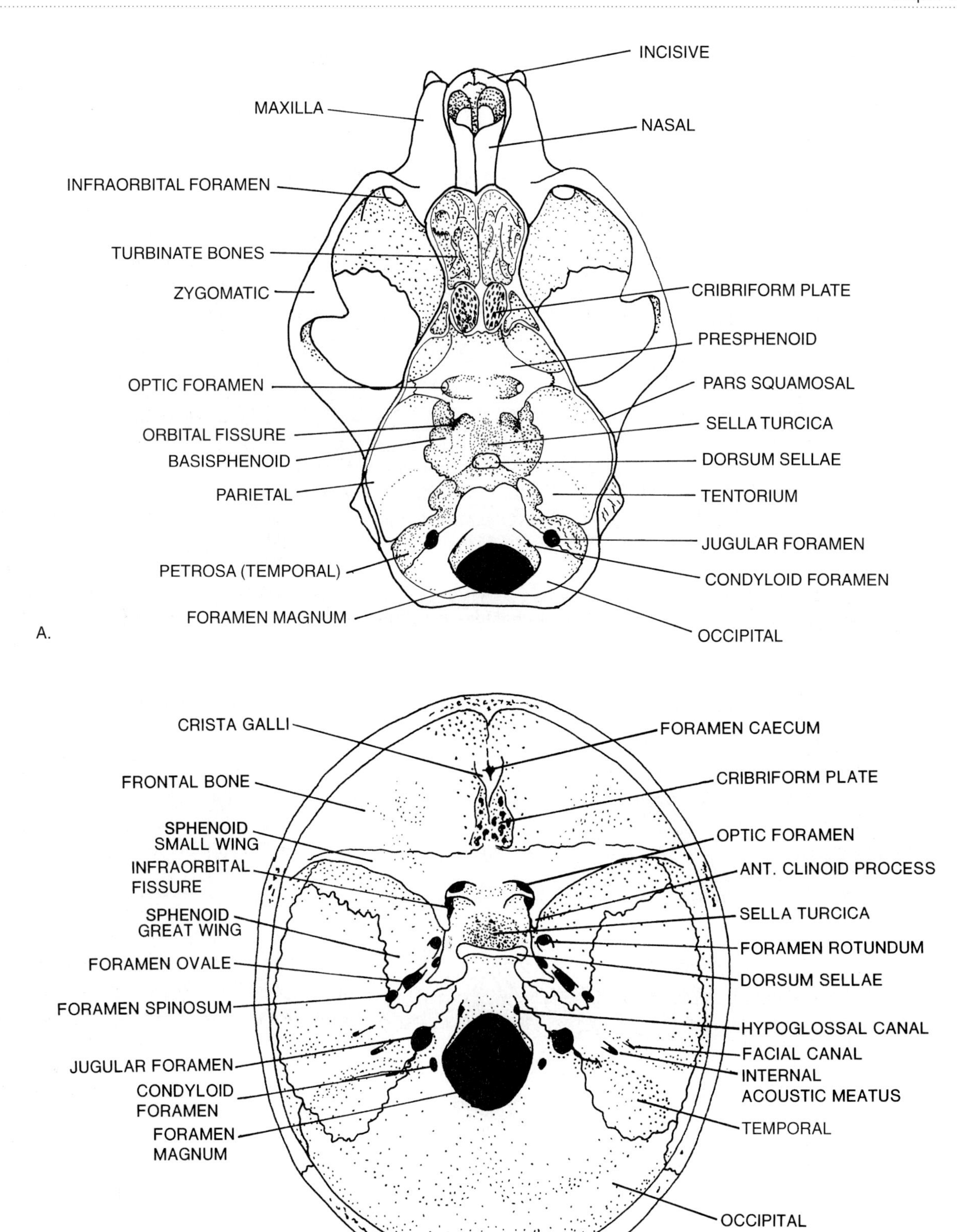

FIGURE 3.8
Coronal sections of the skulls of the cat (A) and human (B) allowing views of the cranial floors.

jugular vein, the large part of the foramen is occupied by the glossopharyngeal, vagus, and spinal accessory nerves.

6. **SPHENOID** (SF*E*-noyd). A single, median, unpaired cartilage bone in the human that forms the floor of the neurocranium. The human sphenoid is comparable to two bones in the more elongated cranium of the cat, a rostral **PRESPHENOID** and a caudal **BASISPHENOID.** The two units are also present in the newborn human but they fuse into a single bone at about eight months of age.

   The central body of the sphenoid (basisphenoid in the cat) has a recess on its internal floor, the **SELLA TURCICA** (SEL-ah TUR-si-ka) (figures 3.7, 3.8), that holds the **pituitary** (pi-t*u*-i-t*a*r-*e*) **gland.** Two wings or **ala** (*A*-la) extend from each side of the body. The **lesser wings** of the sphenoid project from the cranial (presphenoid) portion as **ORBITOSPHENOIDS** (or-bi-t*o*-SF*E*-noyd) in the caudal wall of the orbits. The optic foramen for the passage of the optic nerve is in the base of the lesser **PTERYGOID** (TER-i-goyd). **CRANIAL CLINOID** (KLI-n*o*yd) **PROCESSES** project caudally from the lesser pterygoid wing just above and lateral to the optic foramina. The **greater wings** are **alisphenoids** (al-*e*-SF*E-noyd)* extending laterally from the caudal (basisphenoid) portion of the bone. **CAUDAL CLINOID PROCESSES** project laterally from the superior ridge of the basisphenoid.

   The gap between the lesser and greater wings is the **SUPERIOR ORBITAL FISSURE** (*O*R-bi-tl FISH-*u*r) (figure 3.5B) for the passage of nerves and blood vessels that serve the muscles of the eyeball. The greater wings are each penetrated by a foramen, the **FORAMEN OVALE** (*O*-val), for the passage of the mandibular branch of the trigeminal nerve (figure 3.6).

   The **PTERYGOID PROCESSES** are vertical shelves of bone extending caudally from the palatines to the petrous portion of the temporal bone. The **lateral pterygoid processes** are the site of origin for the external pterygoid muscles; the **medial pterygoid processes** form the lateral wall of the nasopharynx (n*a*-so-FAR-inks) medially and the origin (together with the pterygoid fossa) for the internal pterygoid muscles. The pterygoid fossa is the depression between the lateral and medial pterygoid processes. A large foramen, the **FORAMEN LACERUM** (las-*E*R-um), opens between the petrosal and the caudal border of the basisphenoid wings.

B. **SPLANCHNOCRANIUM** (figures 3.3–3.7). Facial bones surrounding the digestive and respiratory systems:

   1. **NASAL** (NA-zal) **BONES.** Paired dermal bones covering the roof of the nasal chamber.
   2. **INCISIVE** (in-S*I*-siv) **BONES.** Have sockets for the upper **INCISOR** (in-S*I*-zer) **TEETH.** The openings to the nasal chambers, **EXTERNAL NARES** (N*A*-r*e*z), are surrounded by the nasal bones dorsally and the incisive bones laterally and ventrally. Caudally, the incisives articulate with the **MAXILLARY** (MAK-si-ler-*e*) **BONES.**

> The incisive bones are lacking in the adult human, although some anatomists claim they are fused with the maxillae. Embryologists point out that there is no ossification center for the incisive bone and therefore the human incisor teeth are socketed in the maxilla.

   3. **MAXILLA** (mak-SIL-ah). The large, paired dermal bones caudal to the incisive bones and with sockets for the upper canines, premolars, and molars. Laterally, the maxilla have a short but broad **zygomatic process** that articulates with the zygomatic bone. An **INFRAORBITAL** (in-fra-OR-bi-tl) **FORAMEN** opens through the maxilla between the zygomatic process and the border of the orbit. Nerves and blood vessels serving the face pass through the foramen.
   4. **LACRIMALS** (LAK-ri-mal). Small, paired dermal bones at the rostral dorsal corner of the orbit, just dorsal to the arch of the

infraorbital foramen. A **LACRIMAL CANAL** pierces this bone and allows the passage of the lacrimal duct from the eye to the nasal chamber.

5. **ZYGOMA** (z*i*-G*O*-ma) **(malar** (M*A*-lar), **jugal** (JOO-gal) or **zygomatic bone).** The broad, central piece of each zygomatic arch between the maxilla and squamosal. A dorsal postorbital process is directed toward the frontal postorbital process. A small foramen (not in the table of foramina) for blood vessels occurs in the base of the postorbital process.
6. **VOMER** ([V*O*-mer] figures 3.5B, 3.7A). An unpaired endochondral bone on the dorsal surface of the palatine bones and supporting the perpendicular plate of the ethmoid bone (see sagittal view of skull).
7. **ETHMOID** ([ETH-moyd] figure 3.7). Consists of a central **PERPENDICULAR PLATE** set in the vomer and thin **TURBINATE** (TUR-bi-n*a*t) **BONES** on the sides of the perpendicular plate. The caudal, perforated surface is called the **CRIBRIFORM** (KRIB-ri-form) **PLATE.**
8. **PALATINES** ([PAL-a-t*i*nz] figure 3.6). Paired dermal bones forming the rostral roof of the mouth (the hard palate).
9. **MANDIBLES** ([MAN-di-blz] figure 3.7A) of the cat are paired dermal bones that articulate with each other in the rostral midline by the **mandibular symphysis** (SIM-fi-sis), allowing limited movement between the two bones. In the human, the two mandibles are fused together so the two halves move as a single unit.

   All of the lower teeth are socketed in the mandible. Caudally, there is a dorsal **CORONOID** (KOR-*o*-noyd) **PROCESS** for attachment of the temporalis muscle and a **CONDYLOID** (KON-di-loyd) **PROCESS** for articulation with the squamosal (figure 3.2). A single pair of lower dermal jawbones is a primary mammalian characteristic. **MENTAL** (MEN-tl) **FORAMINA** occur on the rostral lateral surface of the mandible and a **MANDIBULAR FORAMEN** is at the medial base of the condyloid process.
10. **HYOID** ([H*I*-oyd] figure 3.9). The hyoid complex of the cat consists of eleven bones, ten paired units, and an unpaired **basihyal** (b*a*-se-H*I*-al). The cranial horn has four pairs of bones: a **ceratohyal** (articulating with the basihyal), **epihyals** (ep-i-H*I*-alz), **stylohyals** (st*i*-lo-H*I*-alz), and **tympanohyals** (tim-pa-no-H*I*-alz) (attached to a socket in the tympanic bulla). The caudal horn consists of a single pair of **thyrohyals** (th*i*-ro-H*I*-alz) between the basihyal and the thyroid cartilage of the larynx (see figure 3.9). The human hyoid is a single bone. The posterior horns are fused with the basihyal and the short, anterior processes are attached by ligaments to the styloid process of the temporal bone.

C. **TEETH** (figure 3.10). Modern mammals have lost some teeth and altered the shape of others, which accommodates their special feeding habits. The cat has **PREMOLAR** (PR*E*-m*o*-lar) and **MOLAR** (M*O*-lar) teeth that are cutting teeth rather than grinders, since their food is primarily animal flesh rather than seeds and other plant material that require grinding.

D. Midsagittal Section of Skull (figure 3.7)

   In the midsagittal section, the skull consists of a caudal **cranial cavity** and a rostral **nasal chamber.** The **HARD PALATE** (PAL-at), consisting of **INCISIVE, MAXILLARY,** and **PALATINE BONES,** forms the floor of the nasal chamber. The **ETHMOID BONE** occupies the greater portion of the nasal chamber and the **PERPENDICULAR PLATE** of the ethmoid divides the chamber into right and left halves. Air passages between the ethmoid and palate connect the rostral external nares with the **CAUDAL CHOANAE** ([ko-*A*-ne] figure 3.6). The choanae are seen as the terminal gap between the palatines and the **presphenoid bone.**

   The foramina (see table, Foramina of the Skull) for the passage of nerves and blood vessels are best seen from a median view of the sagittally sectioned skull. The **CRIBRIFORM PLATE (lamina cribrosa)** of the ethmoid (see 7 under splanchnocranium) allows passage of the olfactory nerves from the nasal epithelium covering the ethmoid bones to the olfactory bulb.

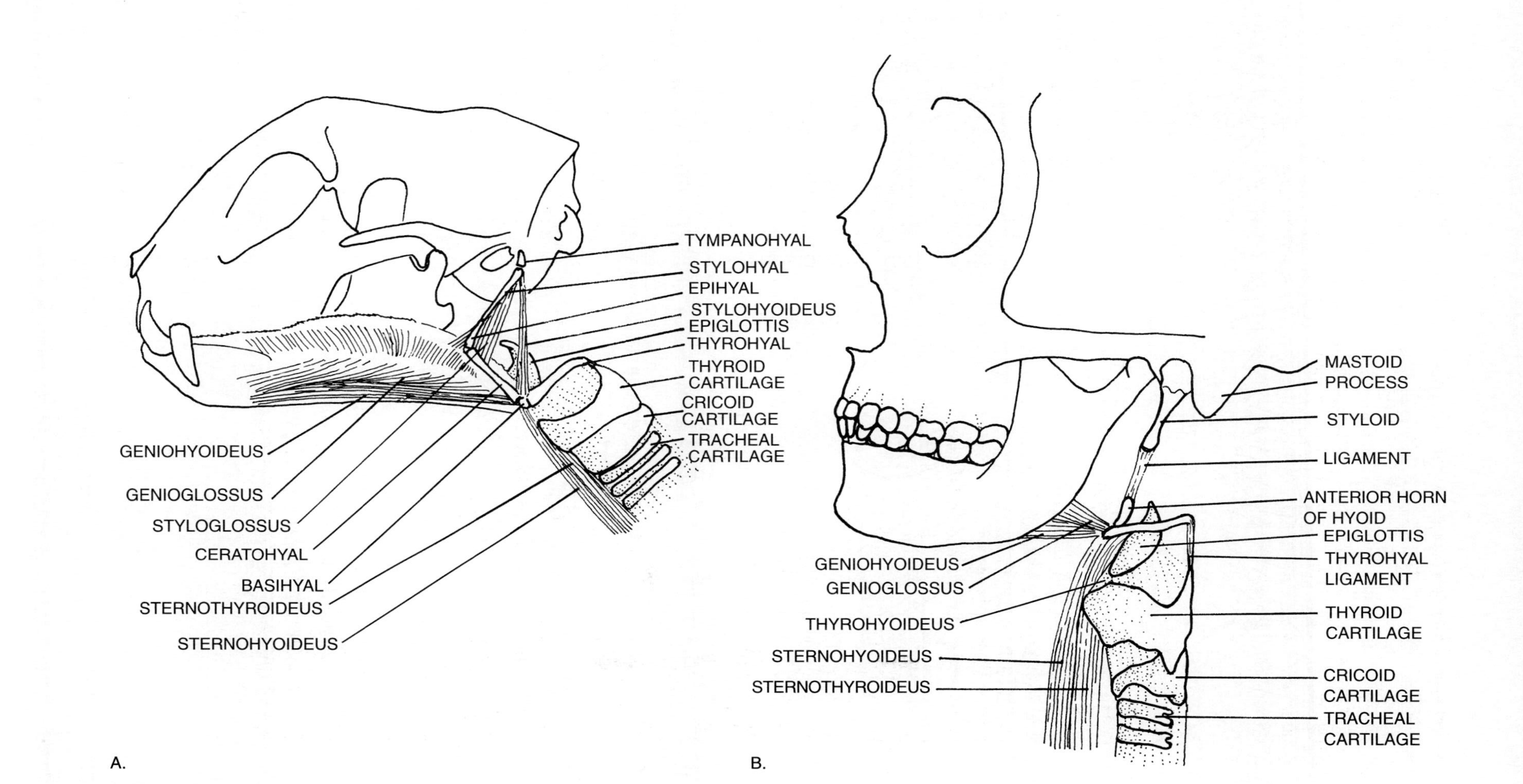

FIGURE 3.9
Lateral views of the cat (A) and human (B) hyoid apparatuses.

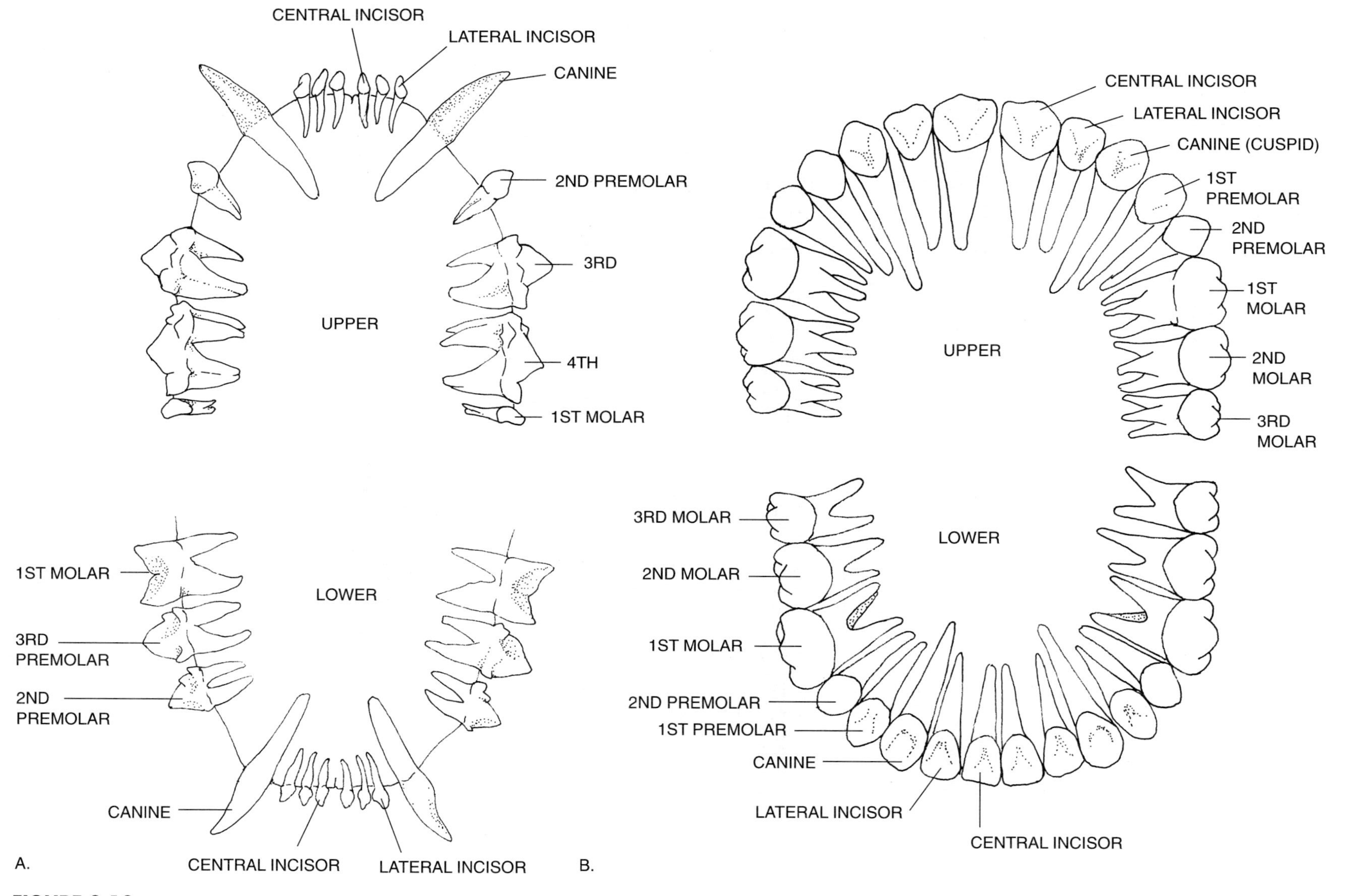

FIGURE 3.10

Dental patterns of the cat (A) and human (B). The primitive condition in mammals is 3/3 incisors, 1/1 canine, 4/4 premolars, and 3/3 molars. Felidae have 3/3, 1/1, 2 or 3/3, and 1/1. In cats the first premolar is absent on the maxilla and the first and fourth on the mandible. The caudal three molars are lost on both jaws. Therefore the premolars are numbered two, three, and four on the upper jaw and two and three on the lower.

Foramina of the Skull

Name	Bone(s) Penetrated	Traversing Structures
Infraorbital	Maxilla, ventromedial to orbit	Trigeminal and orbital nerves and blood vessels
Lacrimal	Lacrimal	Tear duct from orbit to pharynx
Mental	Mandible, usually two; on lateral body (shaft)	Mental nerves and blood vessels
Mandibular	Mandible, medial surface, caudal end	Inferior alveolar nerve and vessels
Optic	Presphenoid, center of orbitosphenoid portion	Optic nerve
Orbital fissure	Presphenoid, basisphenoid, fissure just caudal to optic, posterior wall of orbit	Oculomotor, trochlear, abducens, and ophthalmic (trigeminal) nerves
Incisive	Incisive, maxilla, paired, between canine teeth	Palatine vessels and trigeminal and nasopalatine nerves
Palatine	Palatine, maxilla, horizontal plates	Palatine vessels and nerves (trigeminal)
Sphenopalatine	Palatine, vertical wing, large	Sphenopalatine nerves and blood vessels
Caudal palatine (canal)	Palatine, cranial to sphenopalatine, small	See palatine
Rotundum	Basisphenoid, alisphenoid portion caudal to orbital fissure	Trigeminal (maxillary) nerve
Ovale	Basisphenoid, alisphenoid caudal to rotundum	Trigeminal (mandibular) nerve
Lacerum	Between basisphenoid and bulla of temporal, lateral to auditory tube foramen	Tympanic branches of carotid
Hypoglossal (canal) (h*i*-p*o*-GLOS-al)	Occipital, just lateral to the cranial end of the condyle	Hypoglossal nerve
Stylomastoid (st*i*-l*o*-MAS-toyd)	Temporal, between mastoid and styloid processes	Facial nerve
Condyloid (canal) (KON-di-loyd)	Occipital, caudal to condyles, lateral to foramen magnum	Emissary vein
Magnum	Occipital, center of the bone	Spinal cord joins the medulla, accessory nerves, vertebral arteries
Opening of auditory tube	Bulla of temporal, medial to lacerum	Auditory tube from middle ear
Jugular	Temporal (bulla) and occipital	Glossopharyngeal, vagus, accessory nerves, and internal jugular vein
Olfactory	Ethmoid, cribriform plate	Olfactory nerves
Facial canal	Temporal (petromastoid), caudolateral process of bone, medial surface	Facial nerve
Internal acoustic meatus	Temporal (petromastoid), medial surface	Facial and acoustic nerves and internal acoustic artery

Caudal and ventral to the **cribriform plate** are the **OPTIC FORAMINA** (right and left) for passage of the optic nerves.

The **ORBITAL FISSURE** is lateral and caudal to the optic for the passage of the oculomotor, trochlear, and abducens nerves.

LACERATED and **AUDITORY TUBE FORAMINA** occur between the bulla and the basisphenoid and the **JUGULAR FORAMEN** is between the bulla and the basioccipital bone. The ninth, tenth, and eleventh cranial nerves exit through the jugular foramen. The auditory tube opens through the rostral median corner of the bulla and is bordered caudally by the **PETROSAL BONE.** Between the rostral tips of the auditory bullae is a depression in the basisphenoid, the **SELLA TURCICA,** which holds the pituitary gland.

Dorsal to the bulla, the petrosal bone has two foramina. The **FACIAL** (FASH-*e*-al) **CANAL** is dorsal and caudal to the **INTERNAL ACOUSTIC MEATUS;** the large recess for the paraflocculus is above those two openings. The facial canal for the passage of the facial nerve is a groove in the petrosa covered by the portion of the tympanic bone just caudal to the external auditory meatus. The facial canal is labeled in figure 3.7A. It lies between the paramastoid process and the tympanic bulla. The trigeminal nerve passes through a canal at the rostral medial surface of the petrosa.

Vertebrae, Ribs, and Sternum

A. Vertebral Column

The following describes a typical **VERTEBRAE** (VER-te-br*a*) (see figure 3.14). On one of the **LUMBAR** (LUM-bar) **VERTEBRAE,** locate the **CENTRUM** (SEN-trum) (large, ventral body portion) **NEURAL** (N*U*-ral) **CANAL** (through which the spinal cord passes), **NEURAL SPINE** or **spinous process** (extending dorsally from the neural canal), **TRANSVERSE PROCESSES** (ventral-lateral outgrowths at the base of the neural canal), and the **INTERVERTEBRAL** (in-ter-VER-te-bral) **FORAMINA** (narrowed part of the neural arch near the point of attachment to the centrum), where the spinal nerves branch out from the spinal cord (figures 3.11–3.12).

1. **CERVICAL** (SER-vi-kl) **VERTEBRAE** (figure 3.13). There are seven cervical vertebrae in mammals. The first [**ATLAS** (AT-las)] and second [**AXIS** (AKS-is)] are so distinctive that they have been given separate names. The atlas has a very broad transverse process and large **CRANIAL ARTICULAR PROCESSES** for articulation with the **occipital condyles** of the skull. The atlas lacks a dorsal spinous process and the centrum lacks the central portion. The axis, in turn, appears to compensate for the missing parts of the atlas. The axis has a very large forward projecting spine that extends above the neural arch of the atlas. The centrum of the axis has a forward projecting **ODONTOID** (*o*-DON-toyd) **PROCESS** that fits into the atlas.

   Each cervical vertebra, except the seventh, may be recognized by the **TRANSVERSE FORAMEN,** a small hole on either side of the vertebral canal for the passage of the vertebral artery and vein.
2. **THORACIC** (th*o*-RAS-ik) **VERTEBRAE** (figure 3.14). There are thirteen of these in the cat (twelve in the human), which may be identified by the long dorsal spine and by ribs (costae) that are attached to them. The first thoracic vertebra is similar to the seventh cervical except the thoracic has attachments for the first pair of ribs.

The first seven pairs of ribs (figures 3.11, 3.22) are attached directly to the sternum by costal cartilages (**vertebrosternal** ver-te-br*o*-STER-nl) and are termed **TRUE RIBS.** The next three pairs of ribs are joined to the preceding costal cartilages, rather than to the sternum. These are called **FALSE RIBS** (**vertebrochondral** ver-te-br*o*-KON-dral). The last three pairs (two in the human) of ribs do not join the sternum or the other ribs, and are called **FLOATING RIBS (vertebral).**

Each rib is composed of a vertebral portion (which is ossified or bony) and a cartilaginous portion. The vertebral portion is made up of the head or **CAPUT** (KAP-ut) (articulating with the centrum), a **TUBERCULUM** (tyoo-BER-k*u*-lum) (articulating with the transverse process), a **NECK** (narrow part between the caput and the tuberculum), and the **SHAFT** (longer portion from the tuberculum to the ventral end of the rib). There are neither necks nor tuberculae on the "floating" pairs of ribs.

Usually the head attachment is between the centra so there is a **hemifacet** (hem-*e*-FAS-et) on both borders of the vertebra, but the head of the first rib has its facet entirely

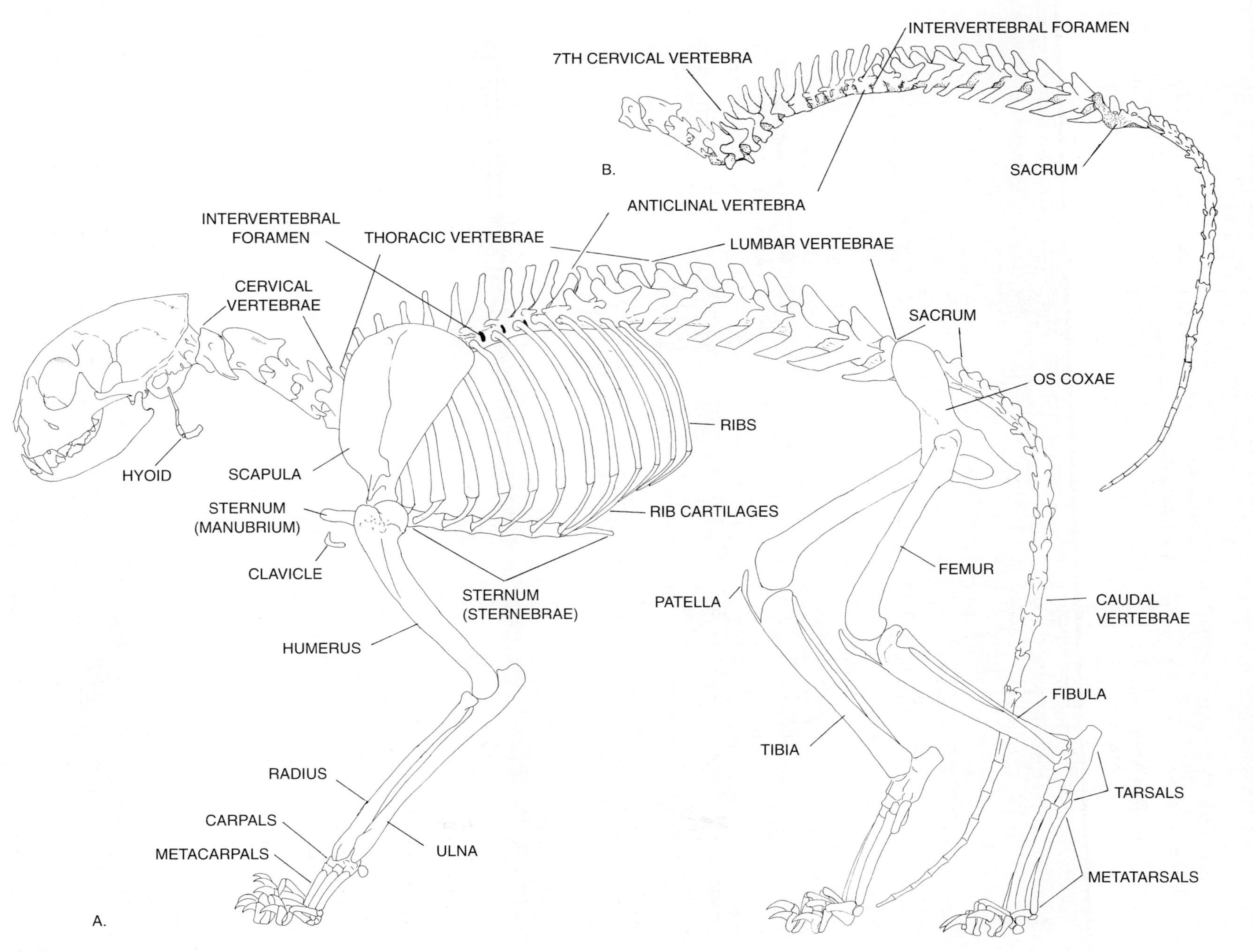

FIGURE 3.11
Lateral views of the cat skeleton (A) and vertebral column (B).

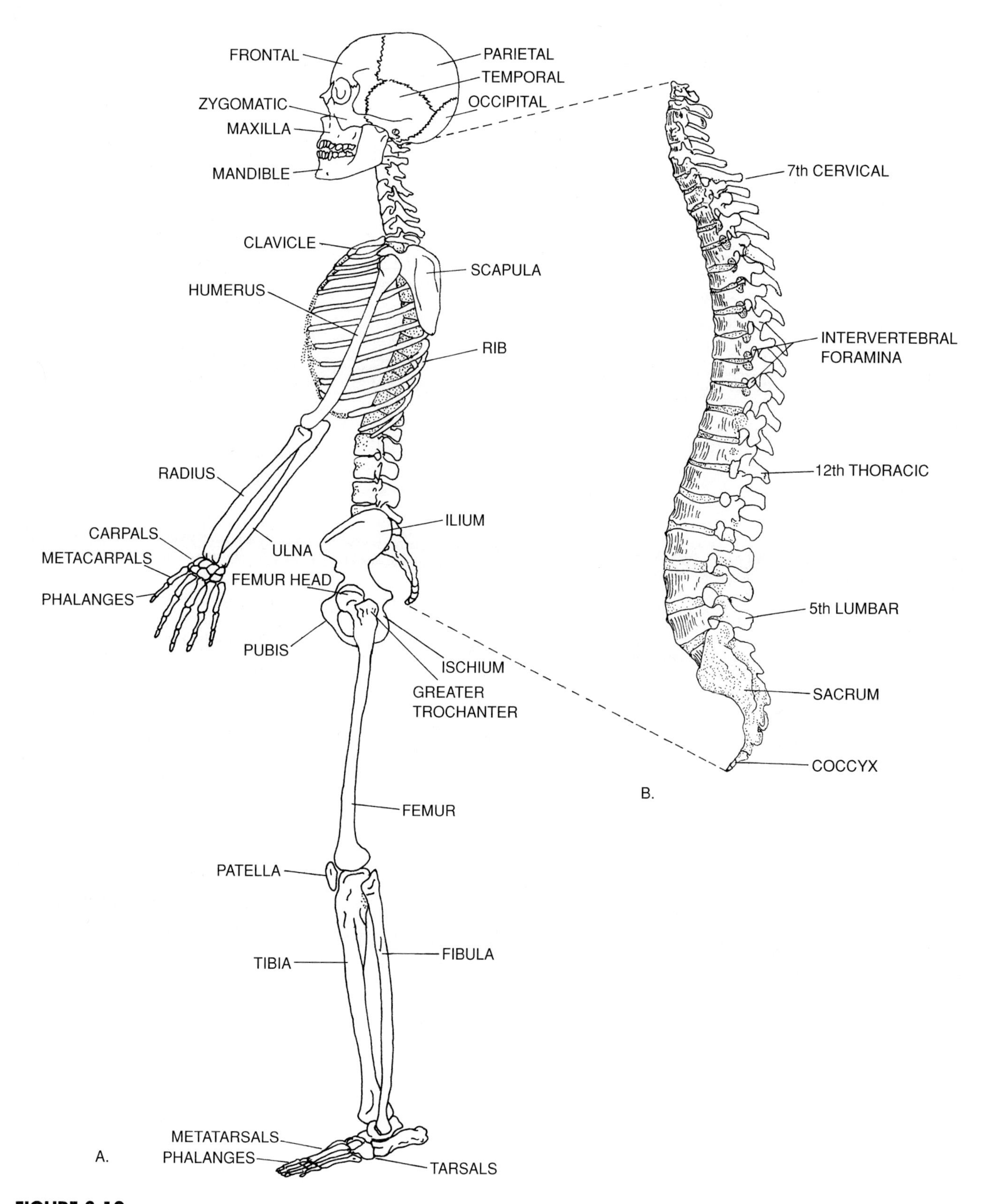

FIGURE 3.12
Lateral views of the human skeleton (A) and vertebral column (B).

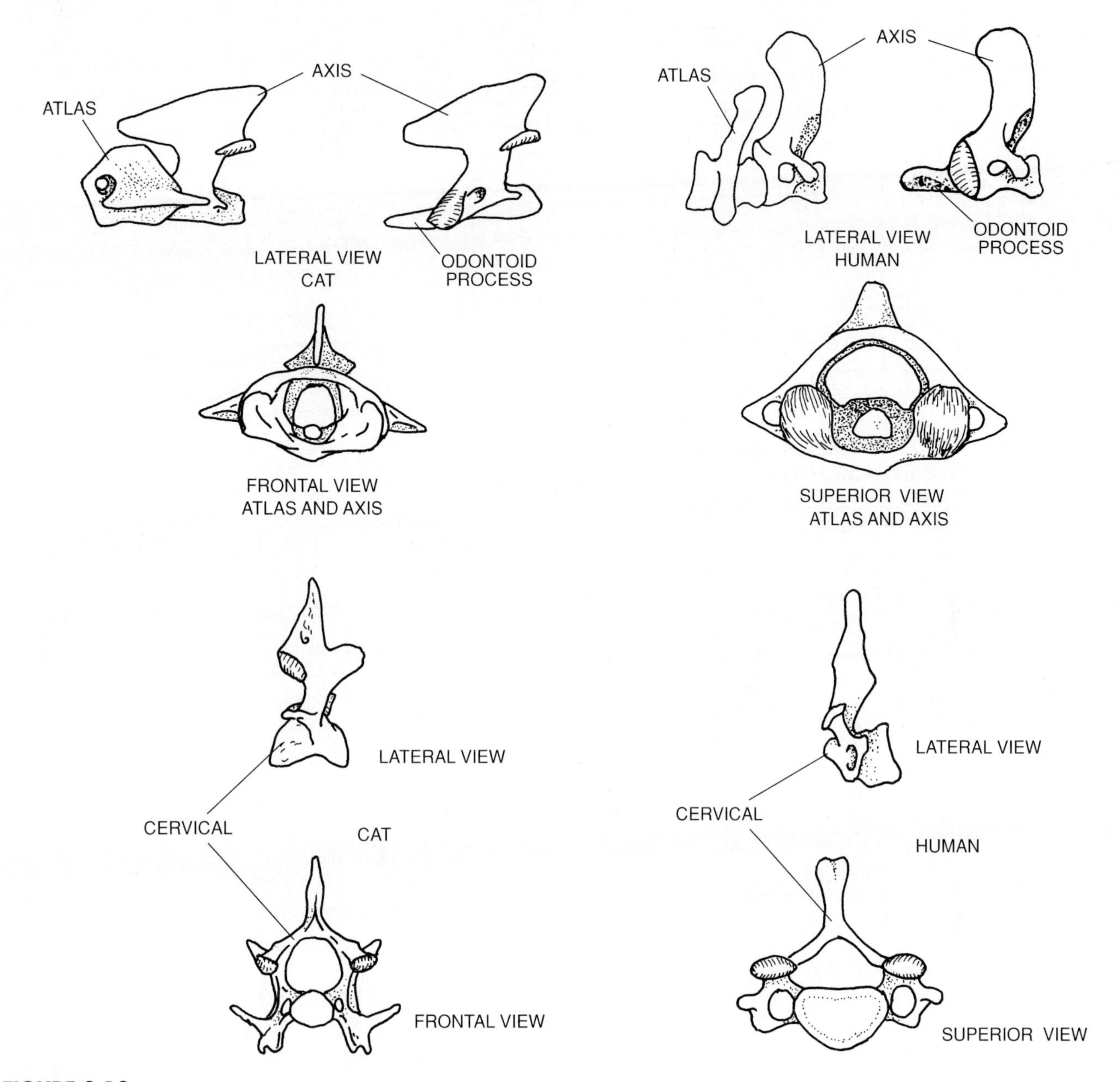

FIGURE 3.13
Atlas and axis and typical cervical vertebrae of the cat and human, lateral (top) and frontal (lower) views.

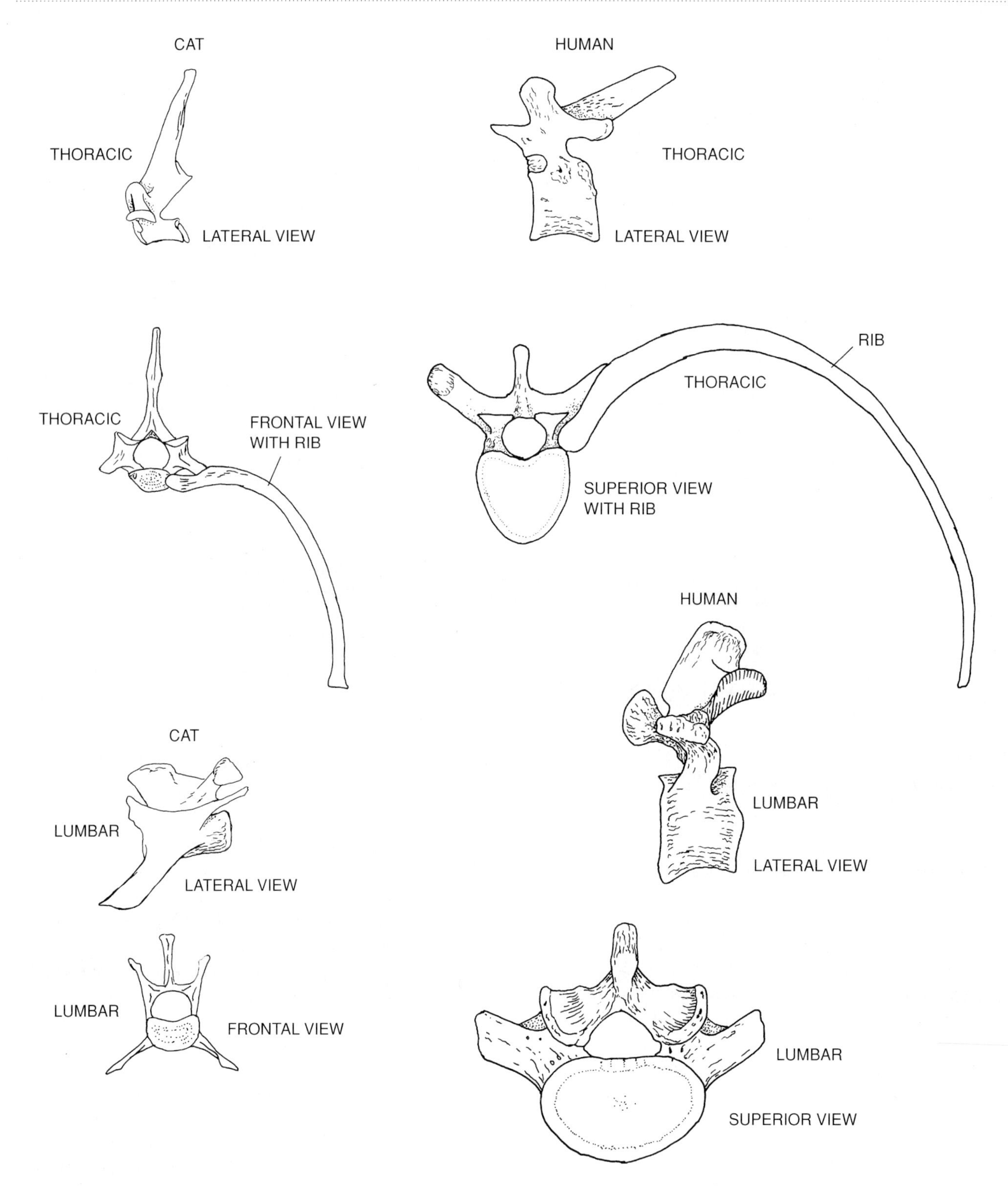

FIGURE 3.14
Typical thoracic and lumbar vertebrae of the cat and human, lateral (top) and frontal superior (lower) views.

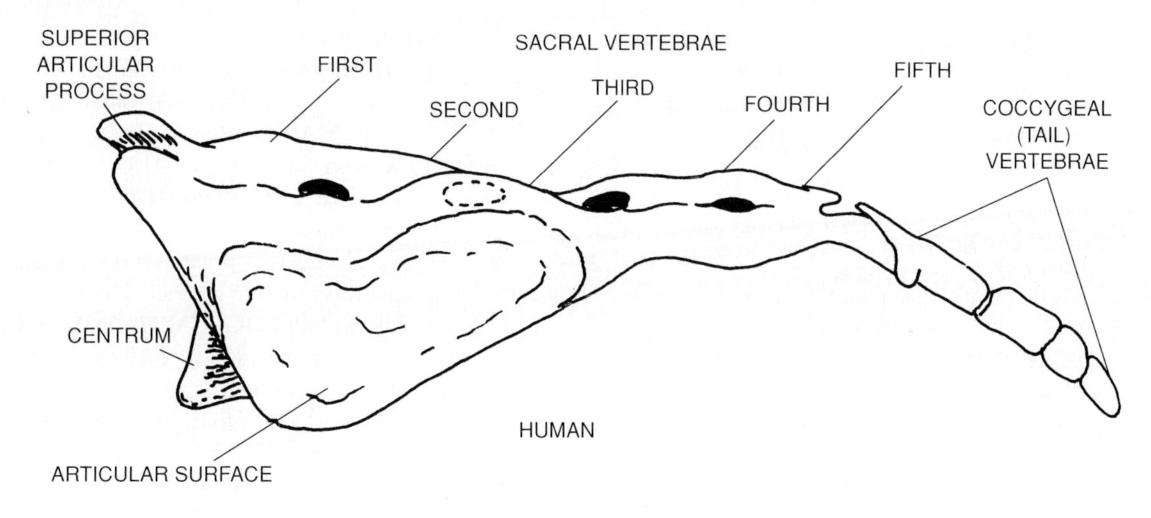

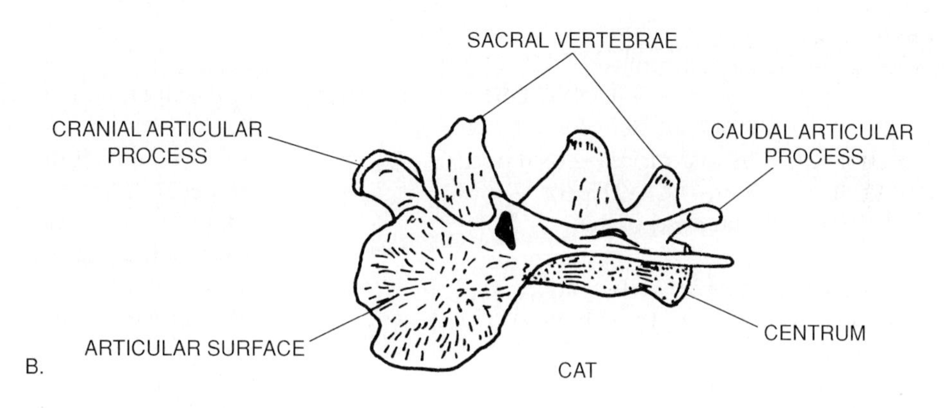

FIGURE 3.15
Lateral views of the cat (A) and human (B) sacral vertebrae.

on the first thoracic vertebra and the last three ribs have their heads attached entirely to their respective vertebrae as well.

3. **STERNUM.** The sternum of the cat (figure 3.11) is composed of six segments called **STERNEBRAE** (STERN-e-br*e*). The cranial sternebrium is the **manubrium** (ma-NYOO-br*e*-um) and the caudal one is the **xiphoid** (ZIF-oyd) **process** (processus xiphoideus). Caudal to the xiphoid is a **xiphisternal** (zif-i-STER-nal) **cartilage** (cartilago xiphoidea). In the human, the sternebrae between the manubrium and xiphoid process are fused into a single **body** (see figure 3.22).
4. **LUMBAR** (LUM-bar) **VERTEBRAE** (figure 3.14). There are six lumbar vertebrae (five in humans), readily identified by the large transverse process which projects ventrally and cranially.
5. **SACRAL** (S*A*-kral) **VERTEBRAE** or **SACRUM** ([S*A*-krum] figure 3.15). A single bone in adults, which is formed by the union of four sacral vertebrae (five in humans).
6. **CAUDAL [COCCYX** (KOK-siks) in the human] **VERTEBRAE.** The number of these varies a great deal in the cat, depending upon the length of the tail, but there are usually 28. When disarticulated, they may be confused with the bones of the sternum and the phalanges unless one is careful to note the exact shape of each of these three types

of bones. Note the dumbbell shape of each caudal vertebra, caused by the wide projecting processes at each end of the bone. There are four fused vertebrae forming the coccyx in the human.

Appendicular Skeleton

A. Pectoral Girdle and Appendage
   1. Pectoral Girdle (figures 3.16–3.19)
      a. **SCAPULAE** (SKAP-*u*-*la*), paired shoulder blades. A spine divides the lateral surface into cranial [**SUPRASPINOUS** (syoo-prah-spi-NA-tus)] and caudal [**INFRASPINOUS** (in-fra-spi-NA-tus)] **FOSSA.** The ventral continuation of the spine is the **ACROMION** (a-KR*O*-m*e*-on) **PROCESS.** Opposite the acromion process, on the medial surface of the scapula, is the hook-shaped **CORACOID** (K*O*R-a-koyd) **PROCESS.** Between the acromion and coracoid processes is the **GLENOID** (GLEN-oyd) **CAVITY** for articulation with the head of the humerus.
      b. **CLAVICLES** (KLAV-i-klz) (clavicula), paired bones. In the cat, these bones do not articulate with any other bones but are embedded between the cleidotrapezius, cleidobrachialis, and cleidomastoideus muscles (see pages 53 and 57). In horses, antelope, deer, seals, and some others, the bone is entirely absent, but it articulates with the sternum and scapula in humans, rodents, and other mammals that need a firm support for the pectoral girdle.
   2. Pectoral Appendage
      a. **HUMERUS** (H*U*-mer-us). The proximal end of the humerus is characterized by a head [**CAPUT** (KAP-ut)] that articulates with the scapula and **GREATER** and **LESSER TUBEROSITIES** (tyoo-be-ROS-i-t*e*s) that serve for muscle insertions. The distal end is characterized by a two-planed articular surface; the rounded, lateral half is the **CAPITULUM** (ka-PIT-*u*-lum) for articulation with the radius, and the concave medial surface, the **TROCHLEA** (TR*O*K-l*e*-a), articulates with the ulnar semilunar notch.

         The cranial edge of the humerus has a large **DELTOID** (DEL-toyd) **TUBEROSITY** for insertion of the deltoid muscles (see p. 57). **EPICONDYLES** (ep-i-KON-d*i* lz) project laterally on each side of the distal end of the humerus.
      b. **RADIUS** (R*A*-d*e*-us). The forearm bone articulating with the capitulum of the humerus.
      c. **ULNA** (UL-na). The forearm bone articulating with the trochlea. The proximal end of the ulna has a **SEMILUNAR** (sem-*e*-LOO-nar) **NOTCH** (circumferentia articularis), which articulates with the humerus and a projection beyond the semilunar notch called the **OLECRANON** (*o*-LEK-ra-non).

In the human, the carpals are in two rows. The proximal row from the thumb side is the **scaphoid** (SK*A*F-oyd), **lunate** (LOO-n*at*), **triquetrum** (tr*i*-KWE-trum), and **pisiform** (pI-si-form). The distal row is the **trapezium** (trah-P*E*-*ze*-um), **trapezoid** (TRAP-e-zoyd) **capitate** (KAP-i-t*a* t), and **hamate** (HAM-*a* t).

Carpal Bones

Cat	Mammalian	Human
Radiale	Scaphoid	Scaphoid
________	Intermediate	Lunate
Ulnare	Triquetrum	Triquetrum
Accessory Carpale	Pisiform	Pisiform
Carpale I	Trapezium	Trapezium
Carpale II	Trapezoid	Trapezoid
Carpale III	Capitate	Capitate
Carpale IV	Hamate	Hamate
Radial Sesamoid	________	________

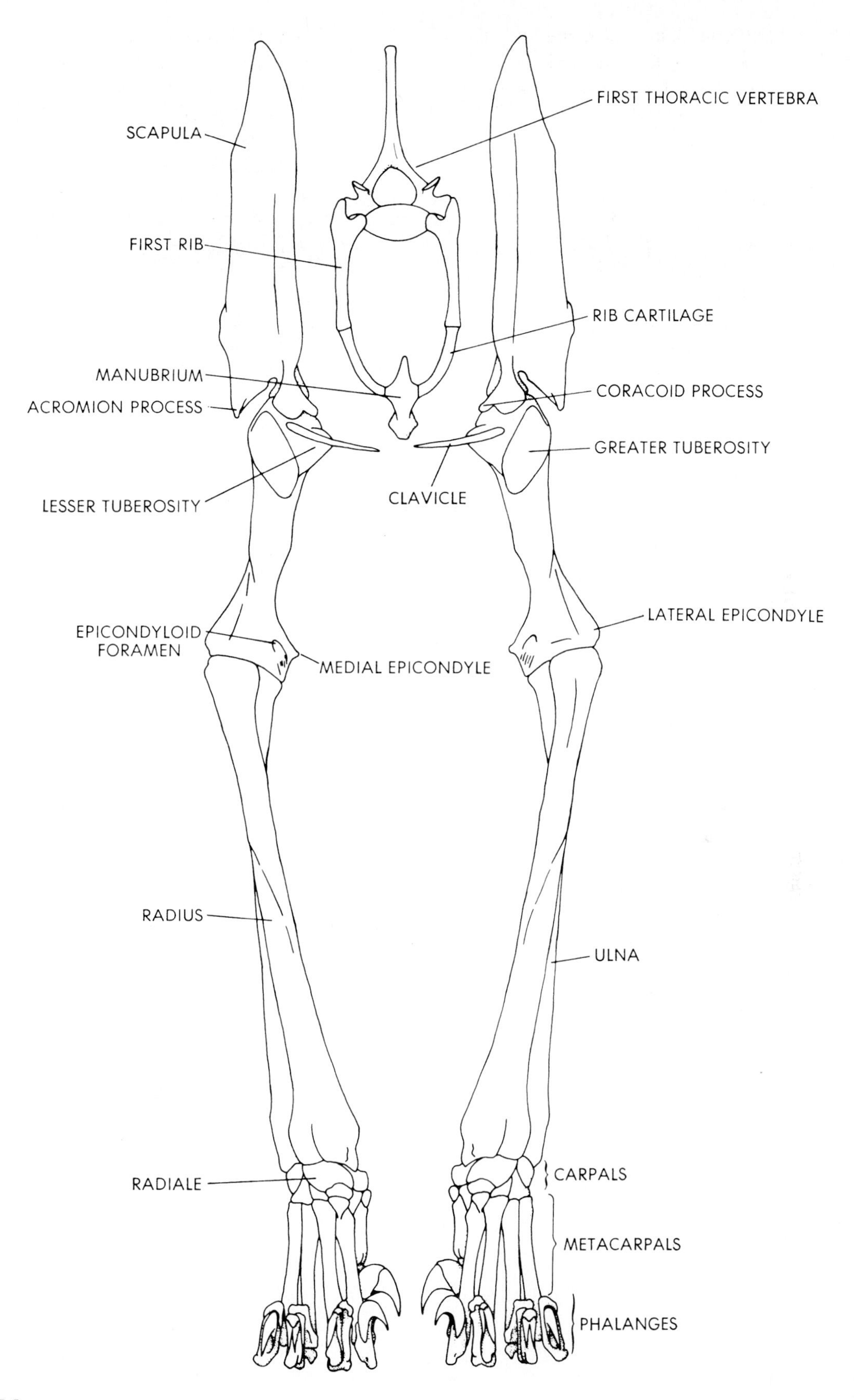

FIGURE 3.16
Pectoral girdle and limbs of the cat, frontal view.

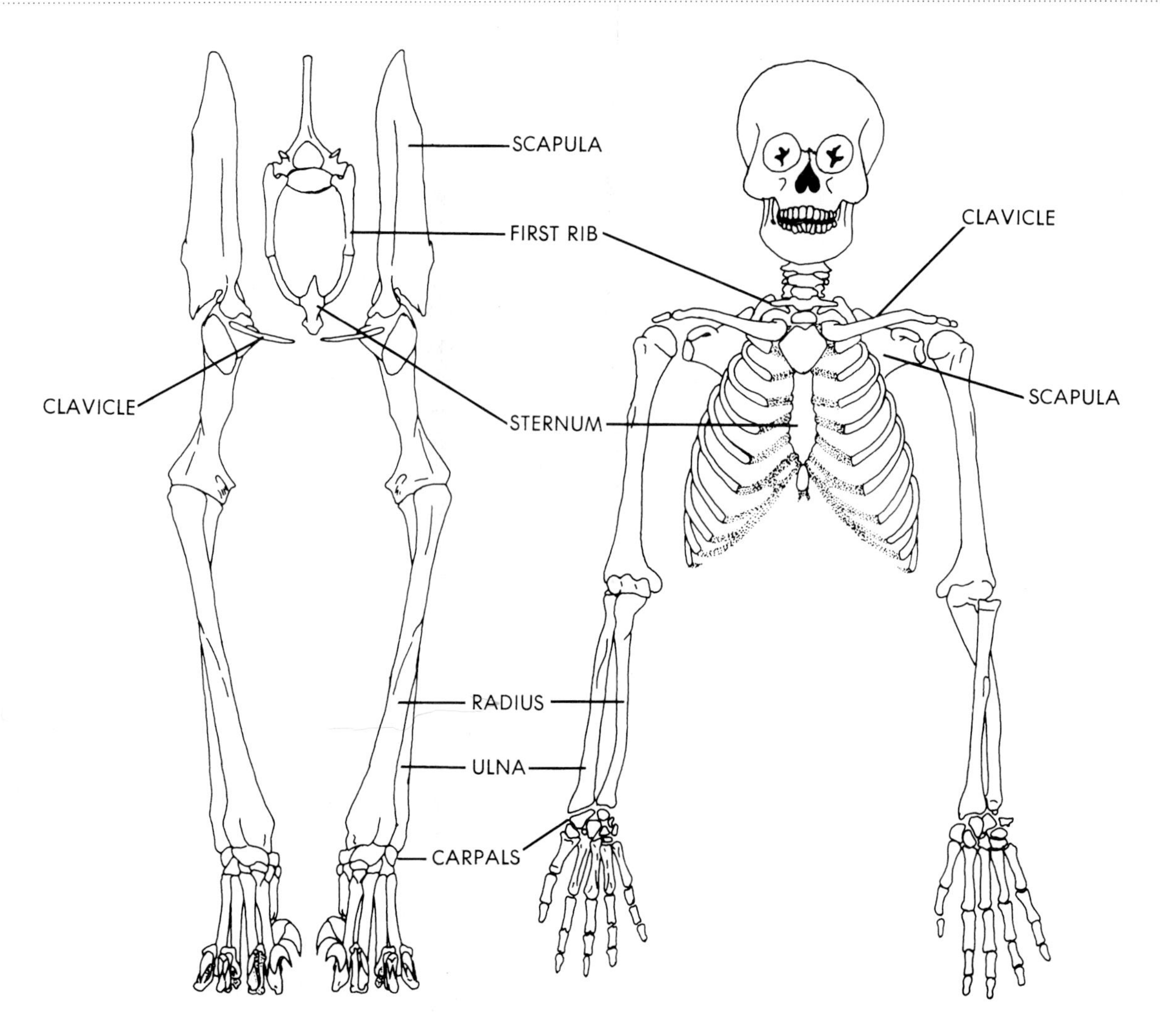

FIGURE 3.17
Comparison of human and cat girdles and limbs, frontal/anterior views.

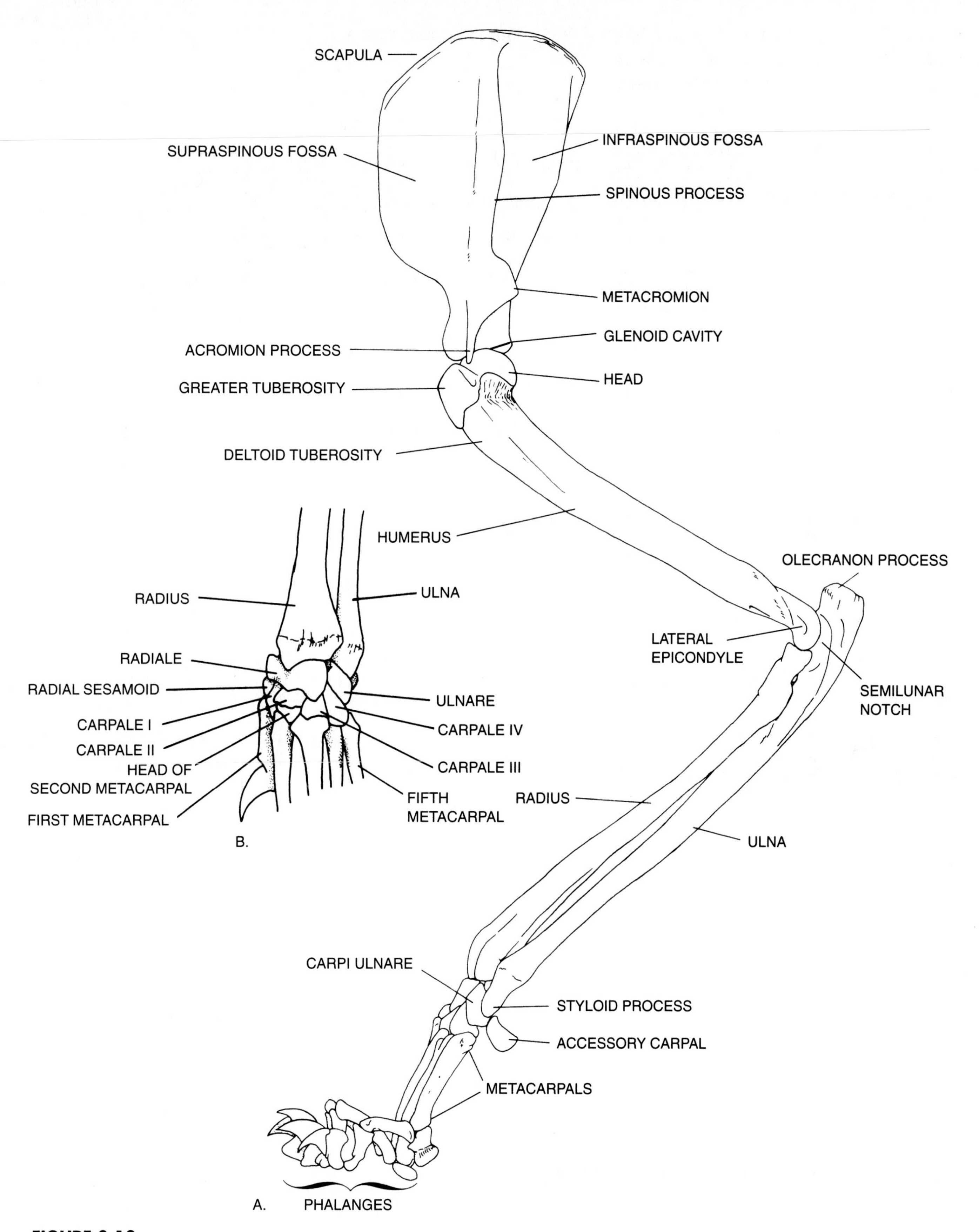

FIGURE 3.18
Pectoral girdle and limb of the cat, lateral view (A) and a frontal view of the cat carpal bones (B).

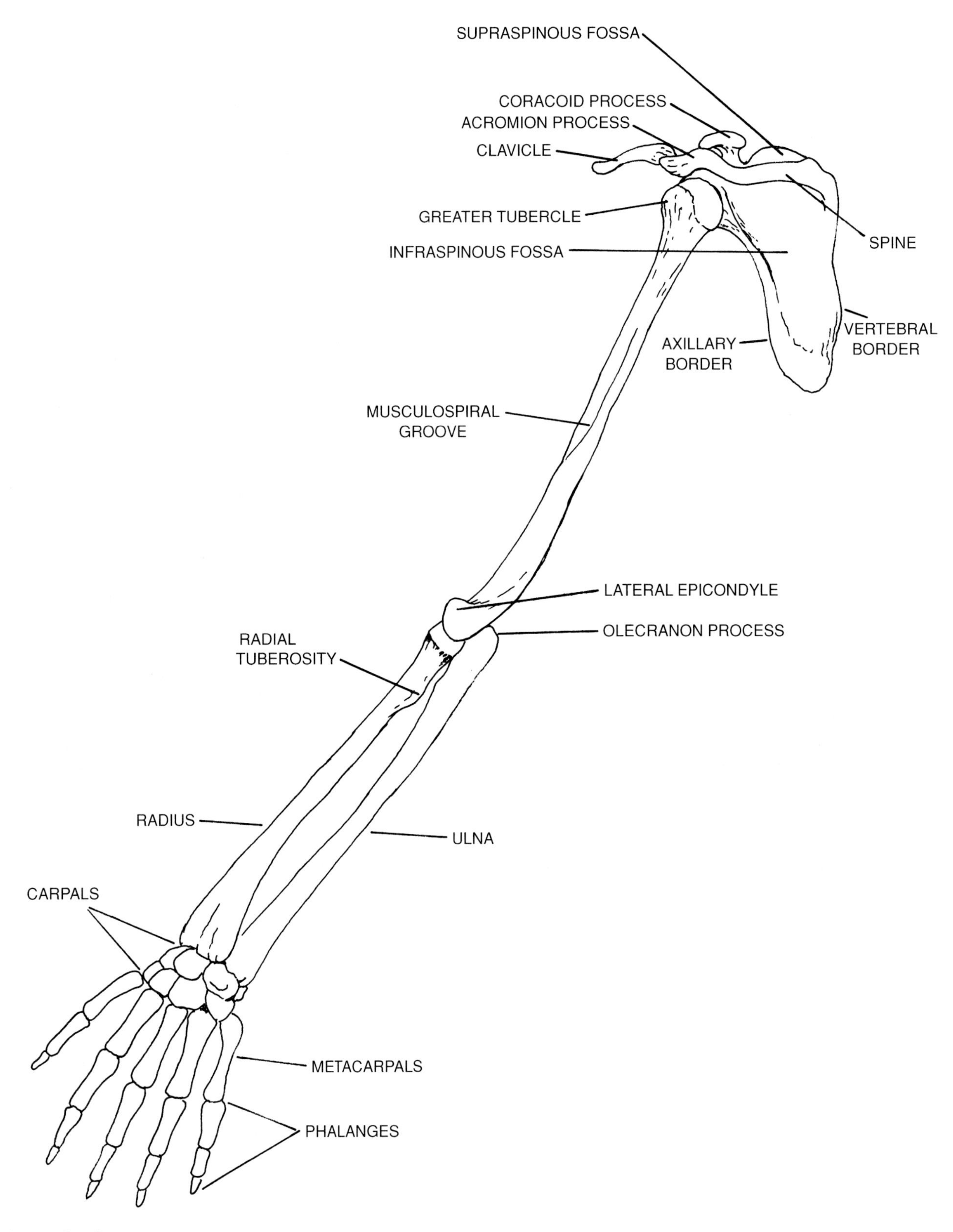

FIGURE 3.19
Pectoral girdle and limb of the human, lateral view.

d. **CARPAL** (KAR-pl) **BONES.** The arrangement of these wrist bones is illustrated in figures 3.18 and 3.19. In the cat, they consist of two proximal bones, the **radiale** (*ra*-de-AL-*e*) and **ulnare** (ul-N*A*-re), based on the radius and ulna respectively, and a **distal carpale series** (I-IV) on each metacarpal except the fifth. The **centrale** (sen-TRA-l*e*) is a small intermediate bone. The **pisiform** (PI-si-form) is a small lateral bone forming a brace for the lateral side of the wrist. See the table of carpal bones on page 34.

e. **METACARPALS** (met-a-KAR-plz). The bones of the **PAWS** (hands). There are five metacarpals in each **MANUS,** one for each digit, and numbered from the first digit (thumb) side

f. **PHALANGES** (fa-LAN-j*e*z). The bones of the digits. The first digit **pollex** (P*O*L-eks) has only two phalanges, but the other four digits each have three phalanges. All five digits of each hand have terminal nails (human) or claws (cat) on the distal phalanges.

B. Pelvic Girdle and Appendage (figures 3.20–3.23)

1. Pelvic Girdle

a. **OS COXAE** (os KOK-s*e*). Composed of four separate units that fuse together to form each bone. The two bones are joined in the ventral midline by the **PUBIC** (PYOO-bik) **SYMPHYSIS** and articulate dorsomedially with the sacrum by the **ILIOSACRAL** (il-*e*-*o*-S*A*-kral) **JOINTS.**

(1). **ILIUM** (IL-*e*-um). The most cranial portion of the os coxae. The most cranial tip of the ilium is flared laterally into a ventral iliac spine. The dorsal border has a **dorsal iliac spine** just above the iliosacral joint.

(2). **ISCHIUM** (IS-k*e*-um). The dorsocaudal portion of the os coxae. The dorsal edge of this bone is raised into an **ischiadic** (is-k*e*-AD-ik) **spine.** The ischium and pubis are separated by the large **obturator** (OB-tyoo-r*a*-tor) **foramen.**

(3). **PUBIS** (PYOO-bis). The ventrocaudal portion of the os coxae. The pubis is joined to its counterpart by the pubic symphysis. The ventral border of the pubis cranial to the symphysis and near the cranial border of the obturator foramen has a prominent **iliopectineal** (il-*e*-*o*-pek-TIN-*e*-al) **eminence** for the origin of the pectineus muscle (see p. 86).

(4). **acetabular** (as-e-TAB-*u*-lar) **bone** (condyloid bone). A small triangular bone within the cat acetabulum.

(5). **ACETABULUM** (as-e-TAB-*u*-lum). The socket for articulation with the head of the femur.

b. **FEMUR** (F*E*-mur). The large bone of the thigh. The proximal end has a head that articulates with the acetabulum and **GREATER** (major) and **LESSER** (minor) **TROCHANTERS** (tr*o*-KAN-terz) for muscle attachments. The head of the femur is set at an angle of about 45° to the shaft of the femur. The distal end has two **articular condyles** separated by an **intercondyloid fossa.**

c. **TIBIA** (TIB-*e*-a). The large medial bone of the leg. The proximal end has two articular condyles separated by a spine. The distal end is fused with the fibula.

d. **PATELLA** (pa-TEL-a). A sesamoid bone cranial to the knee joint; the "knee cap."

e. **FIBULA** (FIB-*u*-lah). A slender, splintlike lateral bone of the leg. This bone is fused distally with the tibia.

f. **TARSALS** (TAR-slz). The ankle bones illustrated in figures 3.20–3.22. There are seven tarsal bones (plus a small sesamoid) arranged roughly in the same pattern as the carpal bones. The two proximal bones are the **talus** (TA-lus) and **calcaneus** (kal-K*A*-n*e*-us), jointed to the ends (fused in the cat) of the tibia and fibula. The central bone is the **central tarsal (navicular** [na-VIK-*u*-lar]) and the distal series are tarsals I–IV (medial, intermediate, and lateral **cuneiform** [K*U*-n*e*-i-form] **bones** and the **cuboideum** [k*u*-BOY-d*e*-um]). See the chart of the tarsal bones.

g. **METATARSALS** (met-a-TAR-slz). The bones of the **FOOT** or **PES.** The cat has five metatarsals. The first, is greatly reduced in size.
h. **PHALANGES.** The bones of the digits. There are three phalanges in each digit. The hallux is absent in the cat. The remaining digits (II-IV) have proximal middle and distal phalanges.

On each foot, man has five digits (figure 3.22) with three phalanges each, except for the first digit [**HALLUX** (HAL-uks)]. The hallux has two phalanges since the distal phalanx is absent.

Articulations

Joints, or **ARTICULATIONS** (ar-tik-*u*-L*A*-shunz), are often thought of as the union of two bones. Actually, the skeleton originates as a single unit in the embryonic mammal and joints are separations between subunits. As pointed out in the introduction to this chapter, skeletal structures are constructed of three tissues: fibrous tissue, bone, and cartilage. Joints utilize all three tissues; resilient cartilage forms the contact surface of the joint and caps the firm, bony, lever arm. The flexible fibrous tissue holds the two contact surfaces together while allowing limited movement between them. The actual shape and construction of the joint surface is dependent on (1) the various muscle attachments that move the parts of the skeleton, (2) the vasculature to the different areas of the joint, and (3) the forces applied to specific regions of the joint by muscle contraction or body weight. Joints that allow movement are **SYNOVIAL** (si-N*O*-v*e*-al) **JOINTS** and classified in six major groups: **spheroidal** (ball and socket), **ellipsoidal, sellar** [(SEL-ar) saddle-shaped], **ginglymus** [(JIN-gli-mus) hinge], **trochoid** [(TR*O*-koyd) pivot], and **arthroidal** (plane) joints. Synovial joints are moved in one (sellar and ginglymus), two (ellipsoidal), or three (spheroidal) directions. Two excellent examples of synovial joints in mammals are (1) the spheroidal hip joint, which allows the maximum planes of movement, and (2) the knee, with special ligaments that limit movement to a single plane. Trochoid and arthroidal joints, which also permit movement in a single plane, are quite specialized. The trochoid is a pivot joint such as that between the atlas and axis, or between the radius and ulna. The arthroidal joint occurs at the wrist or ankle where the radius and ulna glide over the carpals and the tibia glides over the tarsals. Sellar and ginglymus joints allow movements in only one plane and therefore allow only flexion and extension of the joint (for example, between the humerus

Chart of the Tarsal Bones

Cat	Mammalian	Human
Talus	Talus	Talus
Calcaneus	Os calcis	Calcaneus
Central Tarsal	Navicular	Navicular
Tarsal I	Medial cuneiforme	Medial cuneiforme
Tarsal II	Intermediate cuneiforme	Intermediate cuneiforme
Tarsal III	Lateral cuneiforme	Lateral cuneiforme
Tarsal IV	Cuboideum	Cuboideum

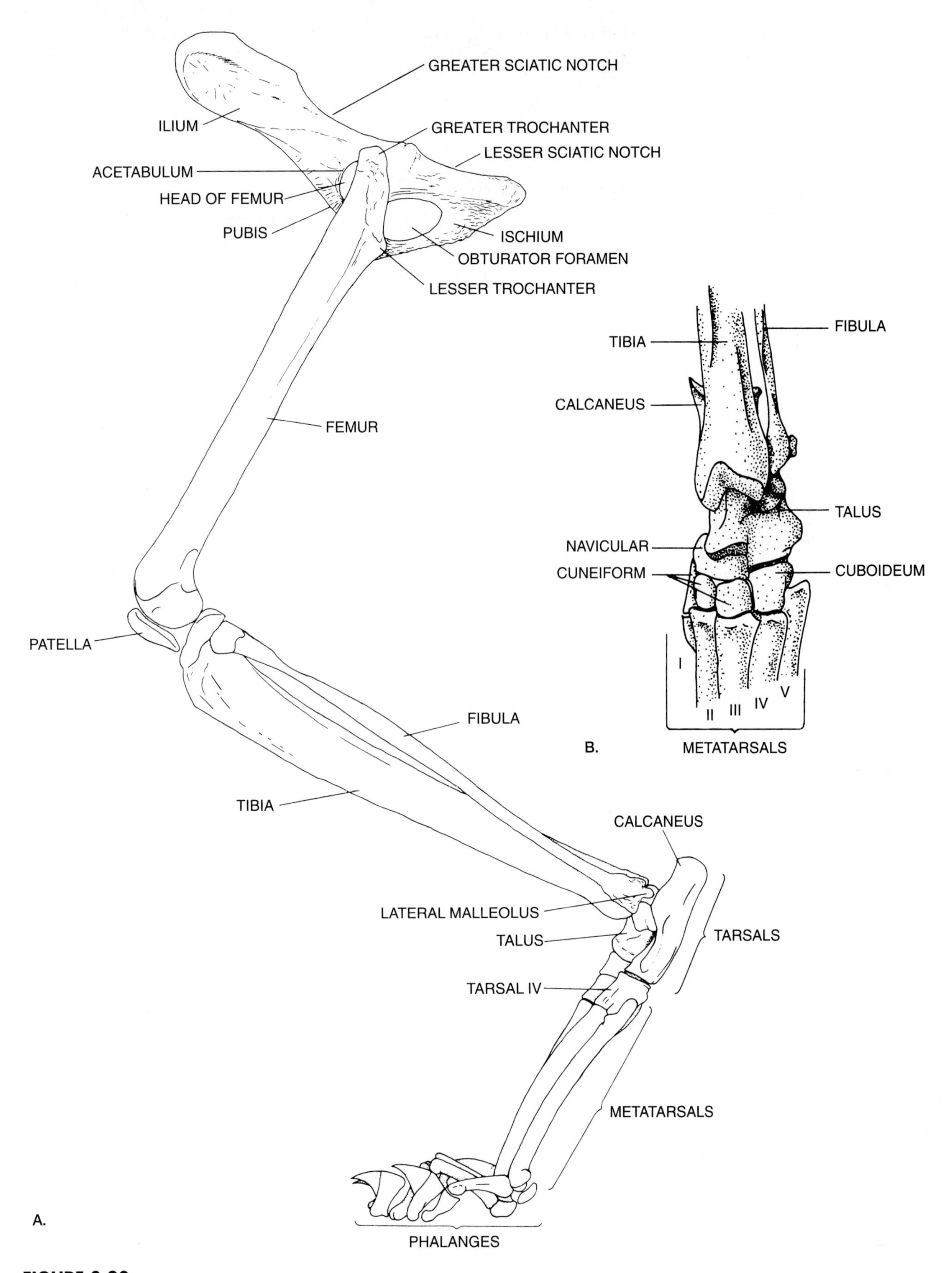

FIGURE 3.20
Lateral view of the cat pelvic girdle and limb (A) and frontal view of the tarsal bones (B).

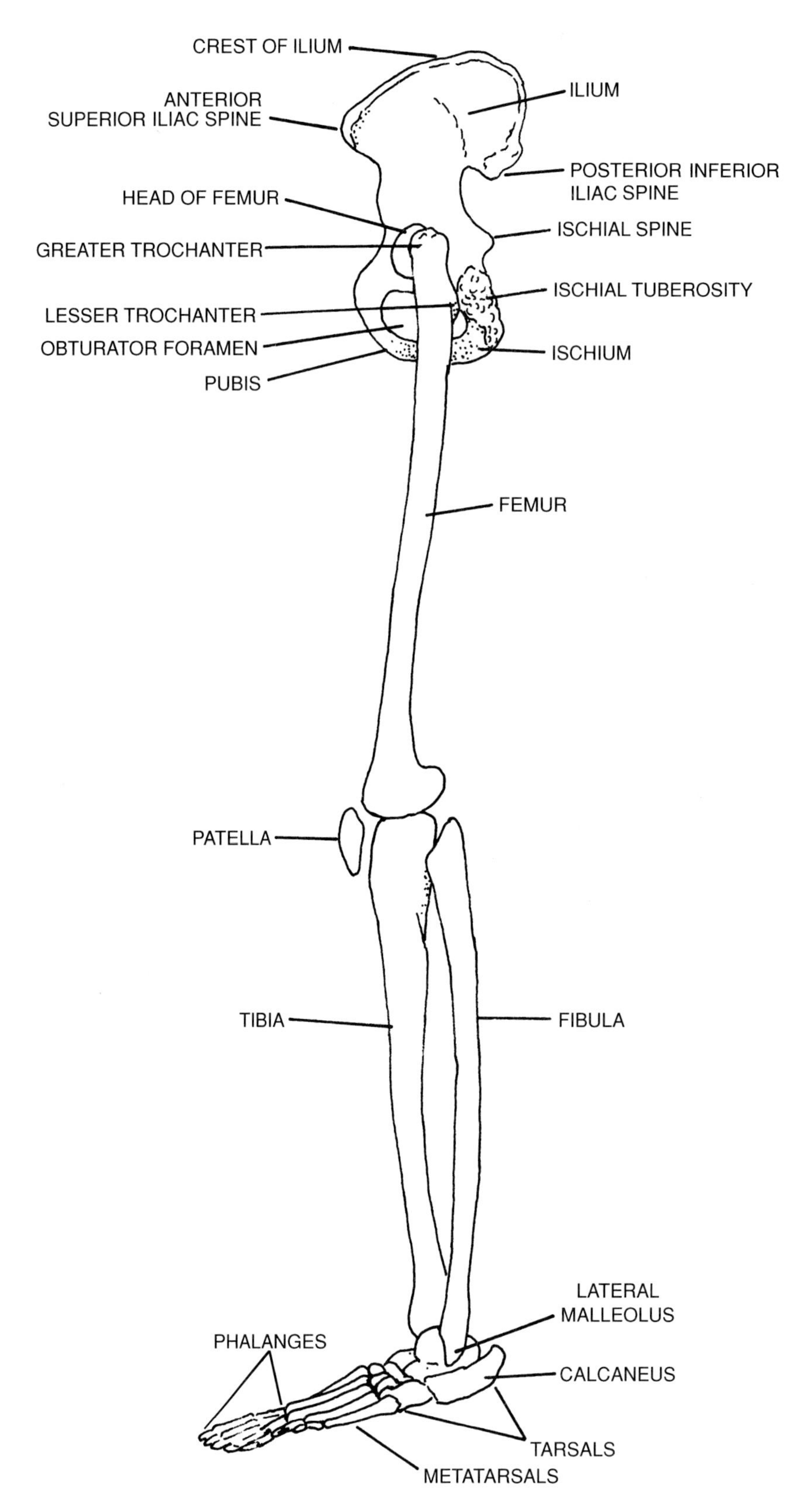

FIGURE 3.21
Lateral view of the human pelvic girdle and limb.

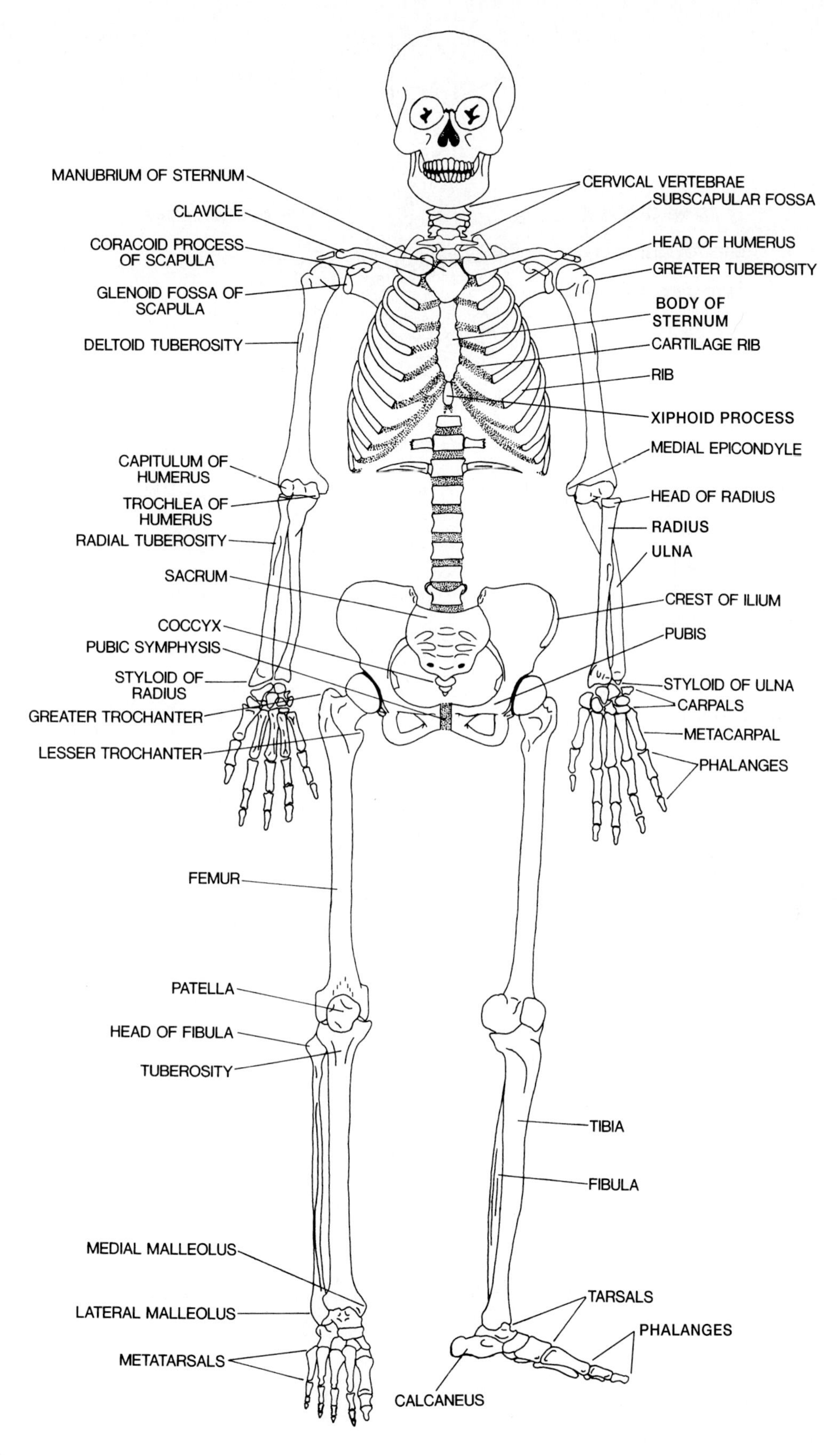

FIGURE 3.22
Anterior view of the human skeleton.

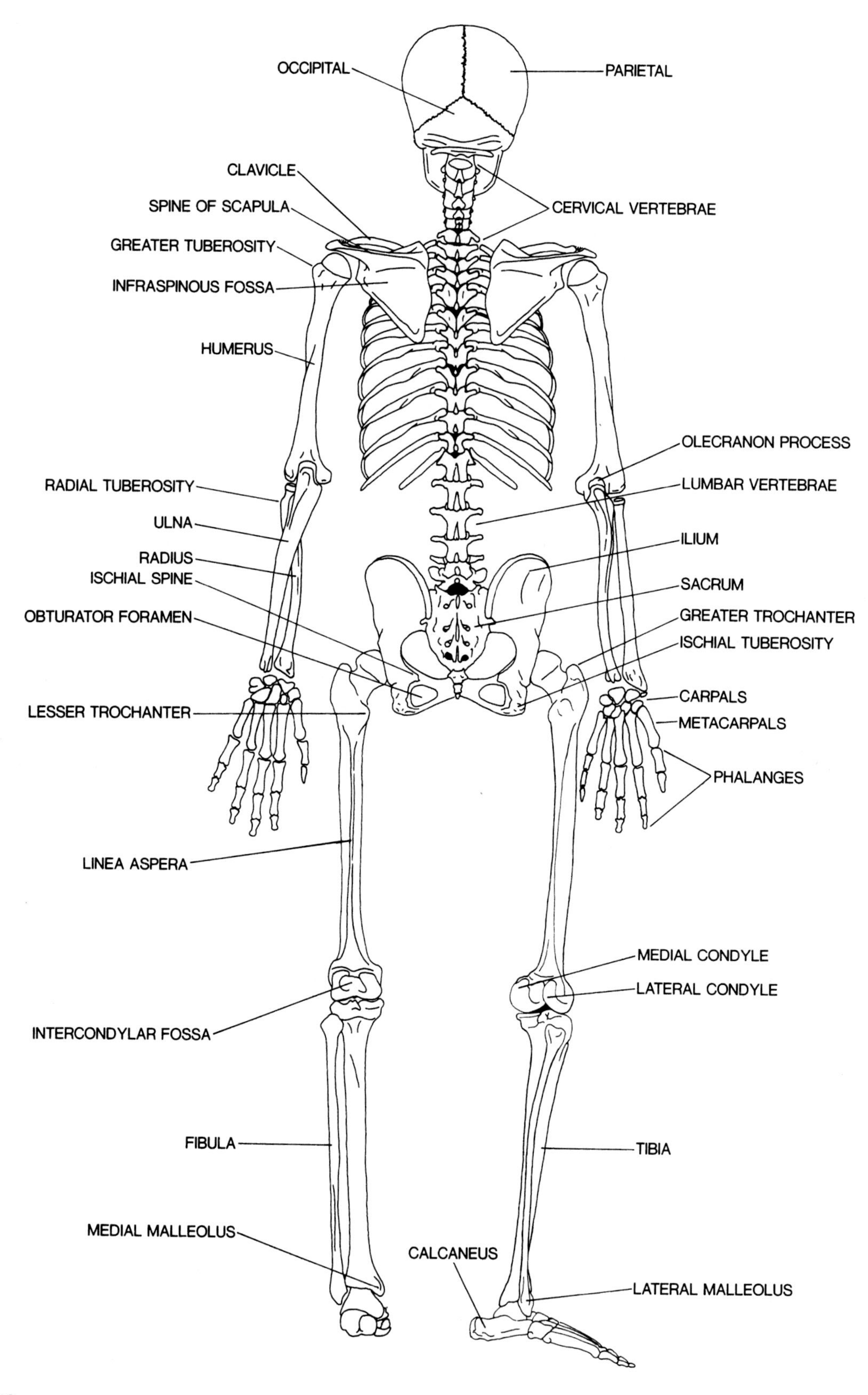

FIGURE 3.23
Posterior view of the human skeleton.

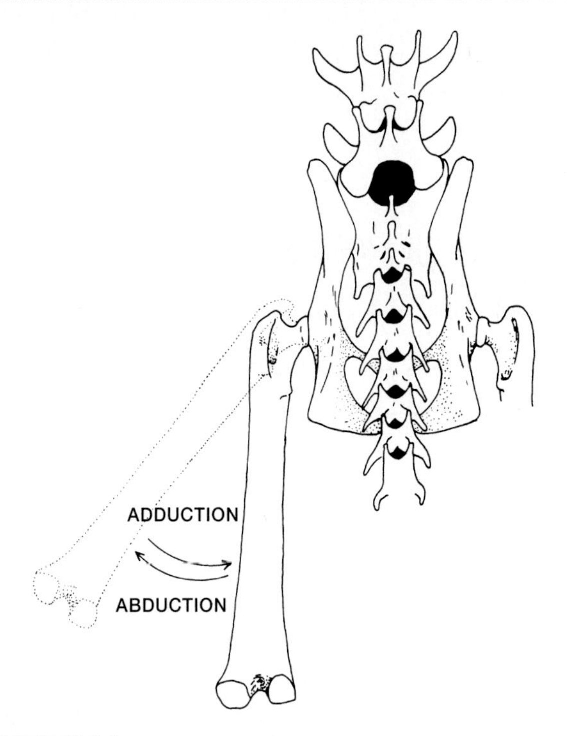

FIGURE 3.24
Dorsal view of the cat pelvic vertebrae, girdle, and femur, illustrating abduction-adduction (arrows).

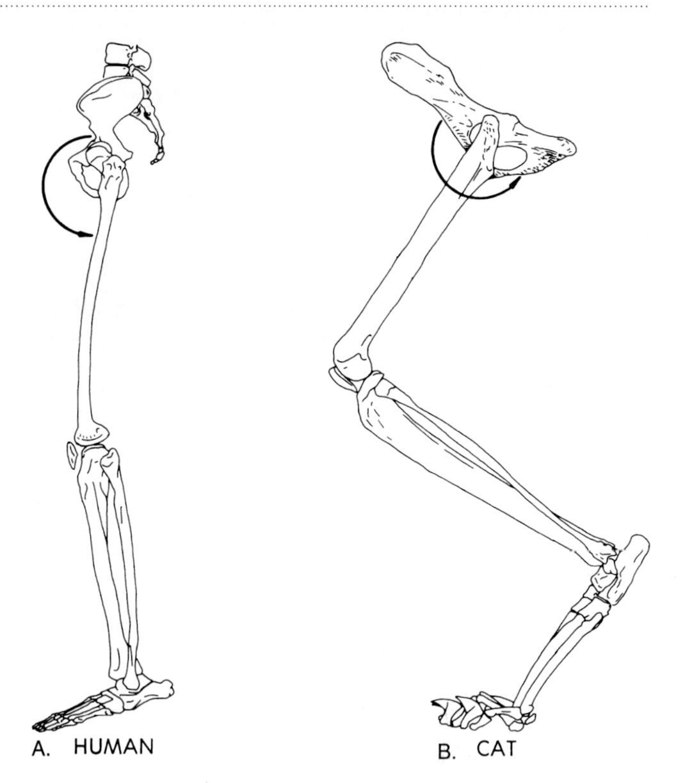

FIGURE 3.25
Lateral views of the pelvic limb and girdle of the human (A) and cat (B) to illustrate thigh movements. In the normal erect position, the human limb is fully extended. The cat femur is held at an angle of approximately 90 degrees to the long axis of the body. The arched arrow indicates the movement of the femur at its joint with the acetabulum.

and ulna). Ellipsoidal joints may be flexed and extended as well as abducted and adducted (as the intercarpal joints). Spheroidal joints flex and extend, abduct and adduct, and may also circumduct (figure 3.24). Synovial joints are encapsulated and filled with fluid (synovium). The synovial fluid is produced by the synovial membrane lining the joint chamber and lubricates the articular surfaces of the joints, thus reducing friction and wear.

The Hip Joint (figures 3.24, 3.25)

The hip joint of the cat allows movement in three planes: flexion-extension, abduction-adduction, and rotation, or cicumduction. This joint and the shoulder are the only joints in the body with three degrees of freedom of movement. The ligament binding the femur head to the acetabulum (pelvic socket) consists of a capsular ligament surrounding the joint between the cartilage lip of the acetabulum and the neck of the femur. Thickened portions of the capsular ligament are distinguished as separate ligaments: the **iliofemoral ligament** on the cranial border of the capsule, the **ischiofemoral ligament** on the caudal border, and the **pubofemoral ligament** on the ventral border. In addition, a ligament, the **transverse acetabular,** extends transversely across the acetabular notch, forming the ventromedial wall of the acetabulum. A **ligamentum teres** (lig-ah-MEN-tum TE-r*ez*) extends from the fovea (F*O*-v*e*-a) in the head of the femur to the acetabular fossa located entirely within the capsular ligament. The ligamentum teres is situated, from a dorsal attachment in the acetabulum to a ventral femoral attachment, so that adduction or flexion compresses the ligament against a fat pad in the acetabular fossa. This movement drives the synovial fluid into the dorsal half of the joint cavity, thus lubricating the dorsal head of the femur and the roof of the acetabulum. The dorsal femur head and the acetabular roof are the greatest pressure-bearing areas of the joint. Since the fluid normally flows into the ventral half of the joint, the ligamentum teres serves to pump the fluid to the areas of greatest wear in the joint's dorsal half. The capsular ligament serves primarily to limit the movement of the joint.

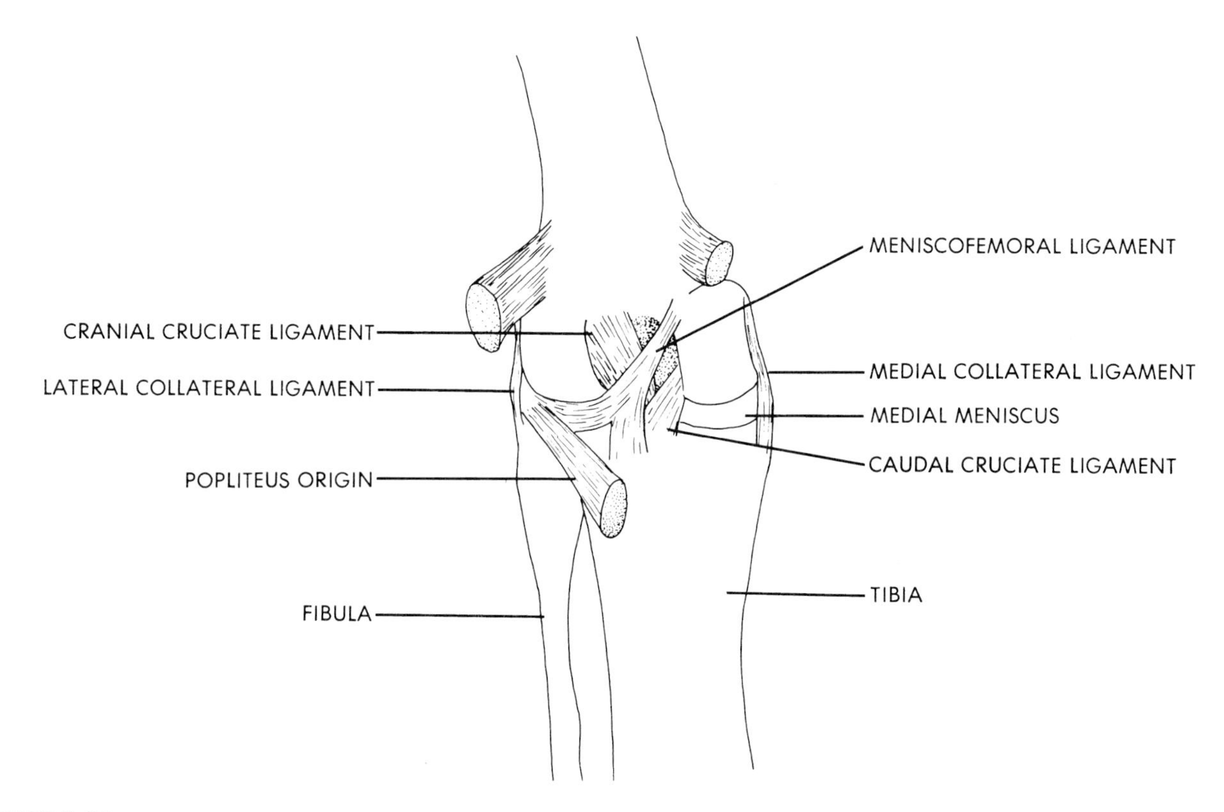

FIGURE 3.26
Caudal view of the cat knee joint.

The Knee Joint (figures 3.26, 3.27)

As the leg moves to a straight position (figure 3.26), the wedgelike **lateral meniscus** (me-NIS-kus) is pulled forward by the **meniscofemoral ligament** and "locks" the joint in the extended position. The popliteal muscle originates on the lateral meniscus. Contraction of this muscle pulls the meniscus back so the now "unlocked" joint may be flexed.

The rounded condyles of the femur roll over the relatively flat tibia much as wheels roll over a flat surface (figure 3.27). A pair of **cruciate** (KROO-she-at) ligaments cross in the midline of the joint between the femur and the tibia. The **cranial cruciate ligament** extends between the menisci of the cranial border of the tibia and the caudal border of the femur. The **caudal cruciate ligament** extends from the caudal border of the tibia to the cranial border of the femur. These ligaments act as elastic bands; the caudal cruciate ligament is stretched as the joint is flexed and the cranial cruciate ligament is stretched as the joint is extended (straightened).

The rolling motion of the femur over the tibia causes less wear than a sliding or pivotal movement would, but this arrangement also requires wedgelike menisci and several ligaments to direct (cruciate) and hold (collateral) the joint in place. Removal of the locking meniscus by contraction of the flexing muscles will allow the elastic cruciate ligament to again direct the movement of the knee.

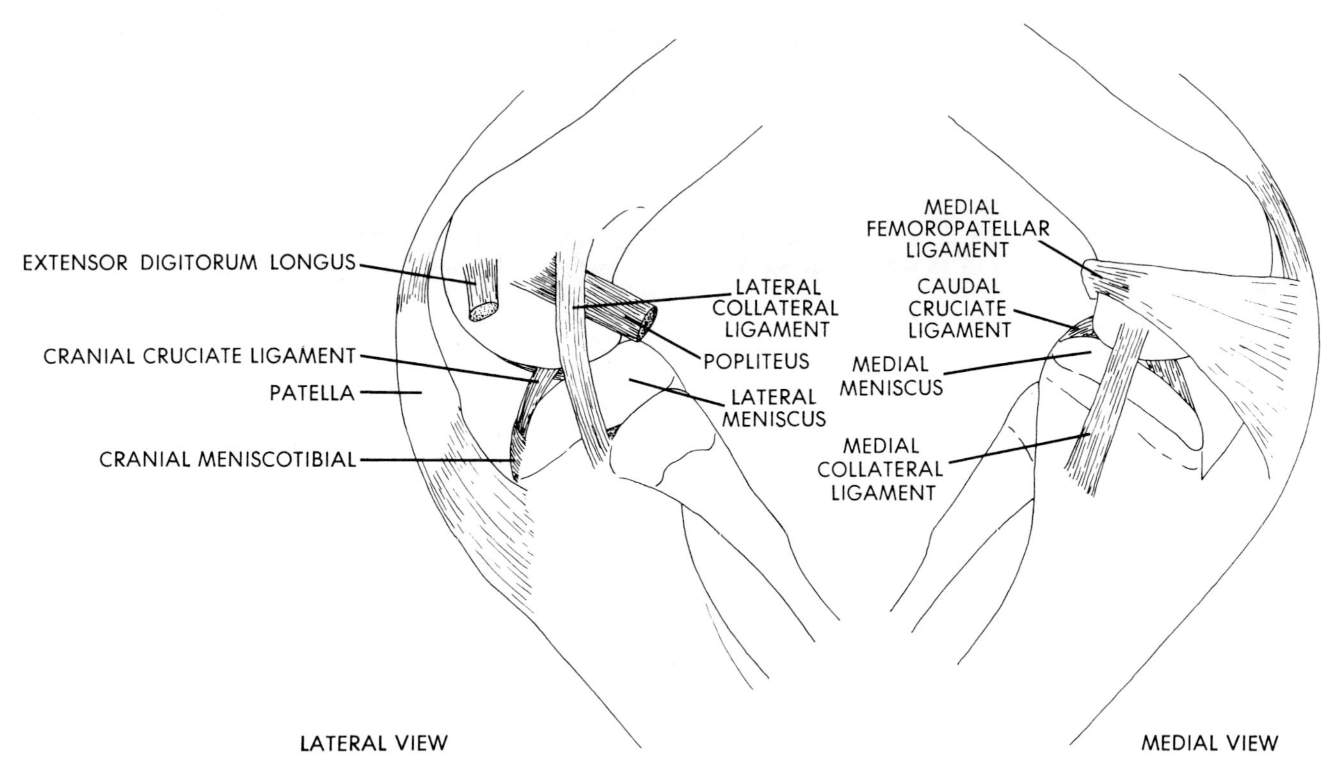

FIGURE 3.27
Medial and lateral views of the cat knee joint.

Suggested Readings

Berman, E. 1986. Fetal and neonatal growth and development. In *Current Therapy in Theriogenology,* edited by D. A. Morrow. 2d ed. Philadelphia: W. B. Saunders Co.

Buckland-Wright, J. C. 1978. Bone structure and the patterns of force transmission in the cat skull (*Felis catus*). *J. Morphol.* 155(1):35–62.

Chiasson, R. B. 1976. Cusp relationships in a carnassial dentition. *Ariz.-Nev. Acad. Sci.* 11(1):30–32.

DuBrul, E. L. 1964. Chapter 10 of *The temporomandibular joint.* 2d ed. Springfield, Ill.: Charles C. Thomas, Publisher.

Freeman, M. A. R., and B. Wyke. 1967. The innervation of the knee joint, an anatomical and histological study in the cat. *J. Anat.* 101(3):505–32.

———. 1967. The innervation of the ankle joint, an anatomical and histological study in the cat. *Acta Anat.* 68:321–33.

Gindhart, P. S. 1972. The effect of seasonal variation on long bone growth. *Hum. Biol.* 44(3):335–50.

Gordon, K. R. 1989. Adaptive nature of skeletal design. *Bioscience.* 39:784–90.

Riach, I. C. F. 1967. Ossification of the sternum as a means of assessing skeletal age. *J. Clin. Pathol.* 20:589–90.

Smith, R. N. 1969. Fusion of ossification centers in the cat. *J. Small Anim. Pract.* 10:523–30.

CHAPTER 4

Muscular System

The ability of mammals to run, fly, jump, or dig results from the contraction of striated, or **SKELETAL** (SKEL-*e*-tal) **MUSCLES** that move segments of the limbs and body. There are more than 450 skeletal muscles in the cat and in the human. Other muscle types found in mammals are smooth, or **VISCERAL** (VIS-er-al) **MUSCLES,** which act on parts of the digestive tract and blood vessels, and **CARDIAC** (KAR-d*e*-ak) **MUSCLE,** which is restricted to the heart (see figure 1.4).

As a muscle contracts, it shortens and performs work. The movement that occurs is called the **ACTION** of the muscle. To accomplish an action, a muscle pulls one attachment of the muscle toward the other. The attachment called the **ORIGIN,** or **HEAD,** is usually on a fixed or immovable bone, and the **INSERTION** is on a movable segment. Often, however, both of the attachments are on movable structures; the actual movement is dependent on "fixing" one attachment so that the other is moved. Midway between the origin and insertion of most muscles is a thickened area called the **BELLY,** or **VENTER** (VEN-ter). Most muscles are attached by thick, connective tissue **tendons** (TEN-duns) at their origin and insertion, but others may have a fleshy attachment. Unless they lie in the midline of the body, all muscles are paired.

Terminology of Muscle Action

FLEXION (FLEK-shun) Reduction of the angle between two bony elements in a joint. Takes place in the sagittal plane. Example: bending the knee.

EXTENSION (ek-STEN-shun) Increase of the angle between two bony elements in the sagittal plane. Example: straightening the knee.

ADDUCTION (ad-DUK-shun) Movement of a structure toward the midline in the frontal plane. Example: moving the arm toward the body in the recovery phase of a butterfly stroke.

ABDUCTION (ab-DUK-shun) Movement away from the midline in the frontal plane. Example: spreading the fingers.

ROTATION Movement of a body part around the long axis. Example: the action of the hand and forearm when using a screwdriver. The parts of this movement are **PRONATION** ([pr*o*-N*A*-shun] [medial rotation]) and **SUPINATION** ([syoo-pi-N*A*-shun] [lateral rotation]).

ELEVATION Lifting or raising a part of the skeleton. Example: shrugging the shoulders.

DEPRESSION Lowering a part of the skeleton. Example: the action of the mandible when opening the mouth.

Naming of Muscles

Muscles are named for their (1) location in the body (gluteus, G. = the rump), (2) shape (rhomboideus, G. = parallelogram with equal opposite sides), (3) origin and insertion (iliocostalis, from the ilium [origin] to the ribs [insertion]), (4) size (major or minor, as in teres major), (5) action (adductor, as in adductor femoris), (6) position (brachialis, G. = the arm), or a combination of these characteristics. For example, flexor digitorum superficialis (flexor = action; digitorum = [location or insertion] on the toes or digits; superficialis = [position] near the surface or external to another muscle with a similar name). The references to shape, location, origin, insertion, action, and size in the names of muscles will be a valuable aid to you in learning the locations and functions of the muscles.

You should learn the name, origin, insertion, action, and location of each muscle of your specimen as listed in the tables on pages 51 to 89. The names, origins, insertions, and actions (or functions) of the cat and human muscles are listed in tabular form and illustrated in the appropriate figures. Locate each muscle in the illustration and compare the muscle in the illustration with your specimen's muscle.

Dissection of the Muscular System

The technique of **BLUNT DISSECTION** should be followed in your study. Use scissors and razor blades as seldom as possible. The use of a scalpel is especially discouraged. The techniques of blunt dissection preserve the specimen and result in a superior dissection.

In blunt dissection, muscles are **ISOLATED** (separated from surrounding tissues) using the probe and forceps. Your instructor will demonstrate the techniques required. It may be necessary in some instances to cut the more superficial muscles in order to locate the deeper ones. Once a muscle is isolated, it may be **REFLECTED** (cut across the belly and the ends turned back so that they remain attached for later study). Never completely remove a muscle or other structure unless instructed to do so. Directions for the dissections precede the muscle tables when appropriate.

Preparation of the Cat for Observation of Musculature (figure 4.1)

Lay the specimen on its back on a dissecting board. If desired, the limbs may be tied in a spread position to the board. If your board has no special means to secure the limbs, use a heavy cord looped over one wrist, pass the cord under the dissecting board, and tie the free end of the cord to the other wrist. Do the same with the ankles. If embalmed and injected cats are used, an incision will be found on the midventral surface of the neck. At this opening, use scissors with one blunt tip and separate the skin from the underlying muscle. Continue the incision (1) caudad to a point just in front of the genitalia. Circumvent the genitalia (2) in making the incision. Likewise, continue the ventral incision (3) rostrad to the mouth, then cut dorsad around the mouth (4) on the left side (of the specimen), caudal to the nose and rostral to the eye. Make lateral incisions (5, 6) from the midventral incision and on the medial surface of the legs to the ankle and wrist. Continue these cuts around the wrist and ankle and separate the skin from the body. BE CAREFUL TO AVOID CUTTING ANY OF THE BODY OPENINGS OR (IF YOUR SPECIMEN IS A FEMALE) MAMMARY GLANDS OR NIPPLES.

The skin is only loosely attached to the body muscles and may be separated from them with your fingers by pulling gently on the skin and inserting your finger under the skin and moving back and forth. Be especially careful along the flank because there is a large, flat, skin muscle (cutaneous maximus) in this area extending from the inner layer of the skin to the armpit. Probe beneath the skin of the shoulder and separate a portion of the cutaneous maximus muscle; then cut it as near its center as possible, leaving one portion with the skin and the other in the armpit.

At the end of each laboratory period, replace the skin and tie it in place on your specimen. (You may, instead, discard the skin and wrap the specimen in wet paper towels after each laboratory period.) THE SPECIMEN SHOULD THEN BE PLACED IN A PLASTIC BAG BETWEEN SESSIONS TO PREVENT DRYING.

Skin Musculature

Skin muscles (not illustrated) are striated muscles that usually lack skeletal attachments. In the cat, the **CUTANEOUS MAXIMUS** (k*u*-T*A*-n*e*-us MAK-si-mus), is a large sheet of muscle (which you cut during skinning) on the side of the body. This sheet originates on each side of the fascia of the **LATISSIMUS DORSI** (la-TIS-i-mus D*O*R-s*i*) and of the ventral pectoral muscles. Fibers of the cutaneous maximus are often confused with those of the latissimus dorsi or with the fibers of the **PECTORALIS PROFUNDUS** (pek-t*o*-RA-lis pr*o*-FUN-dus). These muscles are poorly developed in most humans but some individuals do learn to control them.

The other skin muscles are located on the head and neck. Those of the face are often termed the "muscles of facial expression." The largest of these, the **PLATYSMA** (pla-TIZ-ma), extends over the ventral neck to the face. Skin muscles are usually very thin and diffuse, and are usually removed with the skin. There are more than 30 described for the cat. The muscles of facial expression (21 pair) are well-developed in humans.

The remaining musculature of the cat and human is presented in the following tables. Dissection instructions for the cat are presented when appropriate. Follow the instructions carefully and CONFINE YOUR DISSECTION TO JUST ONE SIDE OF THE ANIMAL. Use the figures for help in locating and identifying the muscles. The cat muscle names, origins, insertions, and actions are provided in the tables. Assume these to be the same for the human unless indicated in parentheses.

Before undertaking further dissection, familiarize yourself with the superficial muscles shown in figure 4.2.

Jaw, Hyoid, and Tongue Musculature

After the skin is removed from the head, there are salivary glands, blood vessels, and fat covering the muscles. *Carefully remove the salivary glands and fat on one side of the head.* If the blood vessels are injected with colored latex, you should be able to leave most of the vessels intact while removing the tissues surrounding the vessels.

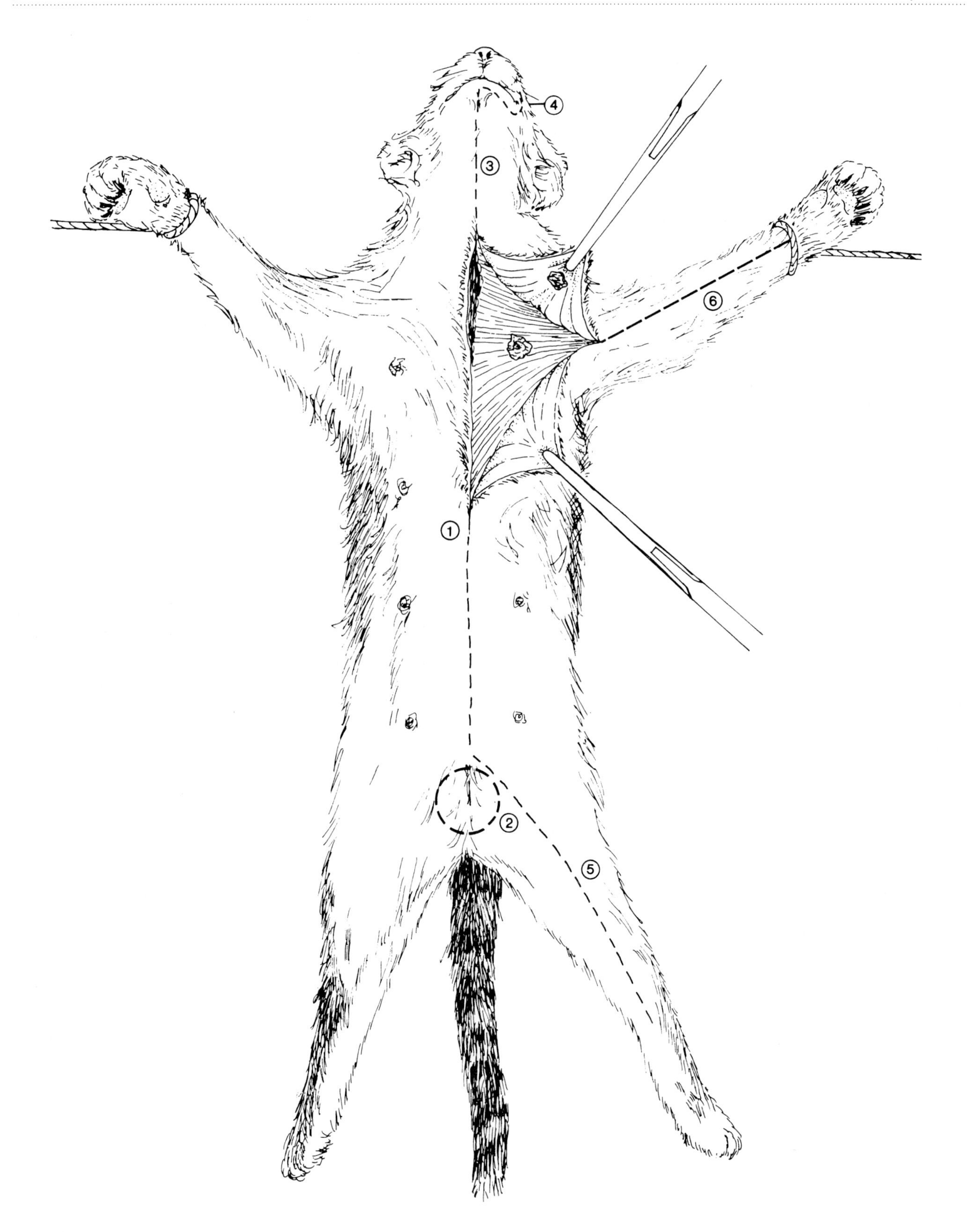

FIGURE 4.1. Dissection incisions for removing the skin of the cat. The limbs are spread in preparation for skinning. Tie the limbs down with heavy string by putting a loop over one wrist, then run the string under the dissecting board to the other side and tie another loop over the other wrist. Repeat this with the ankles and your specimen is ready to skin. Follow the instructions for skinning in the text (page 49) and cut the skin as indicated by the dashed lines in the figure. Use a blunt instrument such as a closed hemostat to separate the connective tissue holding the skin to the musculature.

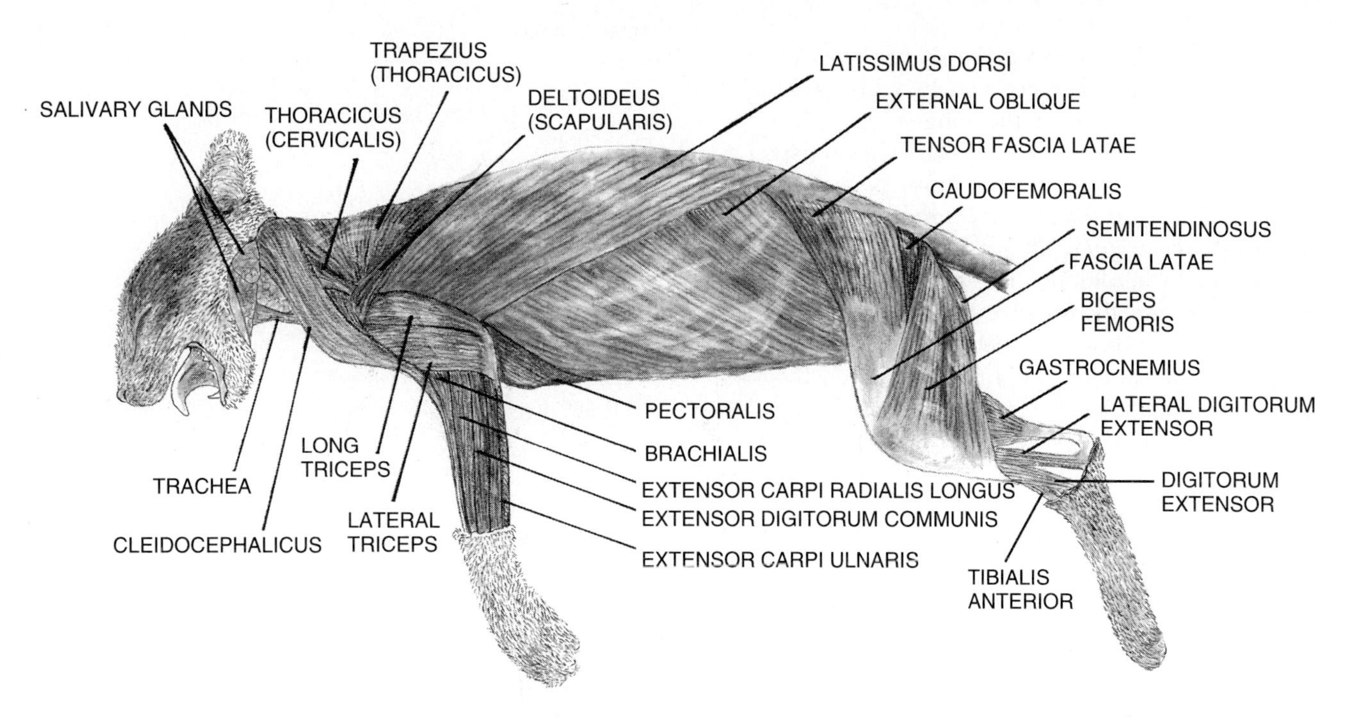

FIGURE 4.2. Superficial lateral muscles of the cat. The skin muscles have been removed.

1. Jaw musculature (figures 4.3, 4.4)

Locate the masseter, temporalis, and digastricus. Then *carefully insert a blunt probe between the rostral and caudal heads of the digastricus and scrape the rostral head loose from its insertion on the mandible. With a blunt probe, scrape the origin and insertion of the masseter loose from the mandible and zygomatic arch. Cut through the two ends of the zygomatic arch with bone cutters and remove the arch. Next, place the blunt probe tip on the inner margin of the mandible and clear the tissue from the bone. Then, cut the mandible with bone cutters at the diastema between the canine and premolars and between the molar and coronoid process. Remove the segment to see the pterygoid muscles.* The remaining muscles of mastication are seen without further dissection.

Name	Origin	Insertion	Action
MASSETER (mas-SE-ter)	Zygomatic arch	Coronoid fossa of mandible	Elevate mandible
TEMPORALIS (tem-po-RA-lis)	Temporal fossa of parietal, squamosal, and frontal	Coronoid fossa of mandible	Elevate mandible
PTERYGOIDEUS LATERALIS (TER-i-goy-de-us)	External pterygoid fossa of palatine	Medial surface of mandible	Elevate mandible
PTERYGOIDEUS MEDIALIS	Internal pterygoid fossa of basisphenoid (hamulus)	Medial surface of mandible	Elevate mandible
DIGASTRICUS (d*i*-GAS-trik-us)	Jugular and mastoid processes of temporal	Lower border of mandible	Depress mandible

2. Hyoid and tongue musculature (figures 3.9, 4.5, 4.15)

The mylohyoideus and stylohyoideus muscles were exposed during the dissection of the digastricus. *Insert a blunt probe beneath the mylohyoideus rostral to the basihyal and separate this muscle from the deeper muscles. Cut the mylohyoideus near the midline. Pull the mylohyoideus toward the mandible (where the digastricus was inserted), thus exposing the deeper muscles.* The geniohyoideus, genioglossus, and hyoglossus muscles will require dissection of the tongue; THIS SHOULD BE POSTPONED UNTIL DISSECTION OF THE ORAL CAVITY (chapter 5, Digestive System). *Transect the stylohyoideus and the sheet of pharyngeal muscle originating from the area just beneath the origin of the stylohyoideus.* This will expose the ceratohyoideus. The sternohyoideus, sternothyroideus, and thyrohyoideus muscles will also be seen with dissection of the neck muscles. The paired sternohyoideus muscles cover the trachea on the ventral midline of the neck. The thyrohyoideus lies beneath them on the ventrolateral surface of the larynx. Locate the following:

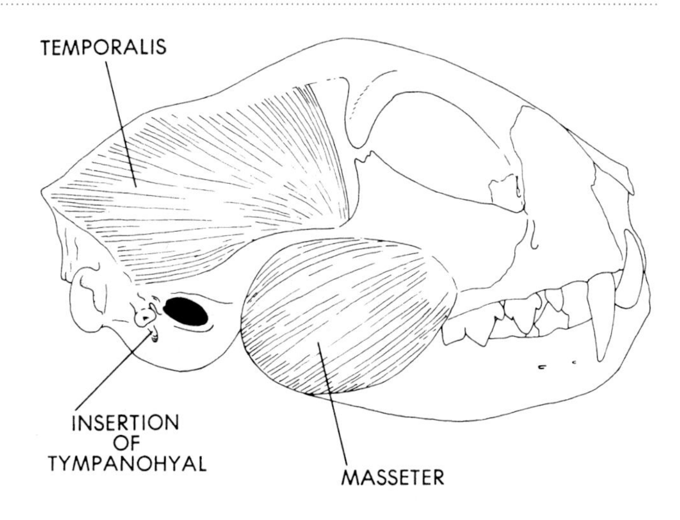

FIGURE 4.3. Lateral view of the cat jaw muscles.

Name	Origin	Insertion	Action
STYLOHYOIDEUS (st*i*-lo-h*i*-OY-d*e*-us) figure 4.15	Stylohyal of hyoid	Basihyal of hyoid	Elevate and draw hyoid rostrad (superior in human)
GENIOHYOIDEUS (je-n*e*-*o*-h*i*-OY-d*e*-us) figure 3.9	Mandible near symphysis	Midline of throat to its opposite	Elevate the floor of mouth, protract hyoid
STERNOTHYROIDEUS (stern-n*o*-th*i*-ROY-d*e*-us) figure 4.5	First costal cartilage	Thyroid cartilage	Draw larynx caudad (inferior in human)
STERNOHYOIDEUS (stern-n*o*-h*i*-OY-d*e*-us)	First rib cartilage (clavicle and manubrium in human)	Basihyal	Draw hyoid caudad (inferior in human)
THYROHYOIDEUS (th*i*-r*o*-h*i*-OY-d*e*-us)	Thyroid cartilage	Basihyal	Elevate larynx
GENIOGLOSSUS (je-n*e*-*o*-GLOS-us) figures 3.9	Mandibular symphysis	Tissues of tongue	Protract tongue
HYOGLOSSUS (h*i*-*o*-GLOS-us)	Ceratohyal and basihyal	Tissues of tongue	Retract tongue
STYLOGLOSSUS (st*i*-l*o*-GLOS-us)	Stylohyal	Tissues of base of the tongue on either side of the glottis	Curl tongue
LINGUALIS PROPRIA (ling-GWA-lis PRO-pr*e*-a) figure 5.6	Tissues of tongue	Tissues of tongue	Intrinsic musculature for complex lingual movements

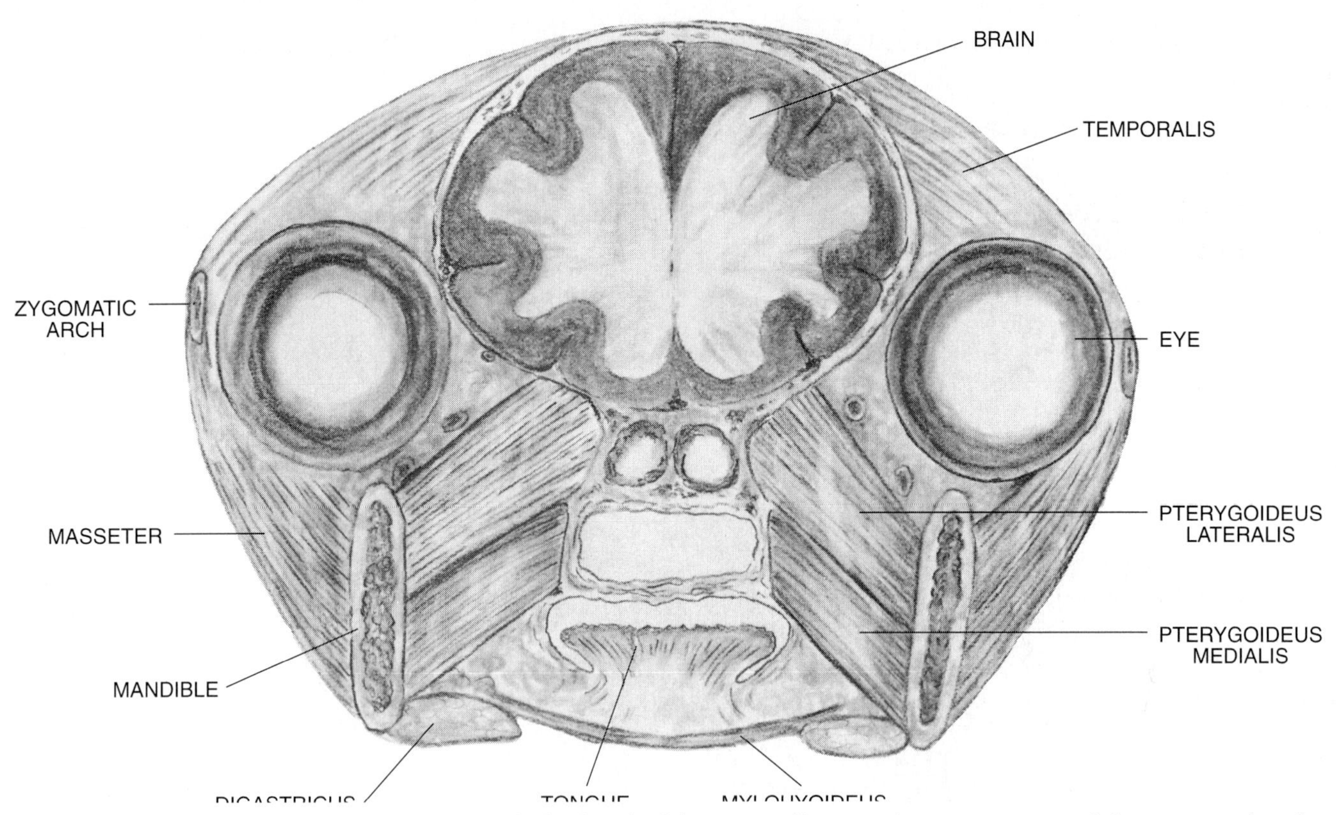

FIGURE 4.4. Diagrammatic section through the head of the cat to illustrate the arrangement of the pterygoid and other muscles of mastication.

Ventral Neck Musculature

Superficial (figure 4.5)

The external jugular vein and anastomosing link should be cut and reflected to dissect the neck muscles. Separate the sternomastoideus from deeper muscles with a blunt probe and transect it halfway between its origin and insertion on the same side as the dissected digastricus and mylohyoideus muscles. Separate the sternohyoideus from the sternothyroideus and transect the sternohyoideus. The thyrohyoideus will be seen beneath the rostral end of the sternohyoideus. DELAY DISSECTION OF DEEP MUSCLES UNTIL YOU HAVE DISSECTED THE PECTORAL GIRDLE AND LIMB MUSCLES (page 71). Locate the following:

Name	Origin	Insertion	Action
STERNOMASTOIDEUS (ster-no-mas-TOY-de-us)	Cranial border of manubrium and fascia of transverse pectoralis	Mastoid and lambdoidal ridge	Rotate head and depress face
CLEIDOMASTOIDEUS (kli-do-mas-TOY-de-us)	Mastoid process	Middle of clavicle	Rotate head and depress snout or draw clavicle forward

In the human, the sternomastoid and cleidomastoid are combined as a single muscle (the **sternocleidomastoideus**) at the insertion. It originates on the clavicle and sternum and inserts on the mastoid process. When the left and right muscles are used in concert, they elevate the face. This is because the fulcrum (the atlas to occipital condyle joint) is anterior to the insertion.

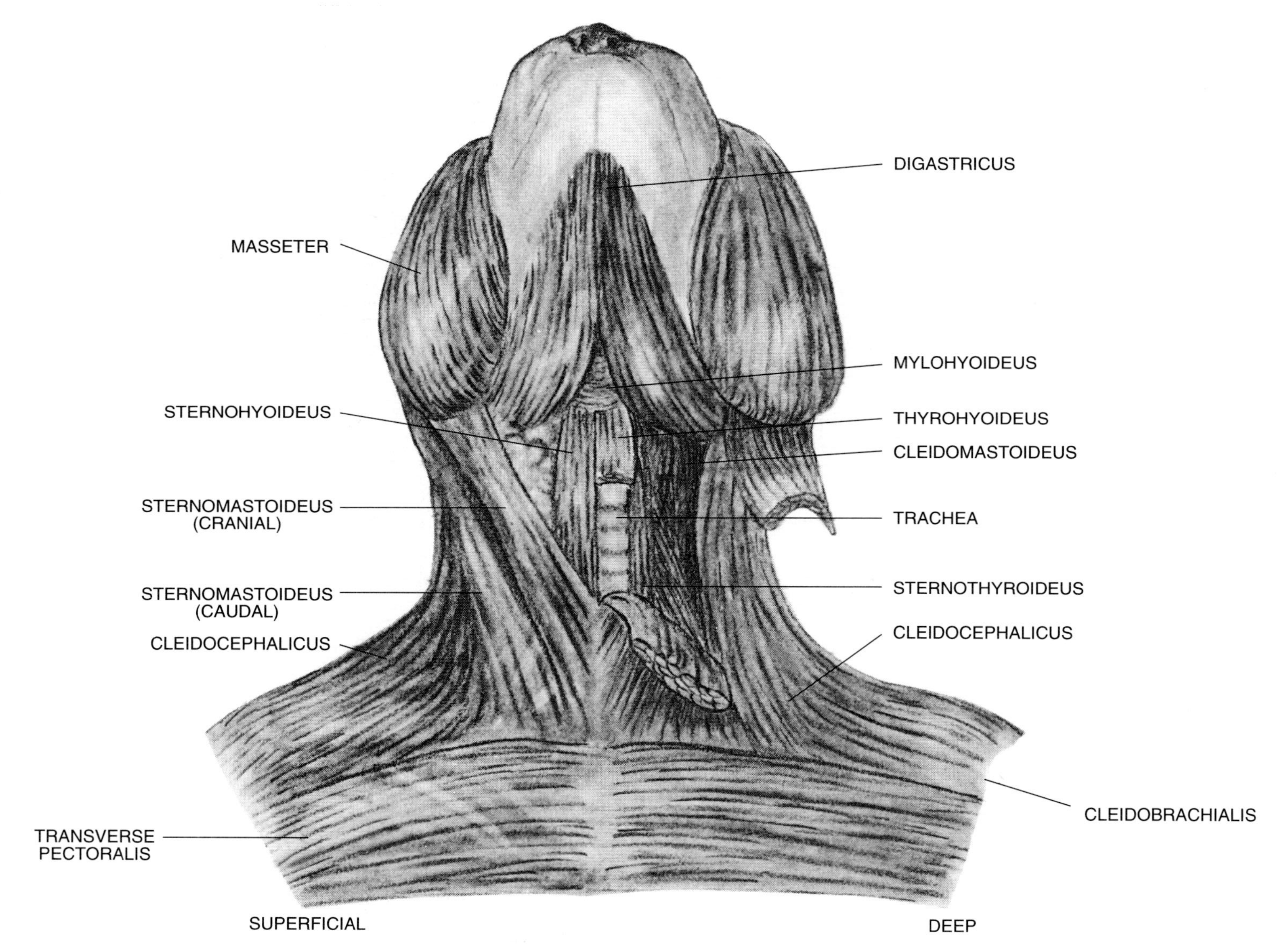

FIGURE 4.5. Ventral view of the throat and neck muscles of the cat.

Pectoral Girdle Musculature

1. Dorsal and lateral (figures 4.5, 4.6 and 4.7)

Identify the following muscles: cleidocephalicus, cleidobrachialis, trapezius, and latissimus dorsi, and *then separate the latissimus dors* confused with it. *As before, separate the muscles with a blunt probe before transecting them.* Note that the cephalic vein crosses over the cleidocephalicus portion of the brachiocephalicus muscle at the shoulder. *Now cut the cleidocephalicus carefully.* A branch of the spinal accessory nerve passes caudad on the neck just beneath the cleidocephalicus and trapezius muscles. DO NOT CUT THIS NERVE. *Next, transect the trapezius. Cut the cleidobrachialis to expose the deltoideus. Loosen and then cut the scapular and acromial portions of the deltoideus. Flex the brachium against the scapula and insert a blunt probe between the scapula and teres major lateral to the insertion of the long head of the triceps. Dissect the breast muscles before completing the dissection of the shoulder muscles on the medial side of the shoulder.*

Locate the following:

2. Ventral and medial (figures 4.8, 4.9)

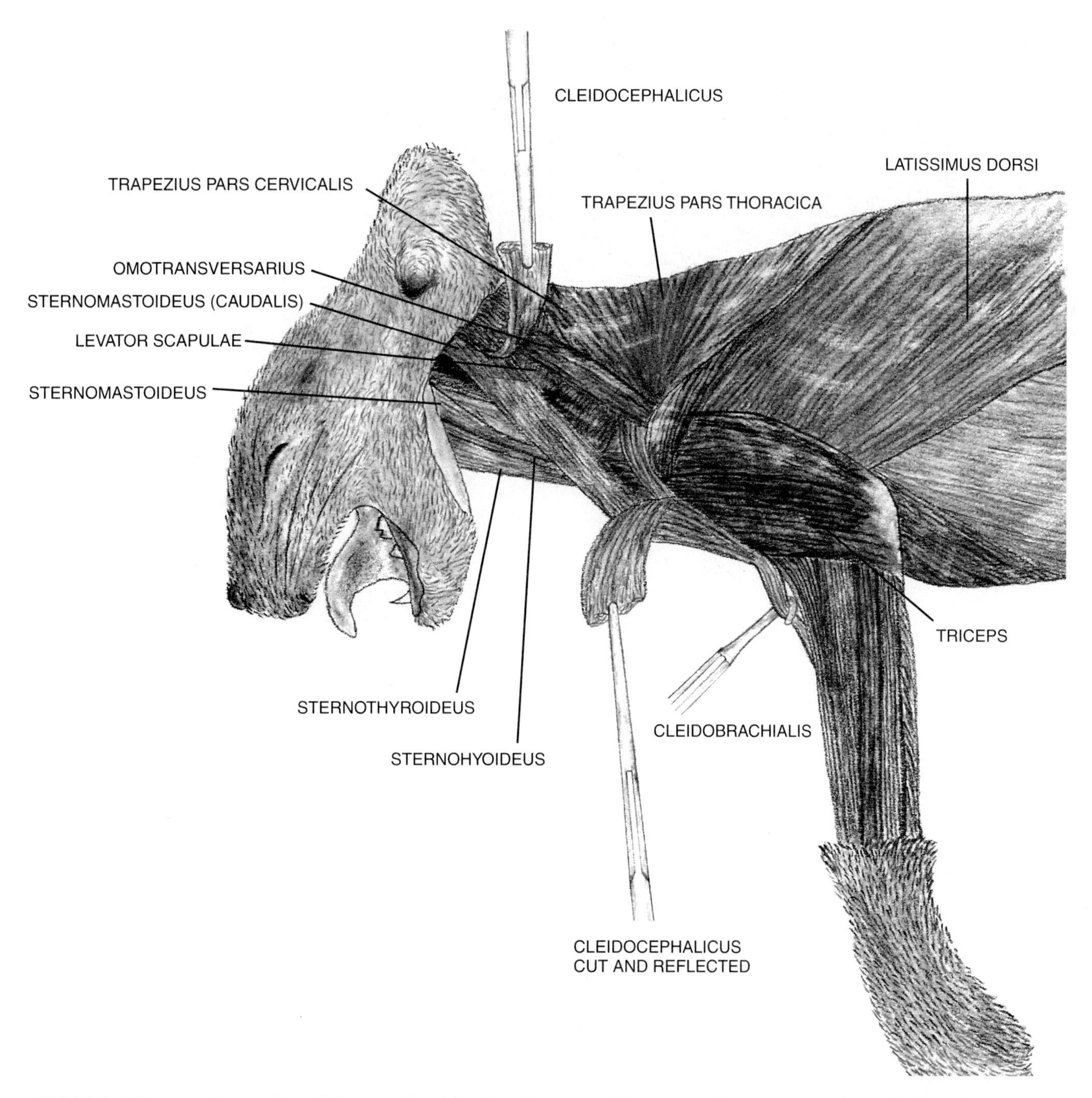

FIGURE 4.6. Lateral muscles of the cat shoulder, brachium, and forearm. The cut ends of the cleidocephalicus are held apart by hemostats and the cleidobrachialis is held by a curved blunt probe.

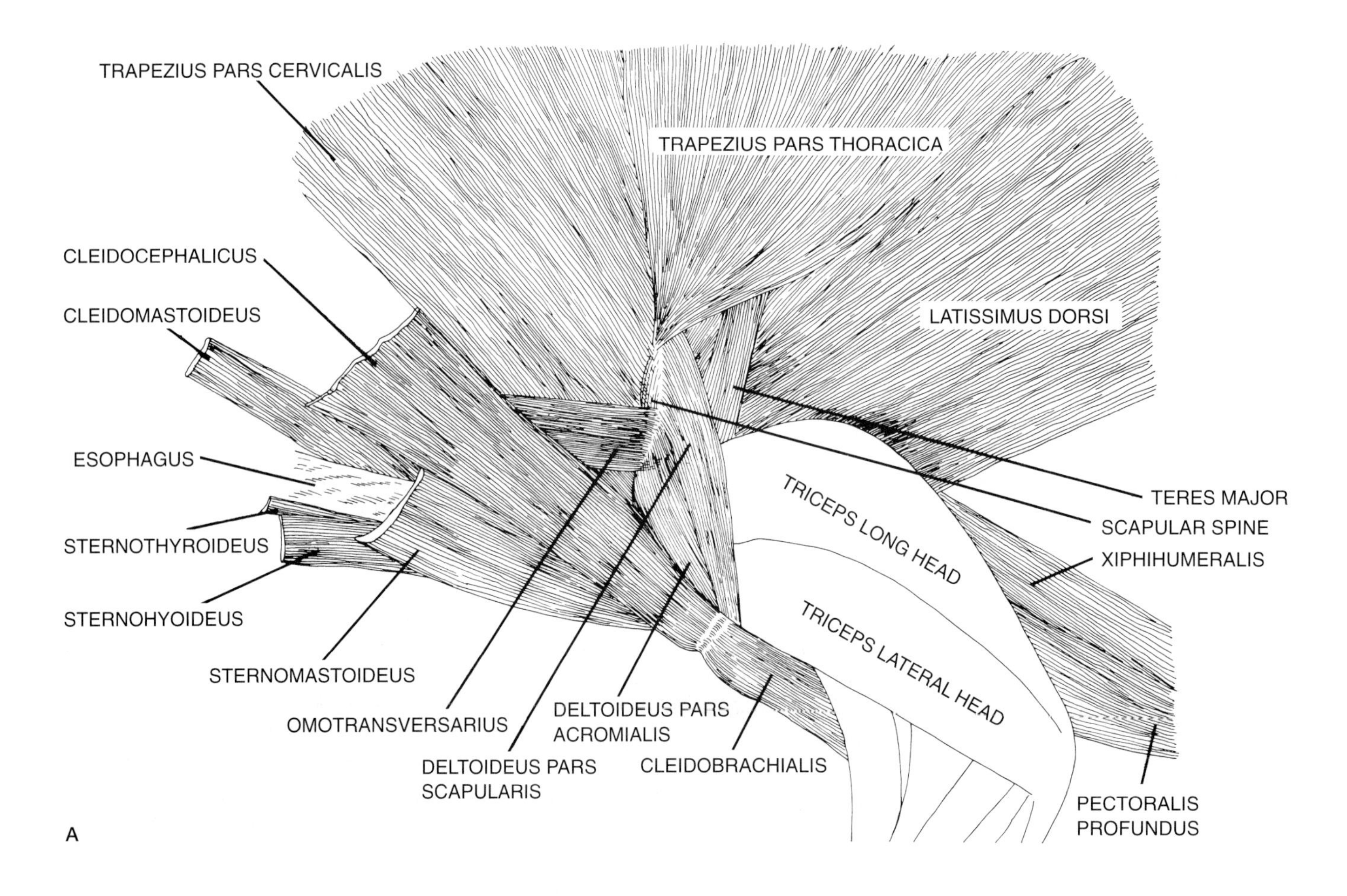

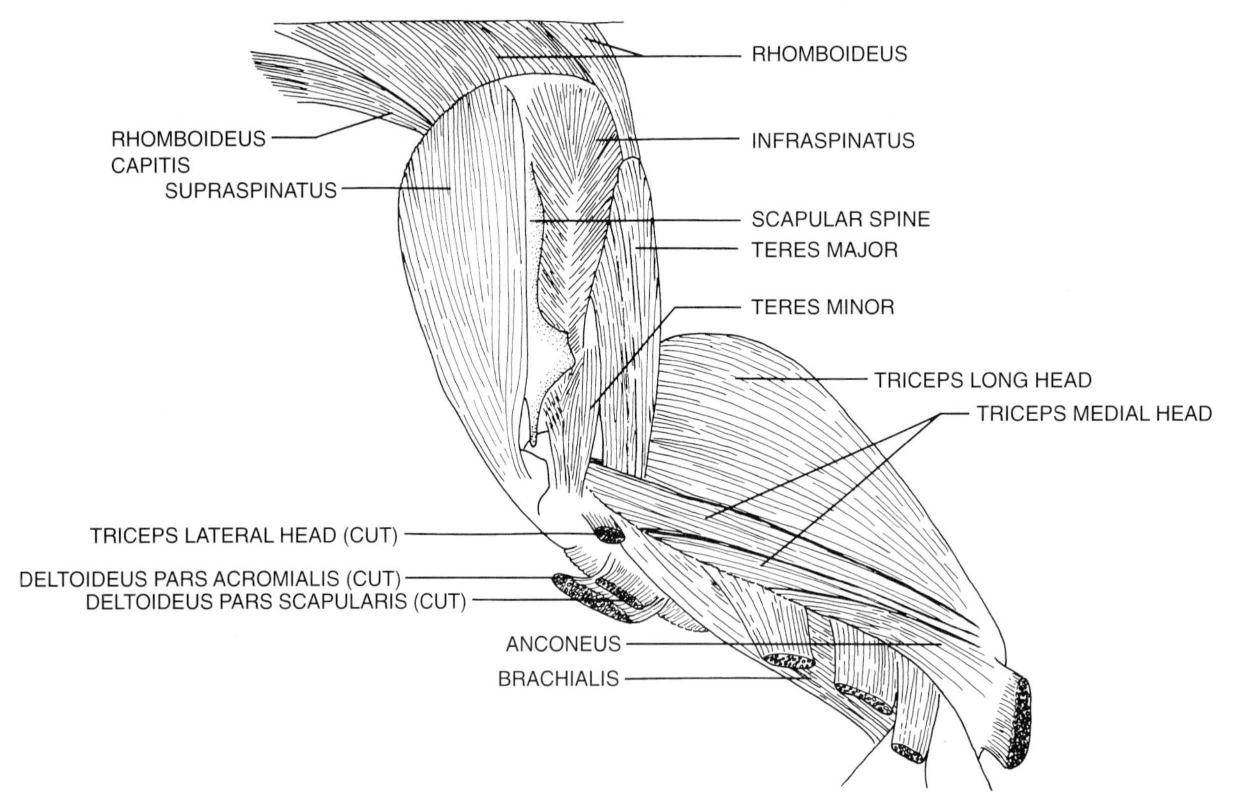

FIGURE 4.7. Pectoral limb muscles of the cat. (A) Superficial lateral muscles of the shoulder and brachium. (B) Deep lateral muscles of the shoulder, brachium, and forearm.

Name	Origin	Insertion	Action
DELTOIDEUS PARS SCAPULARIS (del-TOY-de-us parz skap-*u*-LAR-is)	Scapular spine	Deltoid tuberosity	Extend and rotate humerus laterad (abduct humerus in human)
PARS ACROMIALIS (a-kr*o*-m*e*-AL-is)	Acromion process	Pars scapularis	Extend humerus (abduct humerus in human)
BRACHIOCEPHALICUS (br*a*-k*e*-*o*-se-FAL-i-kus)			
CLEIDOBRACHIALIS (a portion of human deltoid)	Clavicle and cleidocephalicus	Ulna, distal to semilunar notch	Flex antebrachium
CLEIDOCEPHALICUS (a portion of human trapezius)	Lambdoidal ridge of occipital	Clavicle and clavobrachialis	Flex antebrachium

The human has a single deltoid muscle originating on the clavicle and acromion. It inserts on the deltoid tuberosity and abducts the brachium. Because the clavicle of the cat is vestigial, the acromial portion of the deltoid fuses with the clavicular portion of the trapezius (as brachiocephalicus). These are, however, clearly homologous with portions of the deltoid and trapezius of the human and have similar innervation and development. In the cat, the two muscles work synergistically to produce powerful extension of the humerus, a movement used in running.

Name	Origin	Insertion	Action
SUPRASPINATUS (soo-pra-spi-NA-tus)	Supraspinous fossa of scapula	Greater tuberosity of humerus	Extend humerus (abduct humerus in human)
INFRASPINATUS (in-fra-spi-NA-tus)	Infraspinous fossa	Greater tuberosity of humerus	Flex and rotate humerus (adduct humerus in human)
TERES MAJOR (TE-r*e*z)	Glenoid border of scapula near vertebral border	Fuse with latissimus dorsi (in cat) and insert on humerus	Flex humerus (also medial rotation in human)
TERES MINOR	Glenoid border of scapula	Deltoid tuberosity	Rotate humerus laterad
TRAPEZIUS (tra-P*E*-z*e*-us)			
PARS THORACICA	Spines of thoracic vertebrae	Fascia of infraspinatus and supraspinatus and spine of scapula	Draw scapula caudad and mediad
PARS CERVICALIS	Spinous processes of cervical and first three thoracic vertebrae	Metacromion and spine of scapula	Draw scapula mediad
RHOMBOIDEUS CAPITIS (rom-BOY-d*e*-us KAP-i-tis) (deep to trapezius)	Lambdoidal ridge of occipital	Vertebral border of scapula	Draw dorsal border of scapula craniad
RHOMBOIDEUS (deep to trapezius)	Supraspinous ligament of cervical vertebrae and thoracic vertebrae 1–4	Vertebral border of scapula	Draw vertebral border of scapula mediad and dorsad

Name (Continued)	Origin	Insertion	Action

The rhomboideus capitis is usually lost in humans and the rhomboideus is subdivided to form the rhomboideus major and minor. When the scapula is fixed, the rhomboideus capitis elevates the head of the cat. The rhomboideus is used to stabilize the pectoral appendage against the axial skeleton. In humans, major and minor act to adduct the scapula.

Name	Origin	Insertion	Action
OMOTRANSVERSARIUS (o-mo-trans-ver-SA-re-us)	Transverse process of atlas and basioccipital	Metacromion	Draw scapula craniad; a sling muscle

The omotransversarius is not found in the human. The **levator scapulae** of the human elevates the scapula (draws it craniad) in a manner similar to the omotransversarius of the cat. Levator scapulae arises from cervical vertebrae C_1–C_4 and inserts on the vertebral border of the scapula.

Name	Origin	Insertion	Action
LATISSIMUS DORSI	Spines of fourth thoracic to sixth lumbar vertebrae and thoracolumbar fascia (in the human, this origin extends to the ilium)	Proximal humerus with teres major (in the human, the intertubular groove)	Flex brachium (in the human, both an adductor and extensor of the brachium)

Separate the descendens portion of the pectoralis superficialis muscle from the deeper muscles with your probe so that the probe passes completely beneath the muscle. Now separate the transversus portion of the pectoralis superficialis and transect it. The latissimus dorsi attaches, in part, to the pectoralis profundus. *Separate these muscles from one another and transect them.* AVOID CUTTING THE BLOOD VESSELS AND NERVES IN THE AXILLARY AREA.

Note that the pectoralis profundus appears to have three parts. The most caudal has been termed the xiphihumeralis (zif-i-h*u*-mer-AL-is) and appears to go beneath (deep) to the middle portion. Both the middle and xiphihumeralis portions form the caudalis and insert together on the bicipital groove of the humerus. The cranialis portion also inserts in the bicipital groove, but somewhat apart from the other caudalis portions. Its insertion is continuous with a tendon that passes to the coracoid process and fuses with the tendon of insertion of the supraspinatus. These divisions were formerly considered separate muscles, but the divisions are no longer recognized by the Nomina Anatomica Veterinaria. *Transecting the pectoralis will expose the transversus costarum and serratus ventralis. Now turn the cat over and dissect the shoulder muscles. Clean the brachiocephalicus from adjacent muscles and transect it. Separate the thoracic and cervical portions of the trapezius to expose the cervical and thoracic portions of the rhomboideus, which also should be cleaned and transected.* The limb is now attached to the trunk by the sling muscles (sternomastoideus, serratus ventralis, omotransversarius, and the latissimus dorsi) plus blood vessels and nerves.

Look over figures 4.8 and 4.9, identify the muscles, and then read the following detailed description.

Name	Origin	Insertion	Action
SUBSCAPULARIS (sub-skap-*u*-LAR-is) figure 4.8	Subscapular fossa	Lesser tuberosity	Adduct humerus (in the human, it also rotates mediad)
CORACOBRACHIALIS (kor-a-ko-br*a*-ke-AL-is) figure 4.9	Coracoid process of scapula	Medial surface of humerus	Adduct humerus; deep to biceps brachii and brachialis

The coracobrachialis is disproportionally larger in the human and acts to flex and adduct the brachium.

Name (Continued)	Origin	Insertion	Action
SERRATUS VENTRALIS THORACIS (ser-RA-tus) figure 4.8	First 10 ribs	Vertebral border of scapula	Depress scapula; a sling muscle
SERRATUS VENTRALIS CERVICIS (levator scapulae) figure 4.6	Transverse processes of the last five cervical vertebrae	Vertebral border of scapula	Depress scapula and support chest; a sling muscle

The serratus ventralis group is a sling muscle group in the cat. That is, it acts to suspend the body between the pectoral appendages (figure 4.10). This function is not important in humans where the **serratus anterior muscles** abduct the scapula, or when the scapula is fixed, elevate the ribs. Elevation of the ribs is an important respiratory movement.

Name	Origin	Insertion	Action
PECTORALIS SUPERFICIALIS DESCENDENS (pek-to-ra-lis)	Manubrium of sternum	Superficial fascia of elbow	Adduct brachium
PECTORALIS SUPERFICIALIS TRANSVERSUS	Raphe of midventral line, manubrium, and sternebrae	Shaft of humerus	Adduct brachium
PECTORALIS PROFUNDUS			
PARS CRANIALIS	Sternebrae	Bicipital groove of humerus	Adduct brachium
PARS CAUDALIS	Xiphisternum	Bicipital groove of the humerus	Adduct brachium

The pectoralis superficialis decendens has no counterpart in the human. The pectoralis superficialis transversus is the **pectoralis major** of the human and a powerful adductor of the brachium. The pectoralis profundus pars cranialis is the **pectoralis minor** of the human. It is disproportionally larger in the cat, where it adducts the brachium. In the human, this muscle inserts on the scapula and depresses it. Adduction of the brachium is an antigravity response in the cat, which results in the larger size of the pectoralis minor. The pectoralis profundus pars cranialis does not exist in the human.

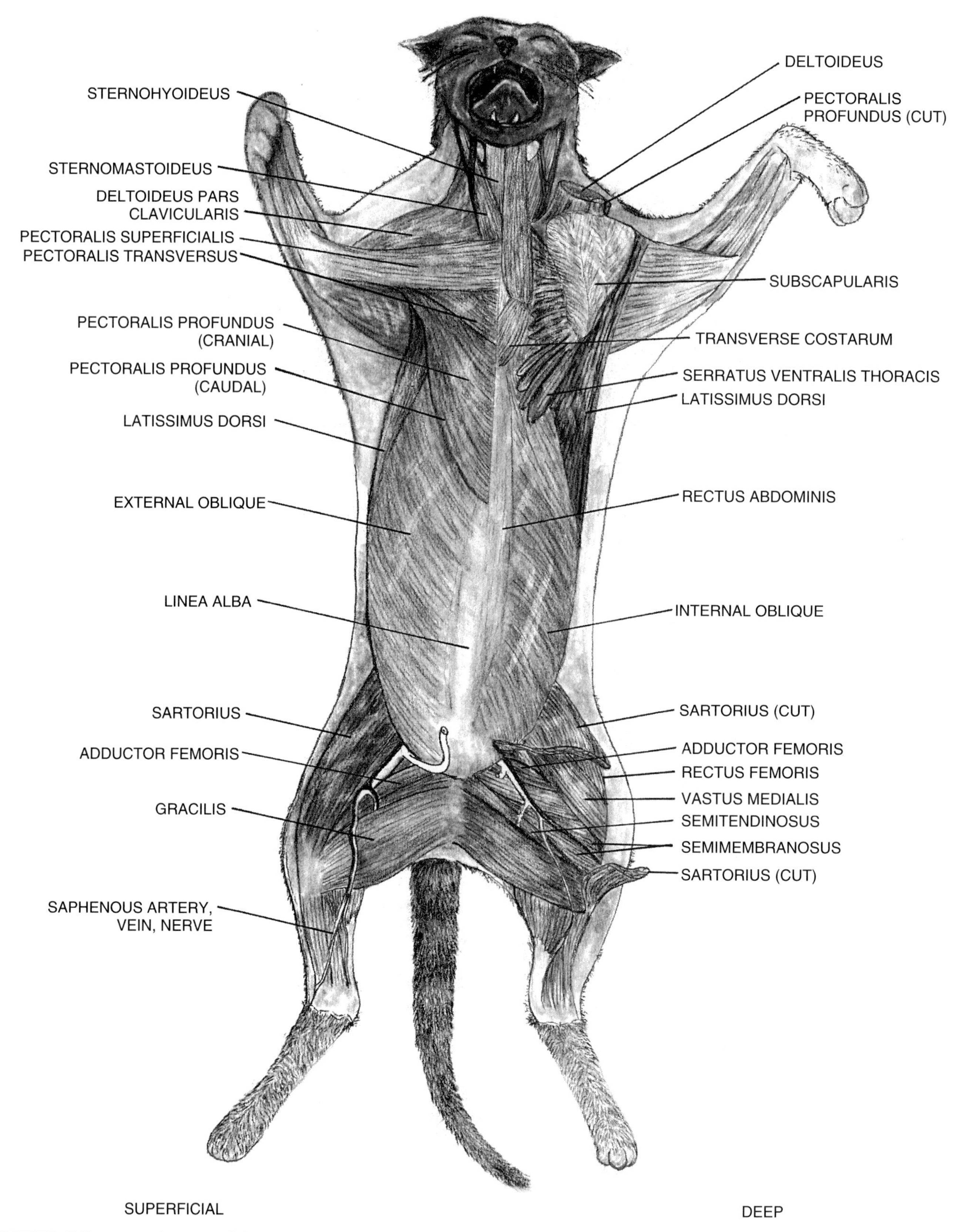

FIGURE 4.8. Ventral view of the superficial and deep muscles of the cat.

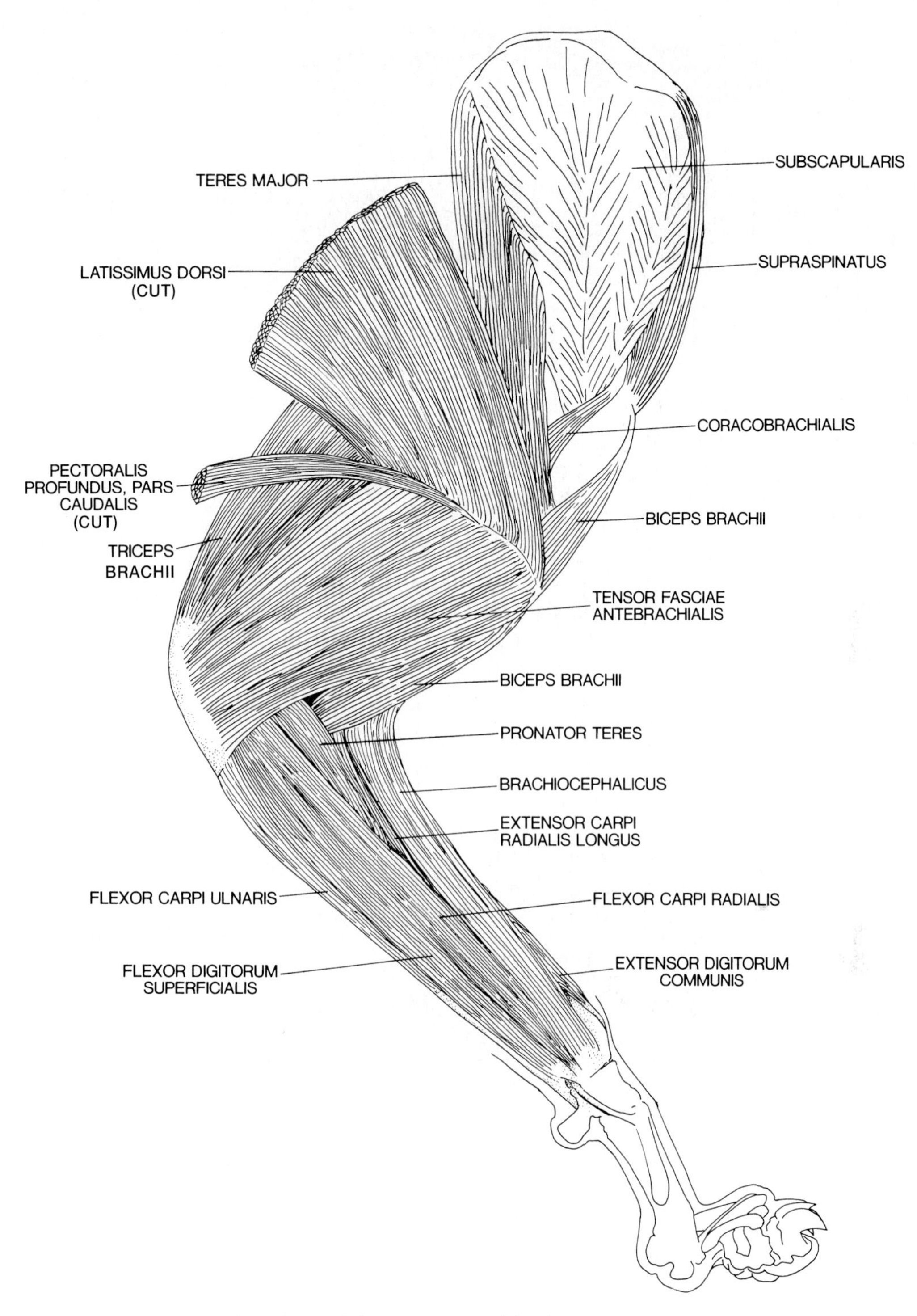

FIGURE 4.9. Muscles of the medial surface of the cat pectoral limb.

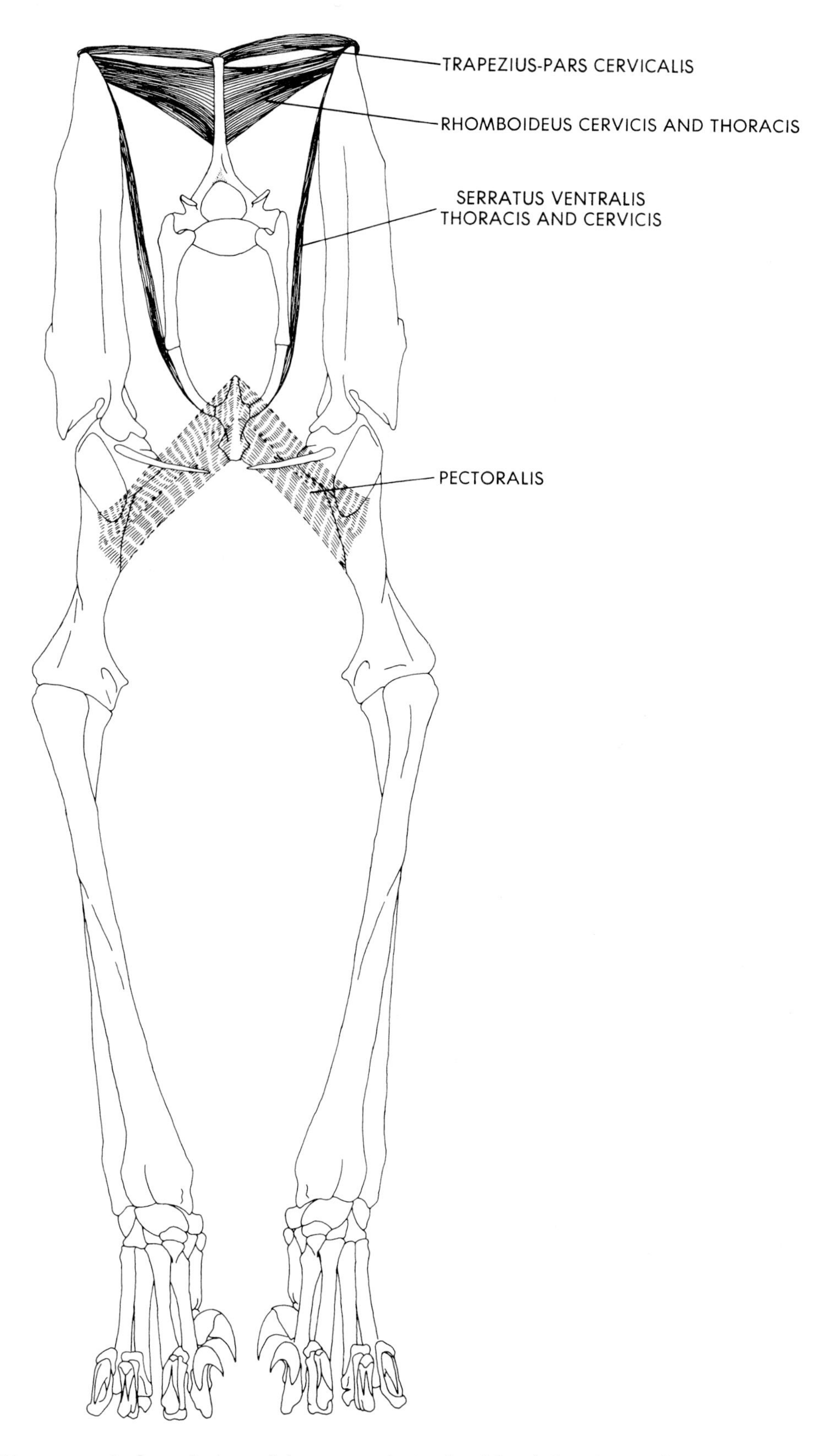

FIGURE 4.10. Diagrammatic frontal view of the cat extrinsic shoulder ("sling") muscles.

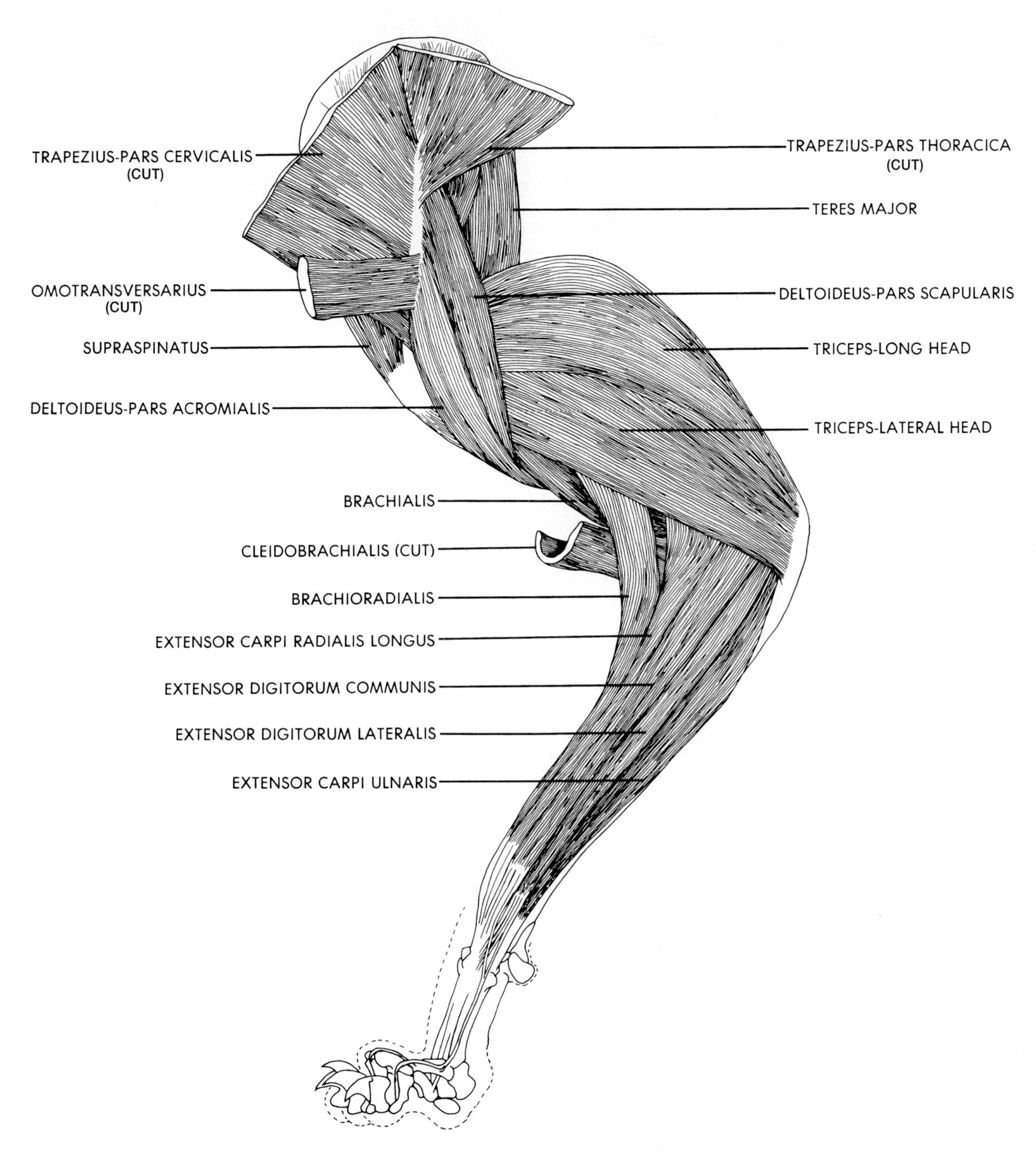

FIGURE 4.11. Deeper lateral muscles of the cat shoulder, brachium, and forearm.

Brachial Musculature

Most of the medial surface of the brachium is covered by the tensor fasciae antebrachii. *Loosen and transect this muscle to observe the muscles of the brachium. Extend the forearm and loosen the triceps brachii laterale with a blunt probe.* This will allow you to observe the other parts of the triceps brachii and anconeus.

Name	Origin	Insertion	Action
TENSOR FASCIAE ANTEBRACHIALIS (TEN-s*or* FASH-*e*-*e* an-t*e*-br*a*-k*e*-AL-is) (epitrochlearis) (figure 4.9)	Tendon of insertion of latissimus dorsi	Olecranon process of ulna	Extend antebrachium

Tensor fasciae antebrachialis has no counterpart in the human and is an extension of the latissimus dorsi muscle in the cat. It is synergistic to the triceps brachii in the cat.

Name	Origin	Insertion	Action
BICEPS BRACHII (B*I*-ceps BR*A*-k*e*-*i*)	Cranial edge of glenoid cavity within the capsule of the shoulder joint (additional origin in man)	Tuberosity of radius	Flex antebrachium
BRACHIALIS (br*a*-k*e*-AL-is) (figure 4.11)	Lateral shaft of humerus from the neck to the	Lateral surface of ulna; distal to supracondyloid ridge	Flex antebrachium semilunar notch

Biceps brachii and brachialis are the main flexors of the antebrachium in both cats and humans. There are two heads of origin for biceps brachii in the human (long and short) and only one in the cat (short). The biceps is larger than the brachialis in man. It is the opposite in the cat because the brachialis is the more powerful flexor when the hand is pronated (the normal posture of cats).

Name	Origin	Insertion	Action
TRICEPS BRACHII (TR*I*-ceps BR*A*-k*e*-*i*) (three heads)	Humerus (two heads) and scapula (one head)	Olecranon of ulna	Extend antebrachium
ANCONEUS figure 4.12	Lateral epicondyle of humerus	Olecranon of ulna	Extend antebrachium

Triceps and anconeus are the extensors of the antebrachium. In quadrupeds, these muscles are very well-developed because they are antigravity muscles, that is, continually supporting the body against the pull of gravity. The anconeus of the human is proportionally quite small, but in the same position as in the cat.

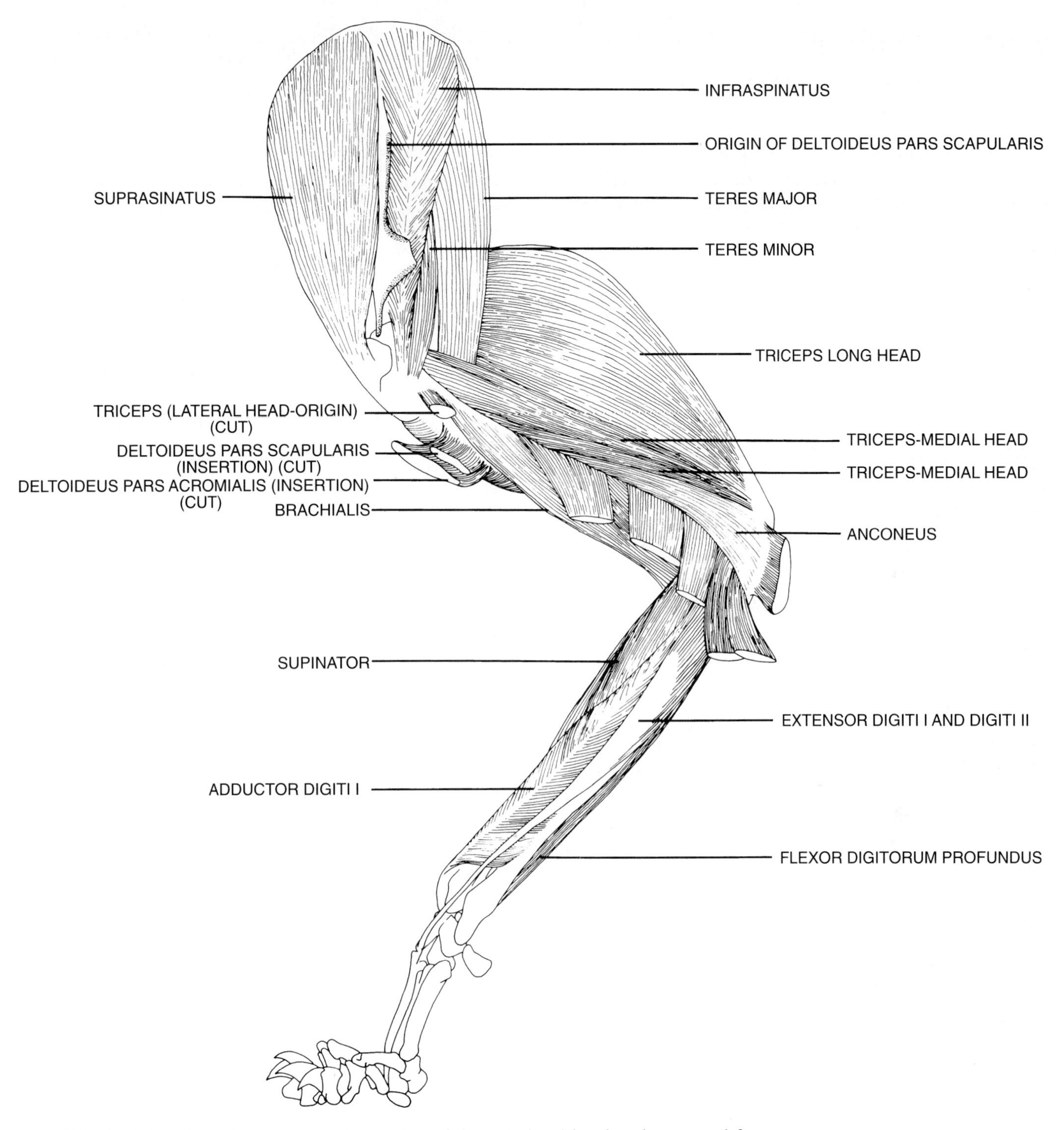

FIGURE 4.12. The deepest lateral muscles of the cat shoulder, brachium, and forearm.

Muscles of the Forearm (Antebrachium)

This dissection is tedious. Check with your instructor to find out if you are to complete this dissection and learn the muscles distal to the elbow.

Cut the brachioradialis, extensor carpi radialis longus, extensor digitorum communis, extensor digitorum lateralis, and extensor carpi ulnaris to observe the deeper forearm muscles.

Now turn to figure 4.14, anterior musculature of the human, and compare the illustrated muscles to those you have just studied on the cat.

Name	Origin	Insertion	Action
BRACHIORADIALIS (br*a*k-*e*-*o*-r*a*-d*e*-AH-lis) figure 4.11	Dorsal border of humerus	Styloid process of radius	Supinate paw (supinate hand, flex forearm in human)
SUPINATOR (SYOO-pi-n*a*-tor) figure 4.12**P**	Lateral ligament between humerus and radius	Proximal and medial shaft of radius	Supinate paw (hand)

The brachioradialis and supinator are the primary supinators of the hand in humans. The muscles of the same name in the cat are not well-developed. A cat in a normal stance holds the paw in a pronated position.

Name	Origin	Insertion	Action
PRONATOR TERES (pro-N*A*-tor) figure 4.9	Medial epicondyle of humerus	Medial shaft of radius	Pronate paw (hand)
EXTENSOR CARPI RADIALIS LONGUS (KAR-p*i* r*a*-d*e*-AH-lis) figure 4.11	Lateral epicondyle of humerus	Second metacarpal	Extend wrist (also abduct hand in human)
EXTENSOR CARPI RADIALIS BREVIS figure 4.13	Lateral epicondyle of humerus	Third metacarpal	Extend paw (extend and abduct wrist in human)
EXTENSOR DIGITORUM COMMUNIS (dig-i-T*O*R-um ko-M*U*-nis) figure 4.11	Lateral epicondyle of humerus	Distal phalanx of digits 2–5	Extend digits (also extend wrist in human)
EXTENSOR DIGITORUM LATERALIS (extensor digiti quinti proprius in human)	Supracondyloid ridge of humerus (extensor tendon in human)	With extensor digitorum communis to middle phalanx of digits except first and second (proximal phalanx of digit 5 in human)	Extend digit(s)
EXTENSOR DIGITI (DIJ-i-t*i*) **I** (Extensor pollicis brevis in human) figure 4.12	Shaft of radius, ulna, and interosseous membrane	Base of first metacarpal (base of proximal phalanx 1 in human)	Extend and abduct pollex
EXTENSOR DIGITI II (Extensor indicis of man)	Ulna near semilunar notch	Middle phalanx of second digit (via tendon of extensor digitorum in human)	Extend second digit

Name (Continued)	Origin	Insertion	Action
EXTENSOR CARPI ULNARIS (KAR-p*i* ul-N*A*-ris) figure 4.11	Lateral epicondyle of humerus and ulna near semilunar notch	Base of fifth metacarpal	Extend wrist (also adduct hand at wrist in human)
FLEXOR CARPI RADIALIS figure 4.13	Medial epicondyle of humerus	Base of second and third metacarpal	Flex paw (flex wrist, abduct hand, and flex forearm)
PALMARIS BREVIS (pal-M*A*-ris) (not illustrated)	Medial epicondyle of humerus (palmar aponeurosis in human)	Palmar fascia to the base of each digit (skin on ulnar side of hand in human)	Flex digits (draw ulnar side of palm mediad in human)
FLEXOR CARPI ULNARIS figure 4.13	Medial epicondyle of humerus and olecranon of ulna	Accessory carpal bone (pisiform in human)	Flex wrist
FLEXOR DIGITI I BREVIS (flexor pollicis brevis of human) (not illustrated)	Annular ligament over radial carpal and carpal III bones (trapezium in human)	Proximal phalanx of digit 1	Flex first digit
FLEXOR DIGITORUM SUPERFICIALIS (not illustrated)	Insertion of tendon of palmaris brevis and flexor digitorum profundus (medial epicondyle of humerus and ulna, radius in human)	Middle phalanx of digits 2–5	Flex digits 2–5 (also flex forearm in human)
FLEXOR DIGITORUM PROFUNDUS figure 4.12	Medial epicondyle of humerus, radius, and ulna (primarily ulna in human)	Distal phalanx of each digit (2–5 in human)	Flex digits (also flex hand at wrist in human)
PRONATOR QUADRATUS (kwod-R*A*-tus) (not illustrated)	Ulna and interosseus membrane	Ventral surface of radius	Pronate paw (hand)
LUMBRICALES (lum-bri-K*A*-lez) (not illustrated)	Ventral surface of flexor digitorum profundus tendon	Base of proximal phalanx of digits (extensor digitorum in human)	Adductor of digits (flex meta-carpophalangeal joints and extend two distal phalanges in human)
ADDUCTOR DIGITI I (adductor pollicis of human) figure 4.12	Carpal III (capitate bone and metacarpals 2, 3 in human)	Proximal phalanx of digit 1	Adduct pollex
INTEROSSEI (in-ter-OS-*e-i*) (not illustrated)	Ventral surface of metacarpals	Proximal phalanx of digits	Flex digits toward pollex (abduct and adduct digits in human)

The interossei of the human consist of a dorsal group and a palmar group. The palmer group adducts the digits, flexes the metacarpophalangeal joint, and extends the middle and distal phalanx. Interossei dorsales abducts the digits, flexes the metacarpophalangeal joint, and extends the middle and distal phalanges.

Name (Continued)	Origin	Insertion	Action
FLEXOR DIGITI II (no homologous muscle in human) (not illustrated)	Ventral surface of second metacarpal	Promimal phalanx of second digit	Flex second digit
ABDUCTOR DIGITI II (no homologous muscle in human) (not illustrated)	Ventral surface of second metacarpal and carpal I	Proximal phalanx of second digit	Abduct second digit
ADDUCTOR DIGITI II (no homologous muscle in human) (not illustrated)	Carpal III	Proximal phalanx of second digit	Adduct second digit
ABDUCTOR DIGITI V (abductor digiti minimi of human)	Accessory carpal bone (pisiform in human)	Proximal phalanx of fifth digit	Abduct fifth digit
FLEXOR DIGITI V BREVIS (flexor digiti minimi brevis of human) (not illustrated)	Fifth metacarpal and carpal IV (hamate in human)	Proximal phalanx of digit 5	Flex fifth digit
ADDUCTOR DIGITI V (similar to opponens digiti minimi of human (not illustrated)	Carpal III bone (hamate of human)	Fifth metacarpal and proximal phalanx of digit 5 (fifth metacarpal only of human)	Adduct fifth digit

Man and cat have very similar musculature distal to the elbow. However, in the human, two extensors work on the thumb, while there is only one extensor in the cat. Three digits of the cat are extended by the extensor digitorum lateralis; in man only the fifth digit is extended. All digits but the thumb are operated by the extensor digitorum communis in the cat, but in the human only the third and fourth digits are affected.

In humans, the palmaris longus is absent in about one out of ten individuals. Flexor digitorum profundus is proportionately larger in the cat than in the human. This is a result of the quadrupedal posture and locomotion of the cat.

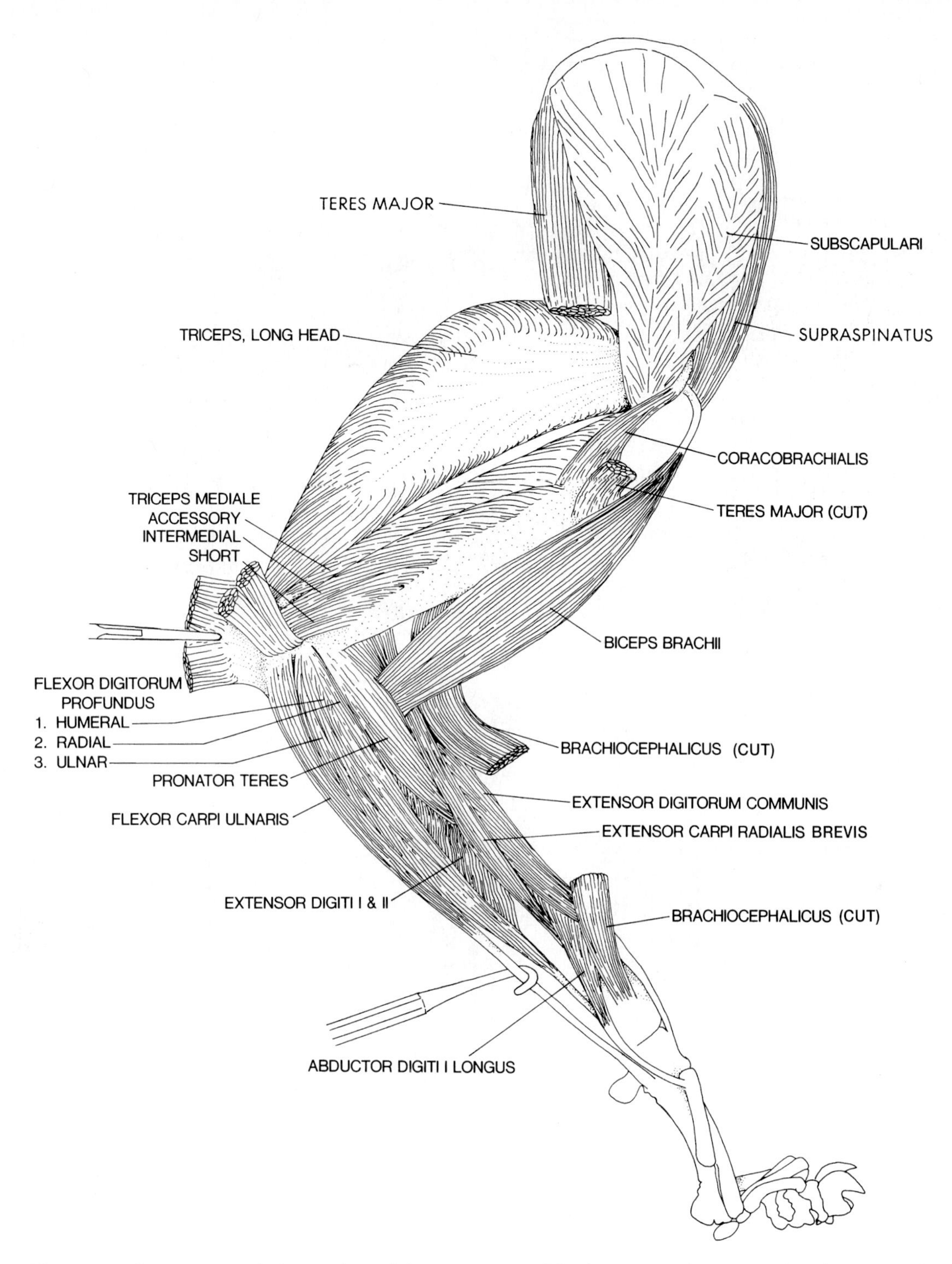

FIGURE 4.13. Deep medial muscles of the cat pectoral limb. Tensor fascia antebrachialis is reflected at the elbow with a hemostat. Flexor digitorum superficialis and flexor carpi radialis are reflected at the elbow.

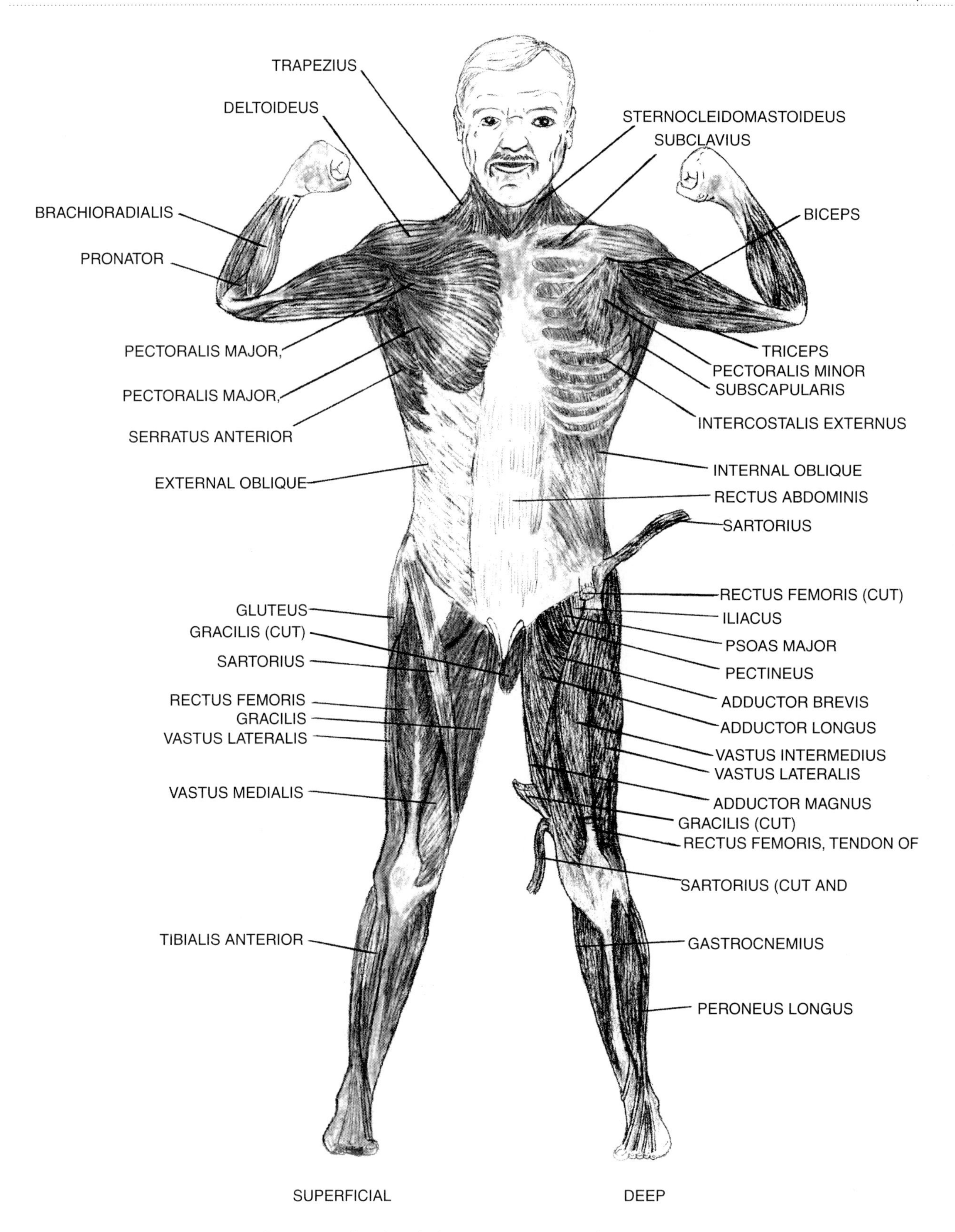

FIGURE 4.14. Anterior view of the superficial and deep musculature of the human.

Neck, Back, and Trunk Musculature

Complete the dissection of the pectoral limb prior to undertaking the following dissection.

The superficial and ventral neck muscles were discussed on page 53, but it was necessary to postpone dissection of the dorsal deep neck muscles until you had dissected the pectoral girdle musculature. Most of the dorsal deep neck muscles are continuations of the deep back and trunk musculature.

The **RECTUS CAPITIS VENTRALIS** fibers run in the same oblique direction as the **LONGUS COLLI** (LONG-gus K*OL-i*) fibers and just rostral to the longus colli, inserting on the atlas. Lateral to the **LONGUS CAPITIS** is the longus colli.

Deep Neck Muscles

1. Ventral (figure 4.15)

The midline neck muscles beneath the trachea and esophagus are the longus colli muscles, running from the thorax to the axis. The rectus capitis ventralis fibers run in the same oblique direction as the longus colli fibers and just rostral to the longus colli, inserting on the atlas. The longus capitis is lateral to the longus colli.

Name	Origin	Insertion	Action
LONGUS COLLI (LONG-gus K*OL-i*) figure 4.15	Transverse processes and centra of cervical and first six thoracic vertebrae	Ventral centrum of all cervical vertebrae	Flex neck
LONGUS CAPITIS (not illustrated)	Ventral transverse processes of second to sixth cervical vertebrae	Rostral portion of basioccipital	Flex head
RECTUS CAPITIS VENTRALIS	Ventral transverse process of atlas	Exoccipital; lateral to condyles	Flex head laterad

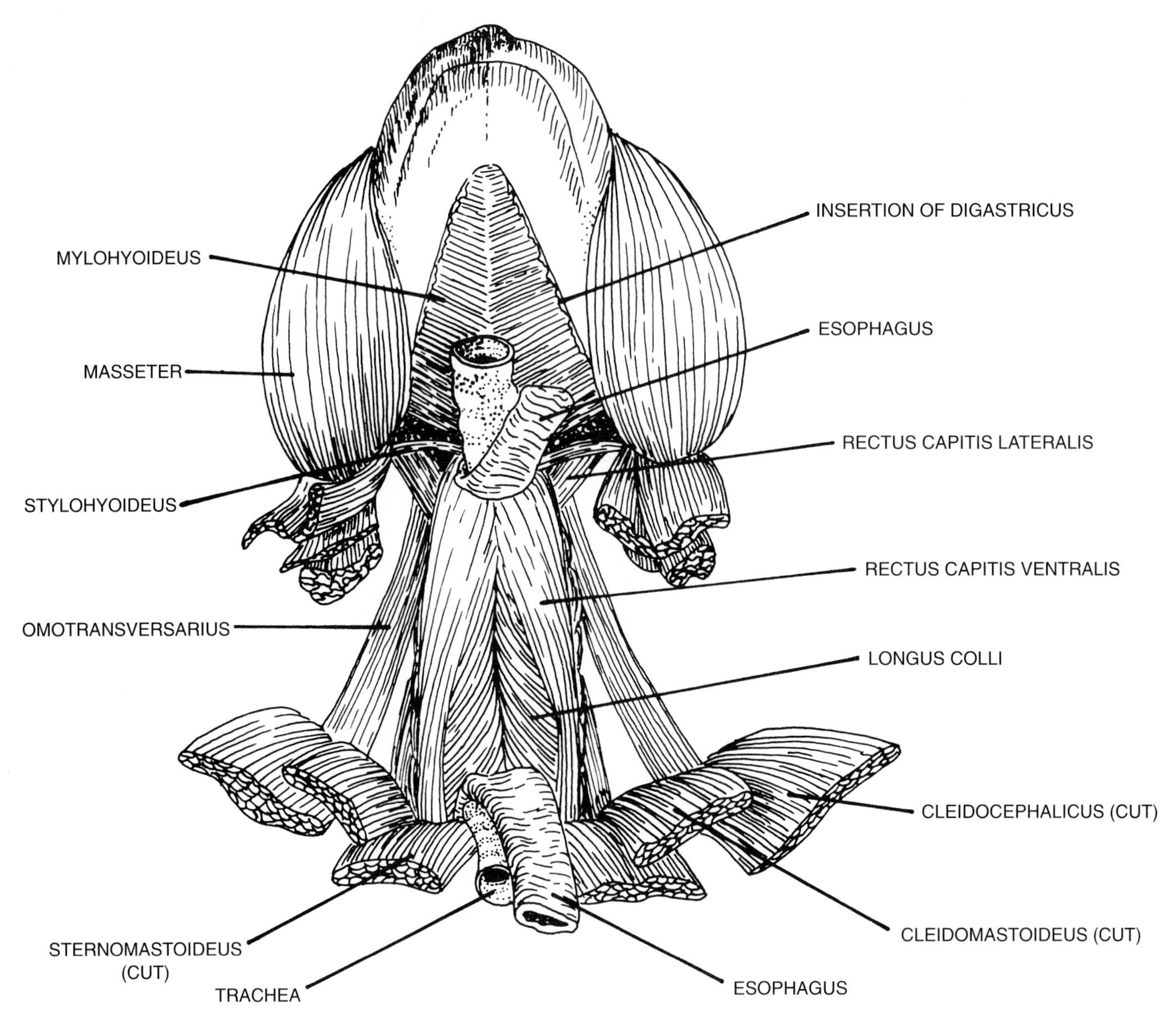

FIGURE 4.15. Ventral view of the deep neck muscles of the cat.

2. Dorsal (figures 4.16, 4.17)

The splenius covers the dorsal half of the neck. *Isolate, cut, and reflect the splenius.* The semispinalis capitis (in two parts, the biventer cervicis and complexus) lies deep to the splenius. After examining the origins and insertions, *cut and reflect both parts of the semispinalis capitis.*

Name	Origin	Insertion	Action
SPLENIUS (SPL*E*-n*e*-us) figure 4.17	Nuchal ligament from first two thoracic vertebra	Lambdoidal crest of occipital bone	Raise (extend in human) and turn head laterad
SEMISPINALIS (sem-e-spi-NA-lis) **CAPITIS** (biventer cervicis and complexus) figure 4.16	Transverse processes of cervical and thoracic vertebrae	Spinous processes of cervical vertebrae and occipital bone	Extend vertebral column and head

3. Lateral (figures 4.16, 4.17)

The scalenus muscles originate on either side of the insertion of the serratus ventralis muscles and insert on the transverse processes of the last four cervical vertebrae. The longissimus cervicis insertion is just dorsal to the insertion of the scalenus. The longissimus capitis extends forward to the head (mastoid process) just dorsal to the insertion of the longissimus cervicis. *Cut the longissimus capitis and reflect toward its origin and insertion.*

Name	Origin	Insertion	Action
SCALENUS DORSALIS (sk*a*-L*E*-nus) (**posterior** in human) figure 4.17	Second and third ribs	Transverse processes of fourth to seventh cervical vertebrae	Flex neck; may draw ribs forward
SCALENUS MEDIUS	Sixth to ninth ribs	Transverse processes of fourth to seventh cervical vertebrae	Flex neck
SCALENUS VENTRALIS (**anterior** in human)	Third or fourth rib	Transverse processes of fourth to seventh cervical vertebrae	Flex neck

In cats and humans, the origins, insertions, and actions of muscles are very similar, for the most part. For the deep neck muscles listed, consult a textbook of human anatomy when precise information is required. The differing postures of cat and human result in slightly different origins and insertions for these muscles.

The scalenus muscles in humans are also active during rested inspiration. Though the scalenus of humans act on only a limited portion of the rib cage, the cephalad motion of the sternum during inspiration likely results from contraction of the scalenus muscles.

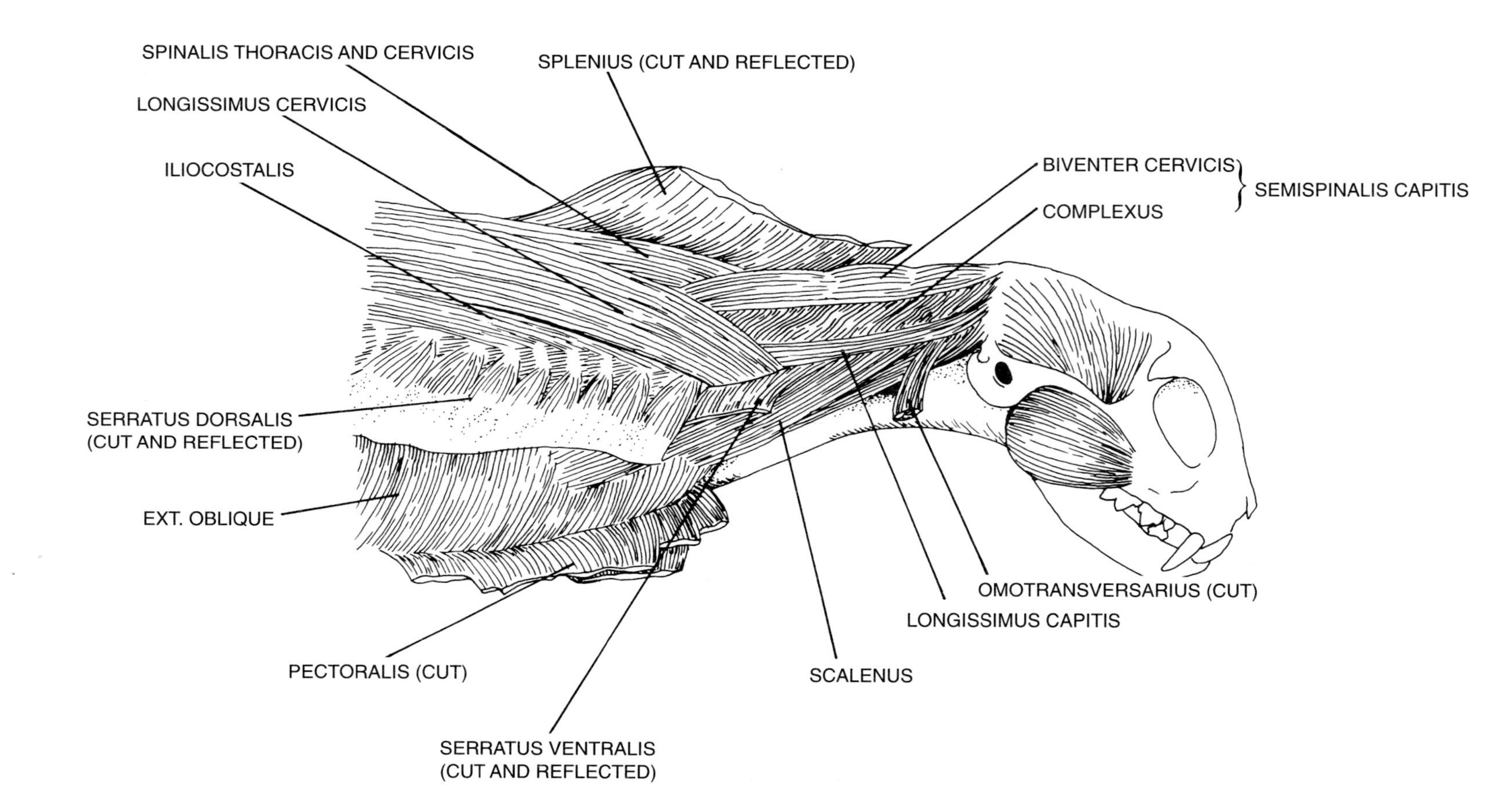

FIGURE 4.16. Lateral view of the deep neck muscles of the cat.

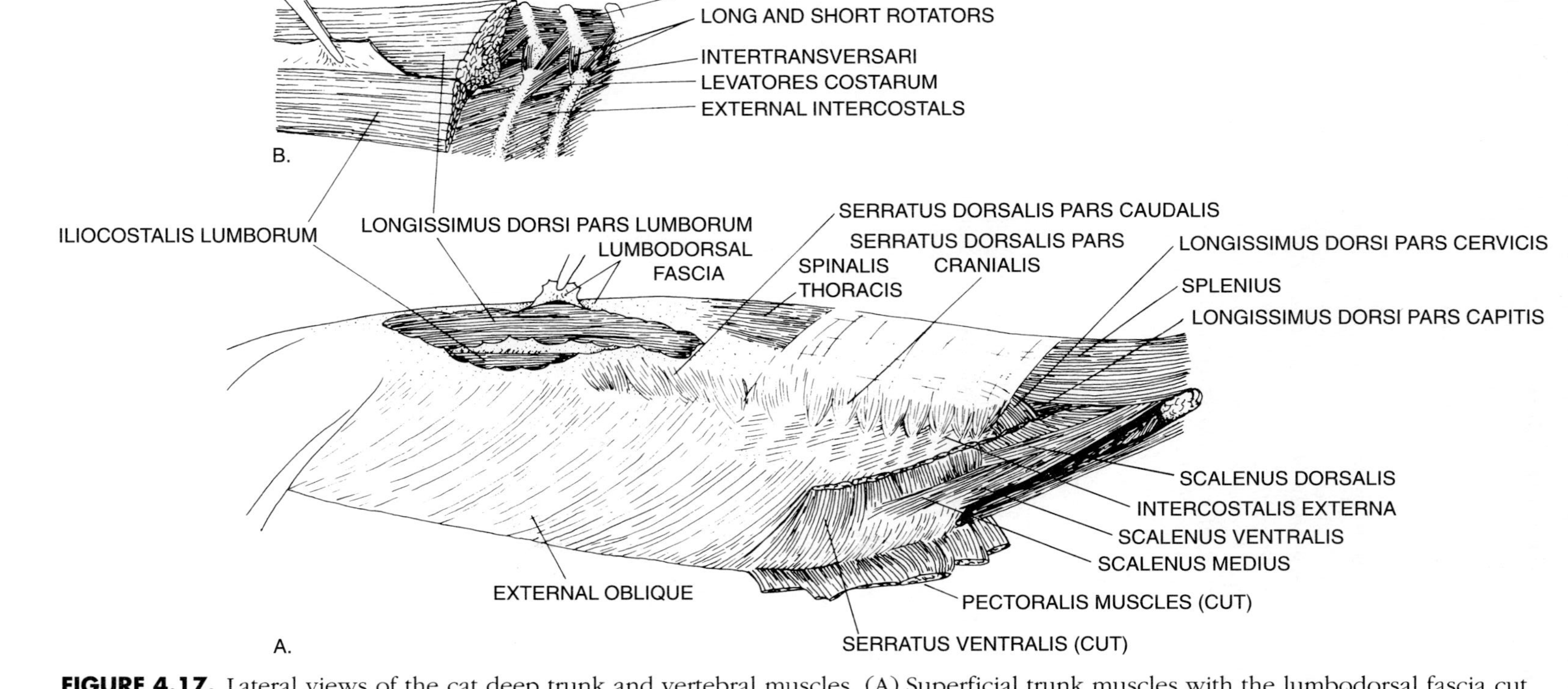

FIGURE 4.17. Lateral views of the cat deep trunk and vertebral muscles. (A) Superficial trunk muscles with the lumbodorsal fascia cut and reflected. (B) Detail of the deeper vertebral muscles.

Back and Trunk Musculature

1. Back (figure 4.17)

Pick away fibers of the intercostales externa to expose the intercostales interna. The transversus thoracis is on the inner surface of the cartilaginous ribs and sternum. As you examine these muscles, BE CAREFUL NOT TO DESTROY THE THORACIC MEMBRANES OR BLOOD VESSELS BENEATH THEM. *Pick away fibers of the external oblique to reveal the internal oblique and continue through the internal oblique to expose the transversus abdominis.*

The caudal back muscles are covered by the heavy lumbodorsal fascia. *Pinch the lumbodorsal fascia with forceps and cut through the fascia lateral to the midline without cutting the muscle beneath it. Continue to cut away the fascia laterad until the fascia passes deep to the lateral border of the longissimus. Grasp the lateral cut border of the fascia with forceps and cut the fascia lateral to this deep fascia to expose the iliocostalis.* Note that the caudal portion of the serratus dorsalis originates on the fascia covering the iliocostalis, and the spinalis originates on the rostral portion of the longissimus fascia. The cranial portion of the serratus dorsalis originates from a sheet of fascia over the dorsal thoracic region. *Cut the sheet of fascia so that the origin of the serratus dorsalis is not disturbed.* The rest of the spinalis and longissimus are exposed by this procedure.

Name	Origin	Insertion	Action
SPINALIS (sp*i*-NA-lis) **DORSI**	Caudal four thoracic vertebrae (in human, from lumbar to cervical vertebrae)	Thoracic and cervical vertebrae (in human, from thoracic to occipital)	Extend vertebral column
LONGISSIMUS (long-JIS-i-mus) **DORSI** (extends from the ilium to the skull; subdivided into four parts)	Iliac crest, spinous processes of lumbar vertebrae, articular processes of thoracic and cervical vertebrae	Lumbar vertebrae, thoracic vertebrae and ribs, transverse processes of cervical vertebrae, mastoid process of temporal bone (in human thoracic vertebrae to mastoid)	Each origin has fibers extending to an insertion immediately cranial to it, thus extending the spine in that region (in human, capitis portion; also extends head)
ILIOCOSTALIS (il-*e*-*o*-kos-TAH-lis)	Ilium, deep lumbar fascia, transverse processes of lumbar vertebrae (and ribs in human)	Ribs of thoracic vertebrae and (not in human) transverse processes of lumbar vertebrae (also transverse processes of seventh cervical in human)	Draw ribs caudad (and extend vertebral column in human)

Name (Continued)	Origin	Insertion	Action
SERRATUS (ser-RA-tus) **DORSALIS** (posterior in human) (divided into cranial and caudal portions)	Fascia of the spinalis and longissimus muscles	Ribs	Cranial portion: draw ribs craniad; caudal portion: draw ribs caudad
INTERSPINALIS	Spinous processes of vertebrae	Spinous processes of vertebrae	Draw vertebral spines together to extend spine
MULTIFIDUS SPINAE (mul-TIF-i-dus SPI-ne) (not illustrated)	Lamina and mammillary processes of sacral, lumbar vertebrae and transverse processes of thoracic vertebrae (to the axis in the human)	Spinous process of third to fifth vertebrae cranial of fiber origin	Extend vertebral column

2. Trunk (figures 4.17–4.19)

Name	Origin	Insertion	Action
INTERNAL INTERCOSTALS (in-ter-KOS-tals) (not illustrated)	Vertebral half of ribs cranial to insertion	Sternal half of ribs caudal to origin	Draw ribs craniad to expand the thoracic cavity during inhalation
EXTERNAL INTERCOSTALS figure 4.17	Sternal half of ribs cranial to insertion	Vertebral half of ribs caudal to insertion	Draw ribs craniad to expand the thoracic cavity during inhalation

There is a continuing discussion about the functions of the intercostals in humans. In stimulation experiments, it has been demonstrated in the dog that the external intercostals are inspiratory and the internal intercostals are expiratory. The anatomy of the chest suggests otherwise, however. The ribs of mammals have two parts: an ossified portion that is jointed to the thoracic vertebrae and a cartilage portion that is jointed to the sternum or to the cartilage rib just cranial to the ones not attached to the sternum. The external intercostal muscles are attached to the bony ribs with the origin of the muscle on the craniad rib, near the joint with the thoracic vertebra. The insertion of the external intercostal is more distal to the thoracic vertebrae on the rib caudal to the rib with the muscle's origin. Thus, when the external intercostal muscle contracts, it moves the caudal rib toward the cranial rib. The internal intercostal muscle fibers are arranged in the opposite direction, but they are attached primarily to the cartilage portion of the rib. Thus, the origins and insertions of the internal intercostal muscles are the reverse of the external intercostals on the cartilage portion of the ribs. Following this reasoning, both internal and external intercostal muscles inflate the lungs when they contract.

It is possible that these muscles exist primarily as a reserve system for breathing in addition to the respiratory movements of the diaphragm. They may also contribute to trunk rotation.

Name (Continued)	Origin	Insertion	Action
TRANSVERSE THORACIC (th*o*-RAS-ik) (not illustrated)	Dorsal sternum	Cartilage ribs	Draw ribs craniad to expand the thoracic cavity during inhalation

The human transverse thoracic is a much different muscle than that of the cat. Called the **triangularis sterni,** it is located on the inner surface of the rib cage and is active during deep expiration, speech, coughing, and laughing. It originates from the sternum and inserts on the costal cartilages of ribs two through six.

Name	Origin	Insertion	Action
DIAPHRAGM (D*I*-a-fram) (not illustrated)	Ventral surface of centra of second to fourth lumbar vertebrae and xiphoid process of sternum	Central semilunar tendon	Depress abdominal viscera and expand thoracic cavity
EXTERNAL ABDOMINAL OBLIQUE (*o*-BL*E*K) figure 4.18	Caudal ribs and lumbodorsal fascia	Aponeurosis to linea alba (also iliac crest in human)	Support abdominal viscera; compress abdomen
INTERNAL ABDOMINAL OBLIQUE figure 4.19	Lumbodorsal fascia and crural arches between crest of ilium and pubis	Aponeurosis to linea alba (also inferior four ribs in human)	Support abdominal viscera; compress abdomen
TRANSVERSUS ABDOMINIS	Inner surface of caudal cartilage ribs, transverse processes of lumbar vertebrae, ilium and crural arches between ilium and pubis (also inferior six ribs in human)	Aponeurosis of linea alba	Support abdominal viscera; compress abdomen
RECTUS ABDOMINIS figure 4.18	Tubercle of pubis	First and second cartilage ribs	Support abdominal viscera; flex spine

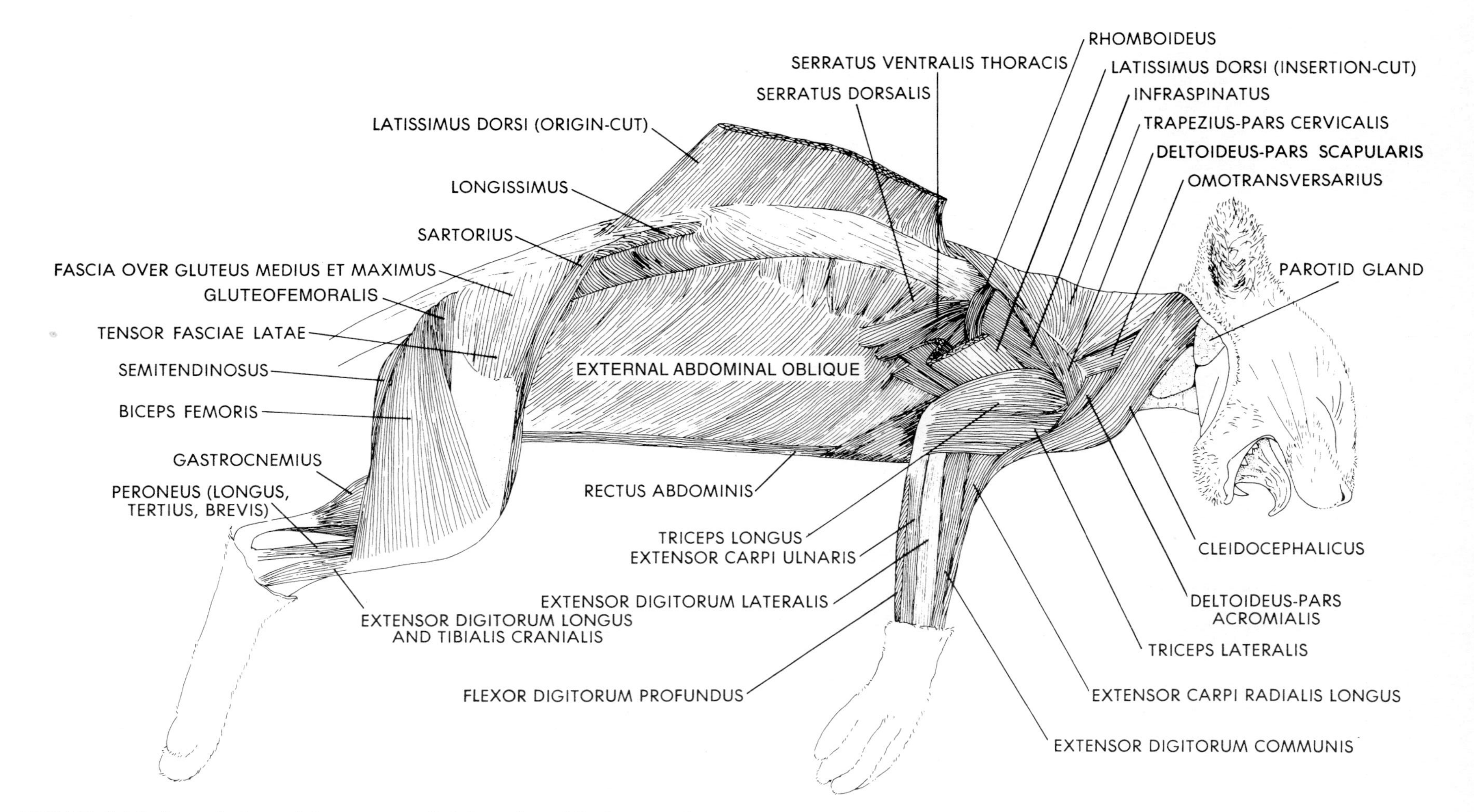

FIGURE 4.18. Lateral view of the cat superficial trunk and limb musculature.

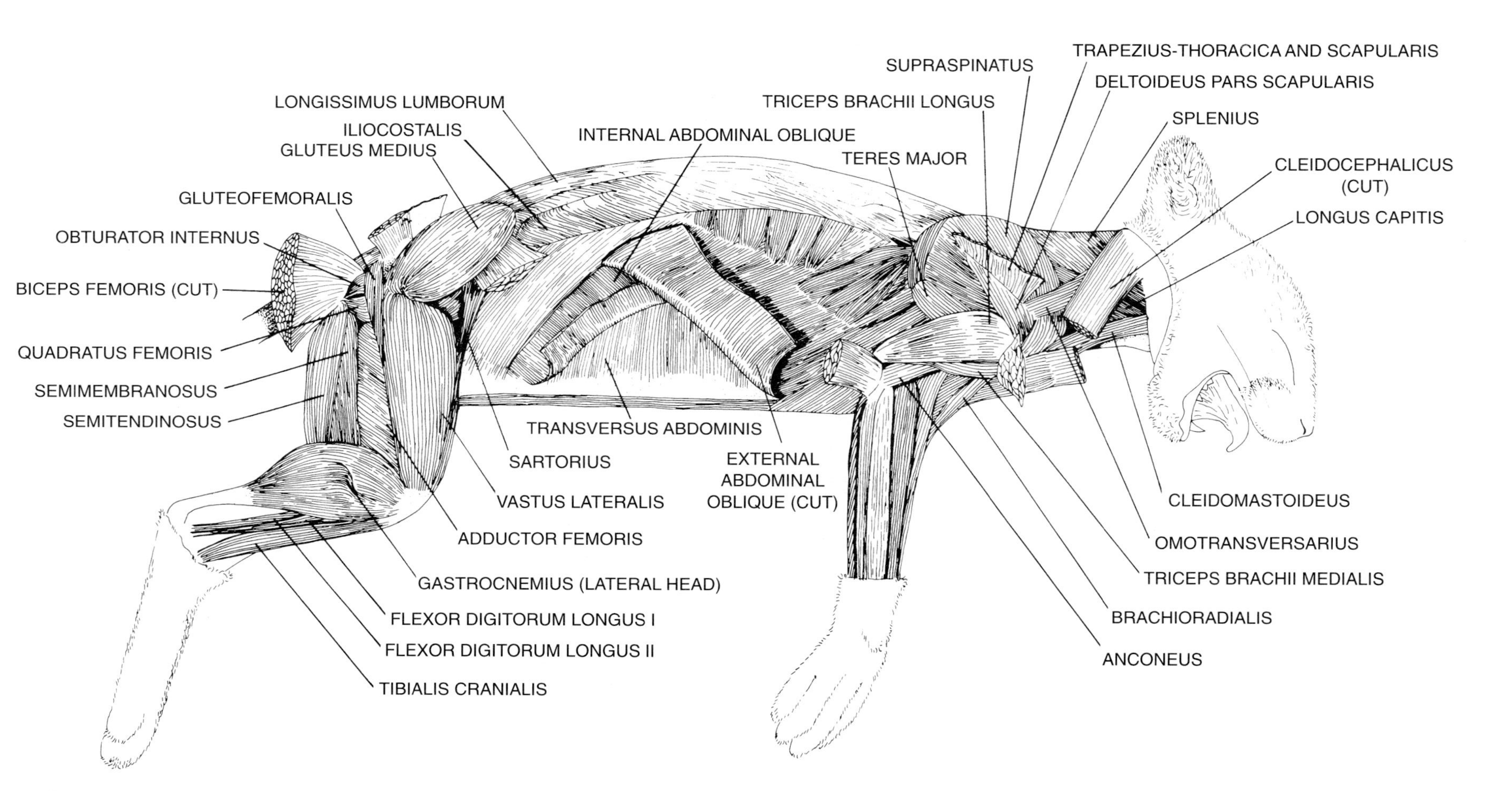

FIGURE 4.19. Lateral view of the cat deep trunk and limb musculature.

Pelvic Girdle, Hip, and Thigh Musculature

1. Lateral, superficial (figure 4.20)

Remove the skin from the thigh and leg and clean away the fat on the surface of the thigh.

Name	Origin	Insertion	Action
TENSOR FASCIA LATAE (TEN-s*or* FASH-*e*-a L*A*-t*e*)	Ilium and fascia of gluteus medius in cat; iliac crest and anterior superior iliac spine in human	Fascia latae (connective tissue sheet covering the gluteus medius and vastus lateralis)	Tense fascia latae and extend leg (flex thigh in human)
GLUTEUS SUPERFICIALIS (GLOO-t*e*-us) (**maximus** in human)	Transverse processes of last sacral and first caudal vertebrae and fascia of hip region	Femur near greater trochanter and fascia latae	Abduct and extend thigh

> The gluteus maximus of the human is proportionately larger than that of the cat. This is related to the posture differences. In the human, the muscle lifts the body from a sitting or squatting position and acts as a high gear muscle during walking and running. The limb of cats is accelerated by the biceps femoris and the gluteus muscle is reduced.

Name	Origin	Insertion	Action
GLUTEUS MEDIUS	Ilium	Greater trochanter	Abduct thigh
BICEPS FEMORIS (B*I*-ceps FEM-*o*-ris)	Tuberosity of ischium	Dorsal border of tibia and patella	Abduct and extend thigh; flex shank

> The biceps femoris is much more powerful in a cat than in a human. In the cat, it extends the femur, propels the body forward, and supports the body against gravity when the knee is flexed. In the human, this muscle is a flexor of the knee joint, though it also extends the thigh when the hip is flexed. Because the insertion is far distal on the appendage, it produces power when initiating extension of a thigh that is already flexed. Thus, in the human, biceps is a low gear muscle used in walking and running. Biceps femoris is part of the hamstring group described on the bottom of page 83.

Name	Origin	Insertion	Action
GLUTEOFEMORALIS (gloo-t*e*-*o*-fem-*o*r-AL-is) (not found in human)	Transverse processes of second and third caudal vertebrae	Patella	Abduct and extend thigh
SARTORIUS (sar-T*O*-r*e*-us)	Crest of ilium	Shaft of tibia	Adduct femur and extend tibia (flex leg and thigh in human)

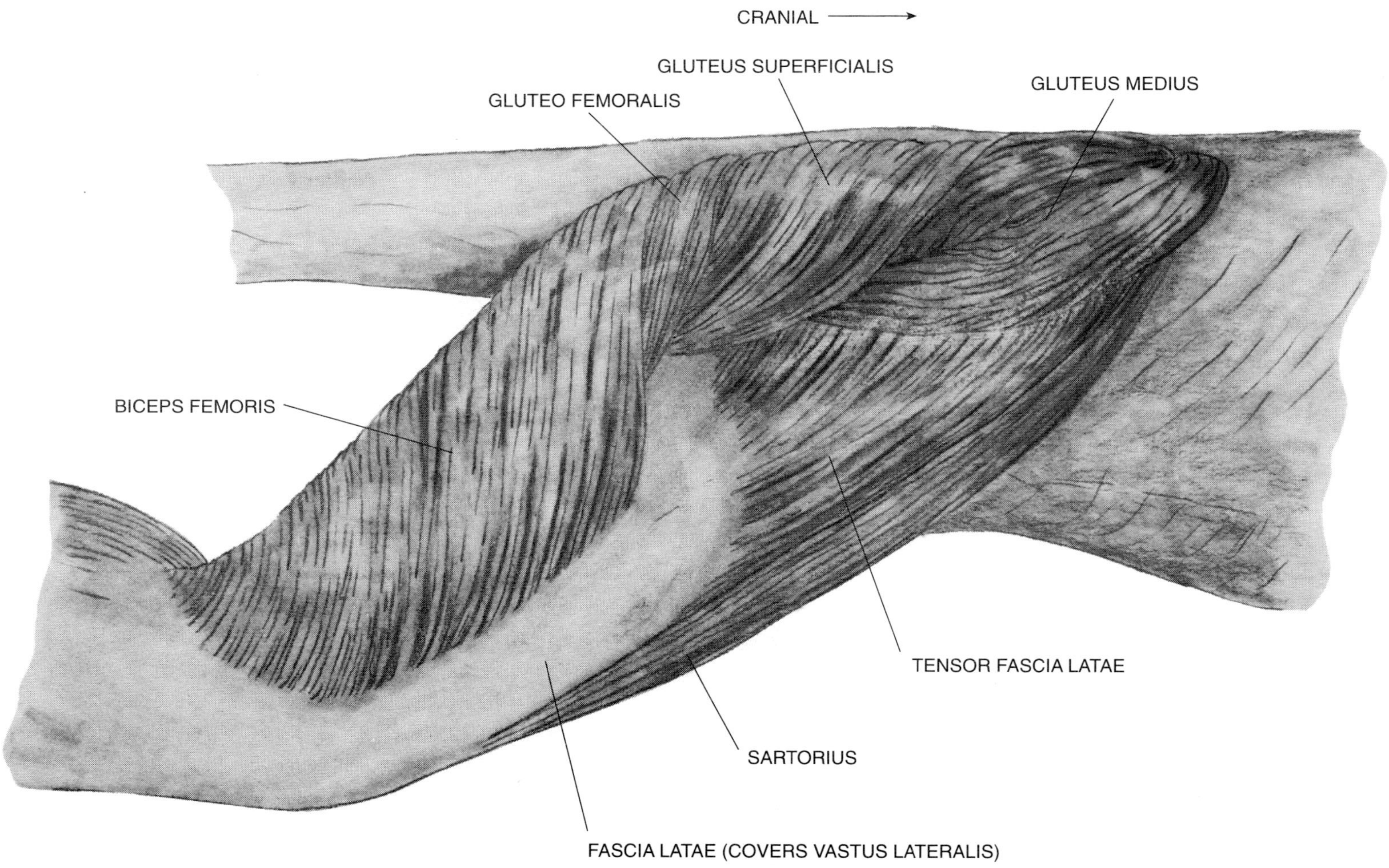

FIGURE 4.20. Superficial muscles of the cat hip and thigh, lateral view.

2. Middle layer (figure 4.21)

Clean and separate the biceps femoris and sartorius from the deeper muscles. BE VERY CAREFUL TO AVOID CUTTING THE SLENDER ABDUCTOR CRURIS CAUDALIS THAT LIES JUST BENEATH THE BICEPS FEMORIS. *Transect the biceps femoris and gluteofemoralis muscles. Separate the tensor fascia latae from the deeper vastus lateralis and gluteofemoralis and cut the tendon (fascia latae) rather than the muscle. Transect the gluteus superficialis muscle.*

Name	Origin	Insertion	Action
ABDUCTOR CRURIS CAUDALIS (KROO-ris) (not found in human)	Transverse process of second caudal vertebra	Insertion tendon of biceps femoris	Extend thigh
GLUTEUS PROFUNDUS (GLOO-*te*-us pro-FUN-dus) (called gluteus minimus in human)	Ventral half of ilium	Greater trochanter	Rotate femur (also abduct femur in human)
OBTURATOR INTERNUS (OB-tu-*ra*-tor)	Medial surface of ischium	Trochanteric fossa	Abduct thigh (also rotate thigh in human)
PIRIFORMIS (pir-i-FOR-mis)	Transverse processes of last two sacral and first caudal vertebrae	Greater trochanter (difficult to separate from gluteus medius)	Abduct and extend thigh (abduct and rotate thigh in in human)
QUADRICEPS FEMORIS (KWOD-ri-ceps FEM-*o*-ris)			
1. **RECTUS FEMORIS**	Ventral border of ilium	Tuberosity of tibia via ligamentum patellae	Extend leg; flex thigh
2. **VASTUS** (VAS-tus) **LATERALIS**	Shaft and greater trochanter of femur	Tuberosity of tibia via ligamentum patellae	Extend leg
3. **VASTUS INTERMEDIUS** figure 4.22	Shaft of femur beneath vastus lateralis and rectus femoris	Capsule of knee joint	Extend leg
4. **VASTUS MEDIALIS** figure 4.23	Medial surface of shaft of femur	Patella and head of tibia	Extend leg

Only the rectus femoris of the quadriceps group crosses both the hip and knee. This is true of the cat and human. Therefore, the action of the quadriceps group is primarily the extension of the leg (the kicking motion of straightening the knee). The patella guides the patellar ligament to the tibial tuberosity on the anterior, proximal tibia. The patella also adds to the leverage that can be exerted on the tibia.

Name	Origin	Insertion	Action
SEMITENDINOSUS (sem-*e*-ten-di-NO-sus) figure 4.22	Tuberosity of ischium	Crest of tibia	Extend thigh; flex leg
SEMIMEMBRANOSUS (sem-*e*-mem-bra-NO-*sus*)	Tuberosity and ramus of ischium	Medial epicondyle of femur (proximal tibia in human)	Extend thigh (also flex leg in human)

The biceps femoris, semitendinosus, and semimembranosus are collectively termed the **hamstrings.** The medial insertion of the semitendinosus and semimembranosus is opposite the lateral insertion of biceps and, thus, prevents rotating forces on the knee joint.

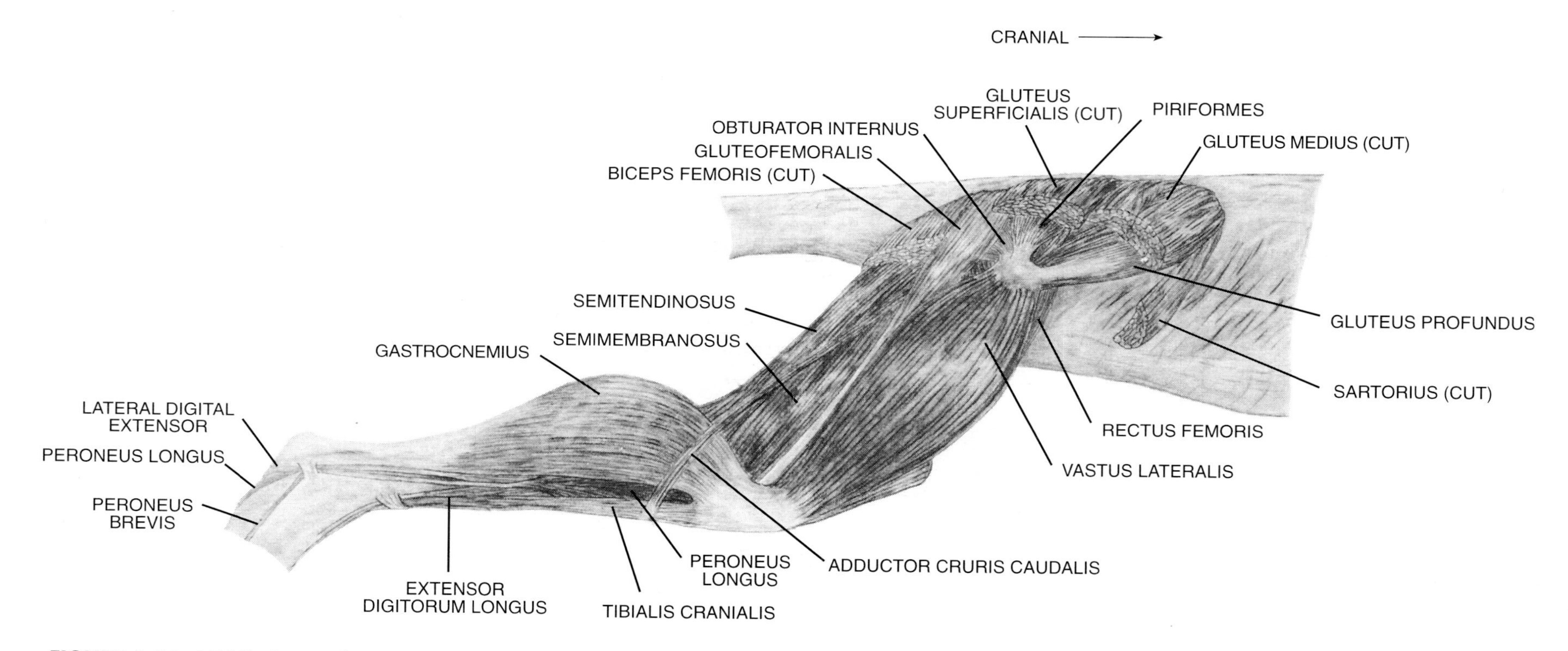

FIGURE 4.21. Middle layer of muscles of the cat hip, thigh, and shank, lateral view.

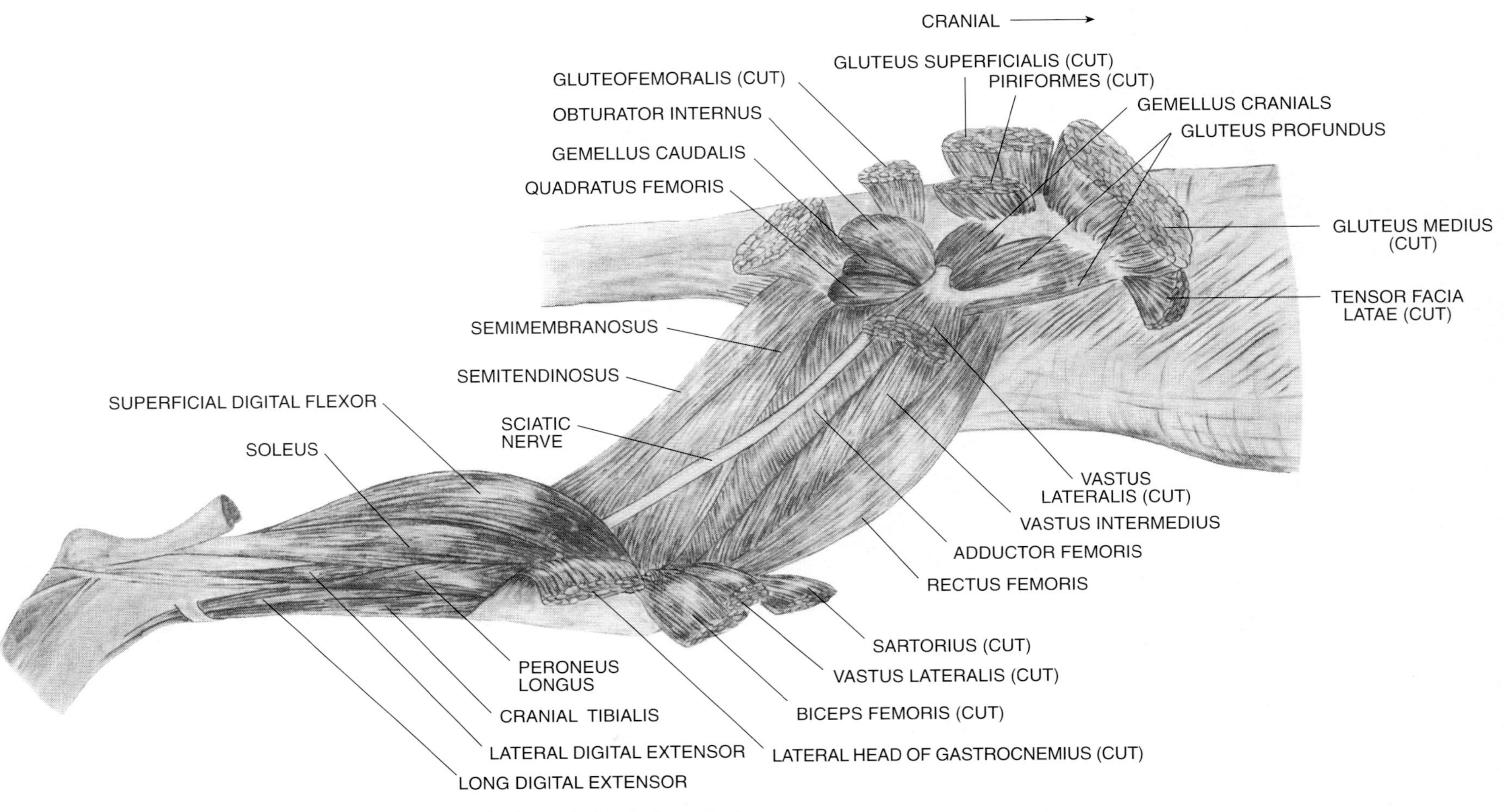

FIGURE 4.22. Deep muscles of the cat hip, thigh, and shank, lateral view.

3. Deep (figure 4.22)

Press the semimembranosus caudad with your probe to expose the quadratus and gemellus muscles. Probe deeper in this area and you will find the obturator muscles. If you have difficulty, cut and reflect the semitendinosus and semimembranosus. This will make the hip joint more flexible and help to expose the obturators. Separate and transect the gluteus medius and piriformis to expose the gluteus profundus.

Name	Origin	Insertion	Action
GEMELLUS (jem-EL-us) **CRANIALIS** (**superior** in human)	Dorsal border of ilium and ischium	Greater trochanteric fossa	Rotate, abduct thigh (rotate thigh laterad in human)
GEMELLUS CAUDALIS (**inferior** in human)	Dorsal border of ischium continuous with gemellus cranialis	Greater trochanter of femur	Rotate, abduct thigh (rotate thigh laterad in human)
QUADRATUS FEMORIS (kwod-R*A*-tus)	Caudal border of ischium near tuberosity	Greater and lesser trochanters of femur (intertrochanteric line of human)	Extend thigh (rotate thigh laterad in human)

4. Medial (figure 4.23)

Note the exit of the femoral artery and vein from the abdominal cavity to the ventromedial surface of the hind limb. *Your probe will fit between the blood vessels and the ventral tendon to which the oblique musculature attaches. Pull your probe forward in this gap and separate the tendon and oblique muscles from the ventral surface of the ilium as shown in figure 4.23B.* BE CAREFUL NOT TO TEAR THE UNDERLYING LOIN MUSCLES.

Name	Origin	Insertion	Action
PSOAS (S*O*-as) **MINOR** (often absent in human) (not illustrated)	Centra of last thoracic and first four lumbar vertebrae (thoracic 12 and lumbar 1 in the human)	Iliopectineal line cranial to acetabulum	Flex vertebral column
ILIOPSOAS (il-*e*-*o*-S*O*-as) (**PSOAS MAJOR** + **ILIACUS** [IL-*e*-ak-us])	Psoas major, transverse processes and centra of lumbar vertebrae and iliacus, ventral border of ilium (the iliac fossa and crest in human) (DO NOT CUT THE NERVE PLEXUS BENEATH THIS MUSCLE)	Lesser trochanter of femur	Rotate thigh laterad (psoas major flexes the thigh and vertebral column; iliacus flexes the thigh)
QUADRATUS LUMBORUM (lum-B*O*-rum) (not illustrated)	Last two thoracic vertebrae and last rib (iliac crest in human)	Transverse processes of lumbar vertebrae and ilium (twelfth rib and transverse processes of L1–L4)	Flex vertebral column (lumbar spine in human) laterad
PECTINEUS (pek-TIN-*e*-us)	Anterior pubis	Shaft of femur near lesser trochanter	Adduct thigh
ADDUCTOR MAGNUS and **BREVIS**	Pubis and ischium	Ventral shaft of femur distal to adductor longus	Extend and adduct thigh
ADDUCTOR LONGUS	Ramus of pubis	Linea aspera on shaft of femur	Extend and adduct thigh
GRACILIS (gras-IL-is)	Symphysis pubis and ischium	Aponeurosis over sartorius and shaft of tibia	Adduct thigh; flex leg

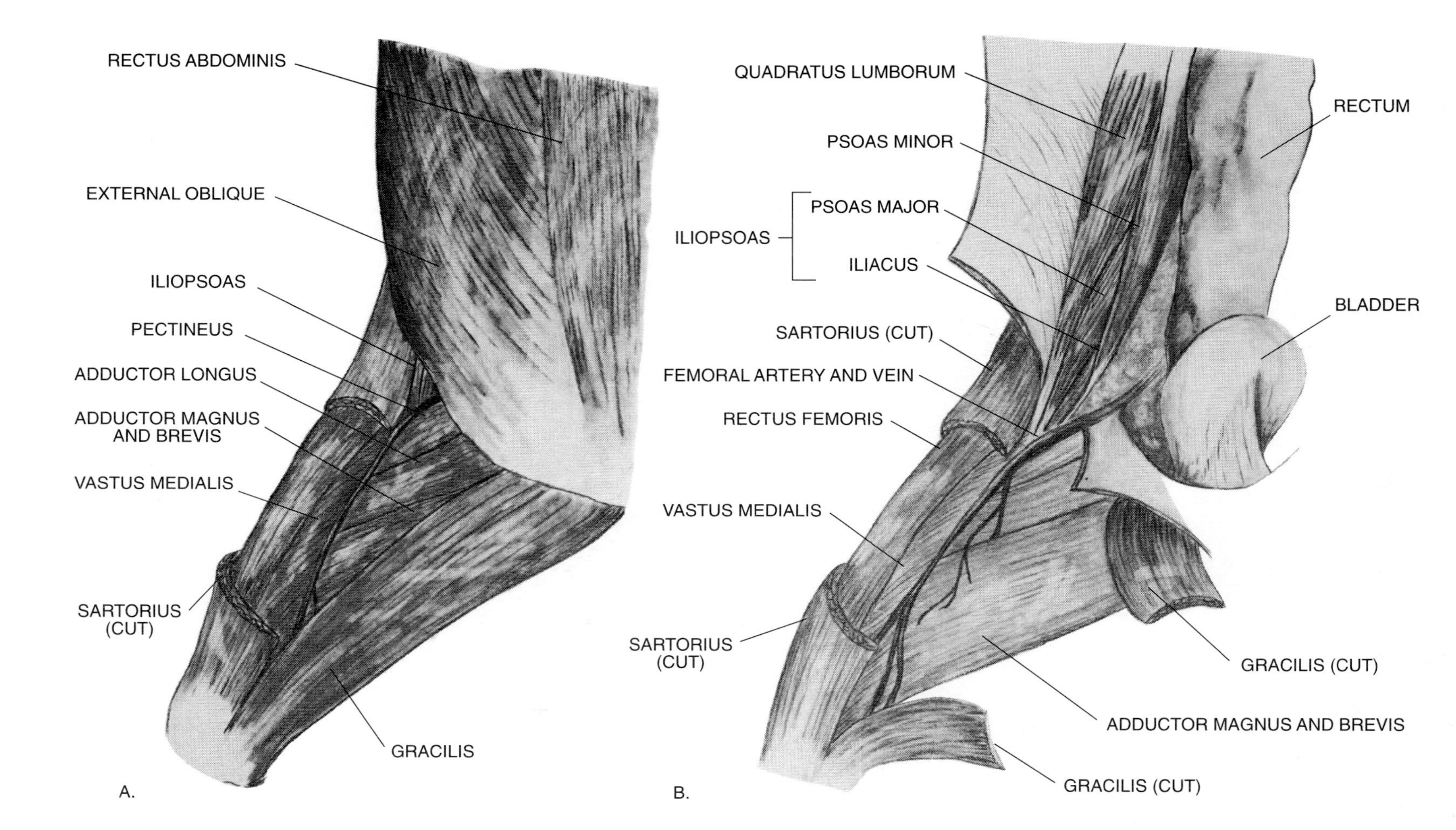

FIGURE 4.23. Medial view of the muscles of the cat hip and thigh. (A) Superficial muscles with a portion of the sartorius removed. (B) Deeper muscles with the body cavity opened to expose the loin muscles.

Leg Musculature (figures 4.21, 4.22)

After identifying the gastrocnemius, isolate and reflect it to see the soleus.

Name	Origin	Insertion	Action
1. Dorsal			
TRICEPS SURAE (TR*I*-ceps SYOO-r*e*) figure 4.21			
GASTROCNEMIUS (gas-tr*o*k-N*E*-me-us)			
LATERALE	Lateral epicondyle and superficial fascia of femur	Tendon of Achilles to calcaneus	Extend foot
MEDIALE	Medial epicondyle of femur and shank	Tendon of Achilles to calcaneus	Extend foot
SOLEUS (S*O*-le-us) figure 4.22	Head of fibula	Tendon of Achilles to calcaneus	Extend foot
FLEXOR DIGITORUM (dig-i-T*O*R-um) **SUPERFICIALIS**	Patella and lateral condyle of femur	Achilles tendon to calcaneus	Extend foot

Flexor digitorum superficialis is very similar in origin, insertion, and action to the **plantaris** of the human. Plantaris originates on the lateral linea aspera and popliteal ligament of the knee joint and inserts on the calcaneus. The action is primarily plantar flexion, though it also flexes the leg. The muscle may not be present at all.

Name	Origin	Insertion	Action
POPLITEUS (pop-LIT-*e*-us) (not illustrated)	Lateral meniscus of knee joint (and lateral condyle of femur in human)	Ventral shaft of tibia (anterior in human)	Flex and rotate leg mediad
FLEXOR DIGITI I LONGUS (**flexor hallicis longus** (HAL-li-cis) of human) figure 4.19	Shaft of fibula	Terminal phalanx of first digit	Flex first digit
FLEXOR DIGITORUM LONGUS	Shaft of tibia	Terminal phalanx of each digit	Flex all digits (not first digit in human)
PERONEUS LONGUS (per-*o*-N*E*-us) figure 4.21	Head of fibula	Base of metatarsals	Flex foot (also pronate foot in human)
2. Ventral			
TIBIALIS CRANIAL (not in human)	Lateral shaft of tibia and head of fibula	First metatarsal	Flex foot
TIBIALIS CAUDALIS (not illustrated)	Near heads of tibia and fibula (interosseous membrane in human)	Lateral tuberosity of central tarsal and tarsal I (tarsals and metatarsal 2–4 in human)	Extend foot (flex foot in human)

Name (Continued)	Origin	Insertion	Action
PERONEUS BREVIS	Fibula and interosseus membrane	Base of fifth metatarsal	Extend foot
PERONEUS TERTIUS figure 4.18	Lateral surface of fibula	Extensor tendon of fifth digit	Extend fifth digit and flex foot

The peroneus brevis and peroneus longus of the human and cat have slightly different insertions and, thus, actions. Peroneus brevis inserts on the base of the fifth metatarsal bone and flexes the foot. Peroneus longus inserts on the first metatarsal and the medial cuneiform where it also flexes the foot. Both pronate the foot.

Name (Continued)	Origin	Insertion	Action
EXTENSOR DIGITORUM LONGUS figure 4.21	Lateral epicondyle of femur	By separate tendons to second and terminal phalanx of each digit	Extend digits and foot

CHAPTER 5

Digestive System

Preparation of the Cat for Observation of the Digestive System, Body Cavities, and Membranes (figure 5.1)

If you have not yet opened the body cavity, do so now. Follow the instructions below.

Open the abdominal and thoracic cavities by (1) cutting from the first rib to the anus with scissors, just to the right (your left) of the midventral line. In the thoracic region you should be cutting through the cartilage ribs. (2) Break or cut each rib on the right side near its vertebral articulation. Make lateral incisions where needed and deflect the body wall to the right to expose the membranes and viscera. Use dissecting needles to pin the skin flaps to the dissecting tray.

Carefully inspect the membranes in the thoracic cavity, and (3) cut laterally to the left side between the diaphragm and the heart. Next (4) cut the cartilage ribs on the left side from the caudal to the rostral ends of the thoracic cavity. Leave the sternum in place and fold back the chest wall on the left side. Again, break or cut the ribs to open the chest wall. Note the internal thoracic artery and vein on either side of the sternum, extending through the cranial mediastinum (the area between the two layers of pleural mesentery) to the cranial vena cava or subclavian artery. The two internal thoracic veins unite at the cranial end of the sternum to form a single vessel (sometimes called the sternal vein) to the vena cava. Now cut the mediastinum loose from the sternum, leaving as much of the vessels as possible attached to the thoracic viscera.

Cat Coelomic Membranes (figures 5.2, 5.3)

Coelomic fluid is both produced and enclosed by thin serous membranes that adhere to the inner body wall **(PARIETAL** [pa-R*I*-e-tal] **MEMBRANE),** adhere to the internal organs **(VISCERAL** [VIS-er-al] **MEMBRANE),** or form a flexible double membrane between the body wall and the internal organs **(MESENTERY** [MES-en-ter-*e*]**).**

The trunk is divided into **THORACIC** and **ABDOMINAL** regions by the muscular **DIAPHRAGM** (D*I*-a-fram). The **THORACIC CAVITY,** contains the lungs and heart. The **ABDOMINAL CAVITY** contains the intestines, liver, pancreas, spleen, kidneys, and reproductive organs.

The actual arrangement of the various membranes is somewhat complex, and the diagrams may help to simplify the explanation. Each membrane or mesentery is usually named in reference to its position or the organ it serves. In the thoracic cavity, the membranes associated with the lungs are termed **PLEURA** (PLOO-ra). Thus, the **PARIETAL PLEURA** lines the thoracic cavity and the **VISCERAL PLEURA** (not illustrated) is on the surface of the lungs. The membranes of the heart are termed **PERICARDIA** (per-i-KAR-d*e*-a) and form a separate chamber between the two pleural cavities. The membrane on the surface of the heart is the **VISCERAL PERICARDIUM (epicardium),** and the coelomic fluid of the heart is between this membrane and the **PARIETAL PERICARDIUM.** The parietal pericardium is located between the right and left membranes of the pleural mesentery. Because of this arrangement, the pericardium of mammalian terminology is said to be a double-walled membrane. It actually consists of a parietal pericardium and portions of the **mediastinum** [m*e*-d*e*-as-T*I*-num] (not illustrated).

The mesenteric portions of the pleural membranes do not contact one another in the midline because the heart, major blood vessels, esophagus, thymus, and fat tissues keep the membranes apart. In this instance, the membranes are referred to as mediastinum rather than mesentery.

The membranes of the abdominal cavity are **PERITONEAL** (per-i-t*o*-N*E*-al) membranes, and the membrane lining the wall of the cavity is the **PARIETAL PERITONEUM.** The membrane on the surface of the abdominal organs is the **VISCERAL PERITONEUM.** Embryonically, there are two mesenteries, dorsal and ventral, but the **ventral mesentery** almost completely disappears during later embryonic development, leaving only the **dorsal mesentery.**

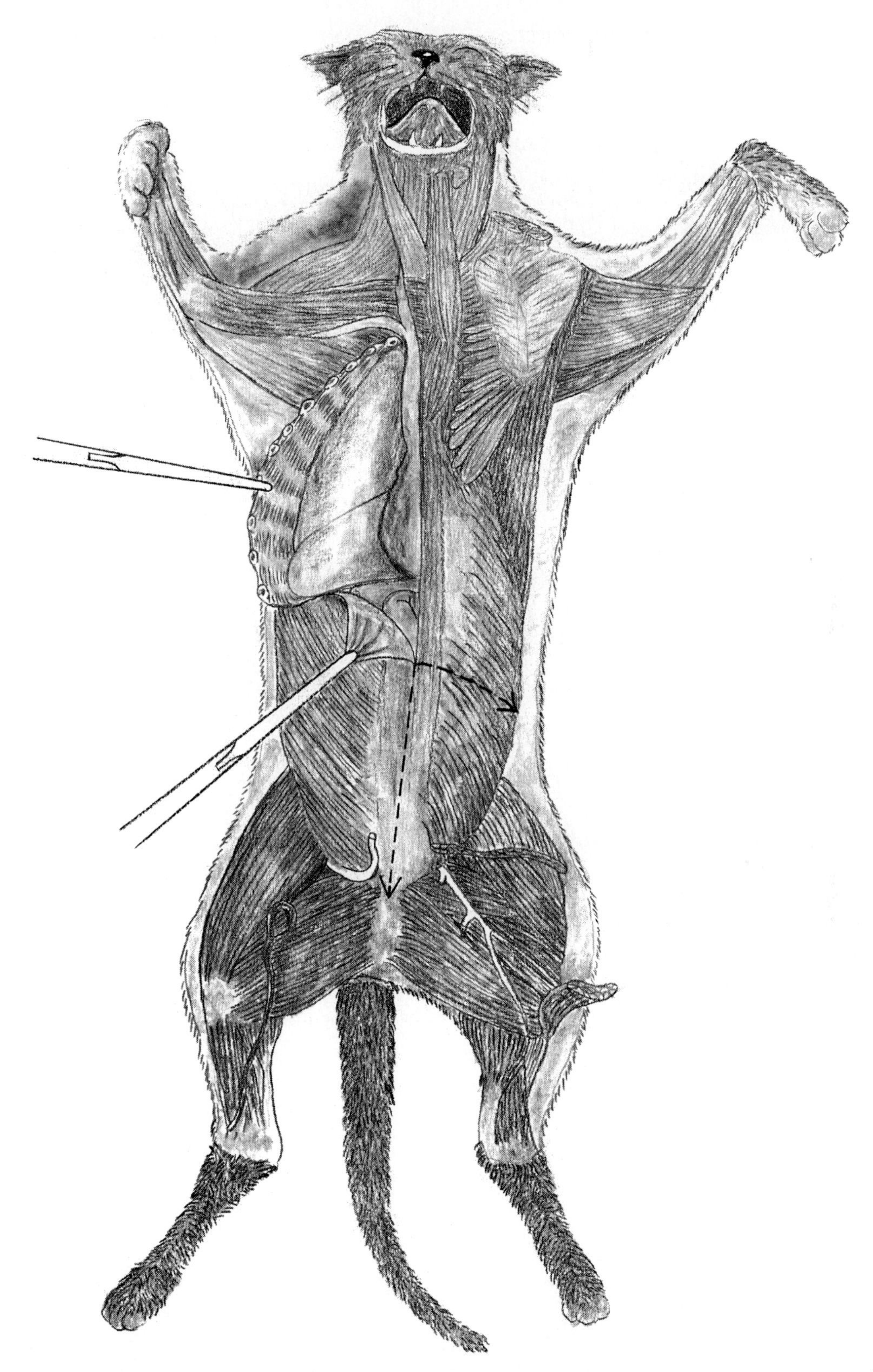

FIGURE 5.1. Dissection of the visceral cavities of the cat. In your earlier dissection of the medial muscles of the pectoral limb you may have separated the limb from the body and therefore you will not be able to tie the specimen down as you did in figure 4.1. Use mounted needles to pin down the skin flaps to the dissecting board. See text (opposite) for instructions.

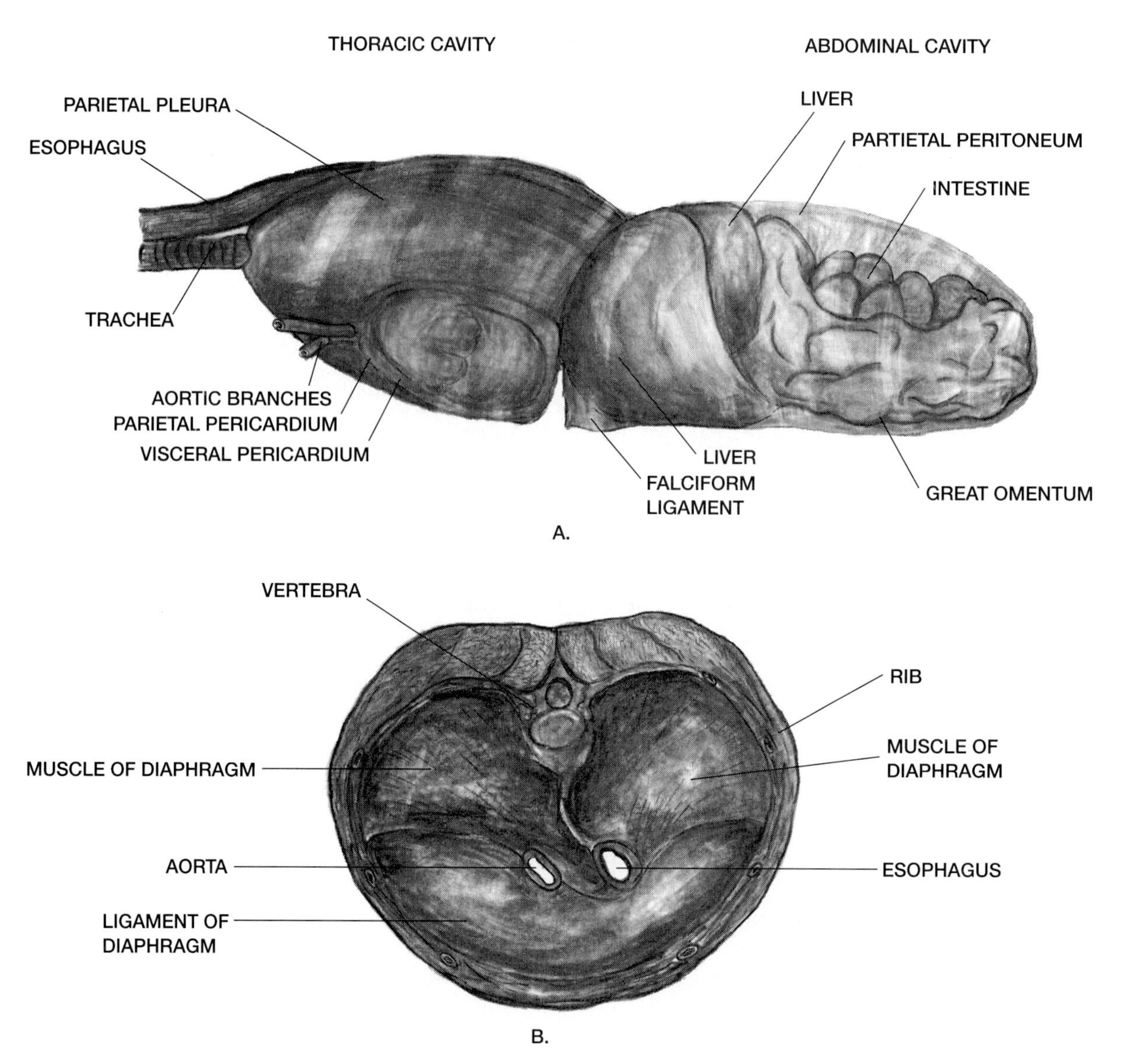

FIGURE 5.2. Diagram of the serous membranes of the cat. (A) In this diagram, all structures except the membranes have been omitted. The cranial (pleural) aspect of the diaphragm is illustrated in (B).

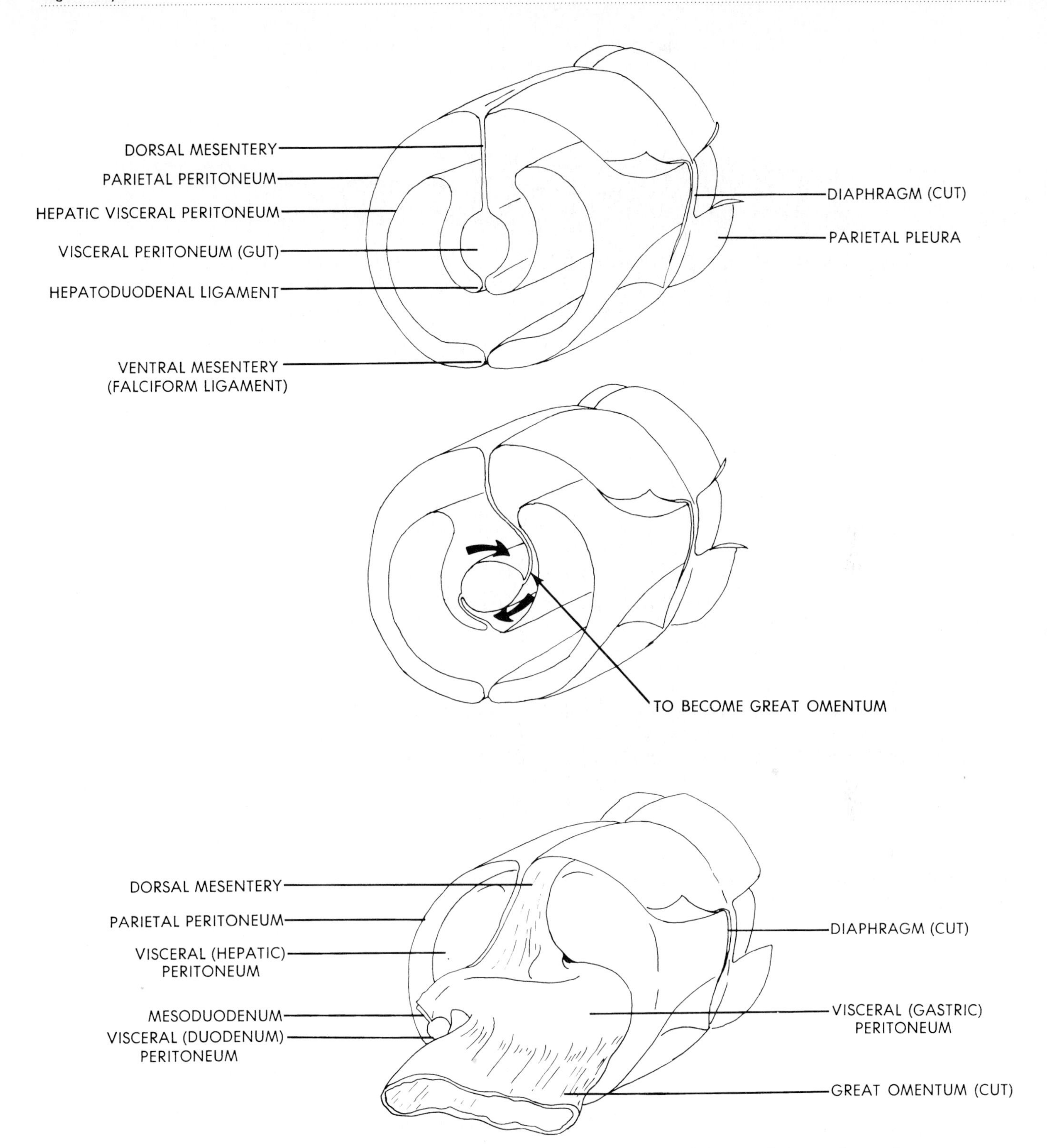

FIGURE 5.3. Stereodiagrams illustrating stages in the development of some of the cat abdominal membranes. Contortions of the embryonic gut in the development of the stomach produces the double folded omentum.

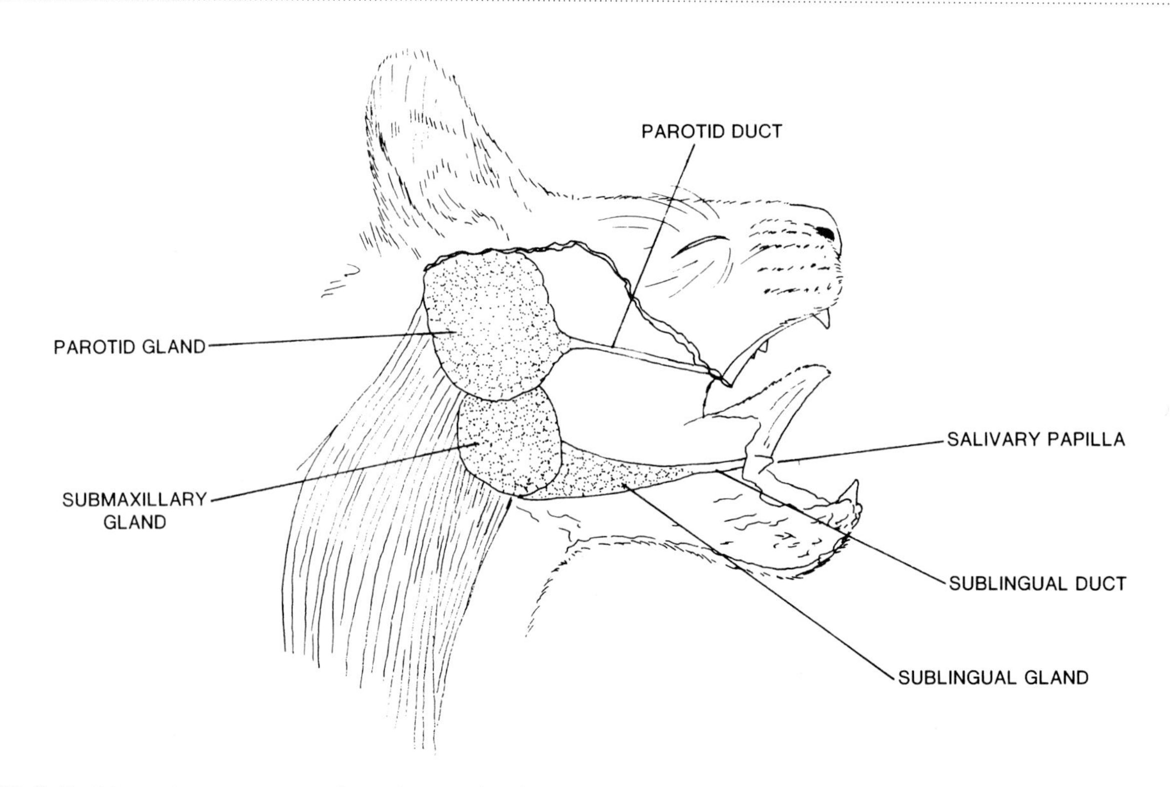

FIGURE 5.4. Dissection to expose the salivary glands on the right side of the head. The right lower jaw is removed.

The only remnants of the ventral mesentery in the adult cat are the **falciform** (FAL-si-form) **ligament** from the ventral body wall and diaphragm to the longitudinal fissure of the liver; the **lesser omentum** (o-MEN-tum) (not illustrated), from the lesser curvature of the stomach (cephalic border) to the liver; and the **suspensory** (sus-PEN-so-re) **ligament** (not illustrated), from the ventral body wall at the extreme caudal end of the abdominal cavity to the urinary bladder.

The abdominal mesenteries are given specific names according to the particular visceral organs that are suspended: **mesogastrium** (mes-o-GAS-tre-um) (not illustrated), to stomach; **mesoduodenum** (mes-o-du-o-DE-num), to duodenum; **true mesentery,** or **mesentery proper,** to all of the small intestine exclusive of the duodenum; **mesocolon** (mes-o-KO-lon) (not illustrated), to the large intestine; **mesorchium** (me-SOR-ki-um) (not illustrated), covers the testicles of the male; **mesovarium** (mes-o-VA-re-um) (not illustrated), to the ovaries of the female; **broad ligament** (not illustrated), to the fallopian tubes and uterus of the female.

The stomach of the mammal undergoes some unusual growth movements during development that affect the appearance of its mesentery. From the entrance of the esophagus, the stomach curves to the right side of the abdominal cavity and rotates, so the dorsal and ventral surfaces are reversed (figure 5.3). During these growth movements, the original dorsal half (to which the mesentery is attached) undergoes a more rapid growth than the ventral portion, thus producing the large, saclike fundic portion of the stomach. These movements also produce an elongated fold of the mesentery, so this membrane lies like a folded apron between the stomach and the intestines. The apronlike mesentery is termed the **GREATER OMENTUM.** A portion of the greater omentum holding the spleen is called the **LIENOGASTRIC** (li-e-no-GAS-trik) mesentery (not illustrated).

Preparation for Observation of the Oral Cavity

With a pair of heavy scissors or bone cutters, cut the muscles, skin, and mandible at the angle of the mouth on both sides and depress the lower jaws with your fingers **(CAUTION: THE TEETH ARE SHARP).**

Salivary Glands (figures 5.4, 5.5)

If the glands or their ducts on the right side have been destroyed during the previous dissection, then skin the left side and study these glands. There are five pairs of salivary glands located in the head. The histology of these glands is illustrated in figure 5.5.

1. **PAROTID** (pa-ROT-id) **GLAND.** Lies superficially beneath the skin and just ventral to the ear. The

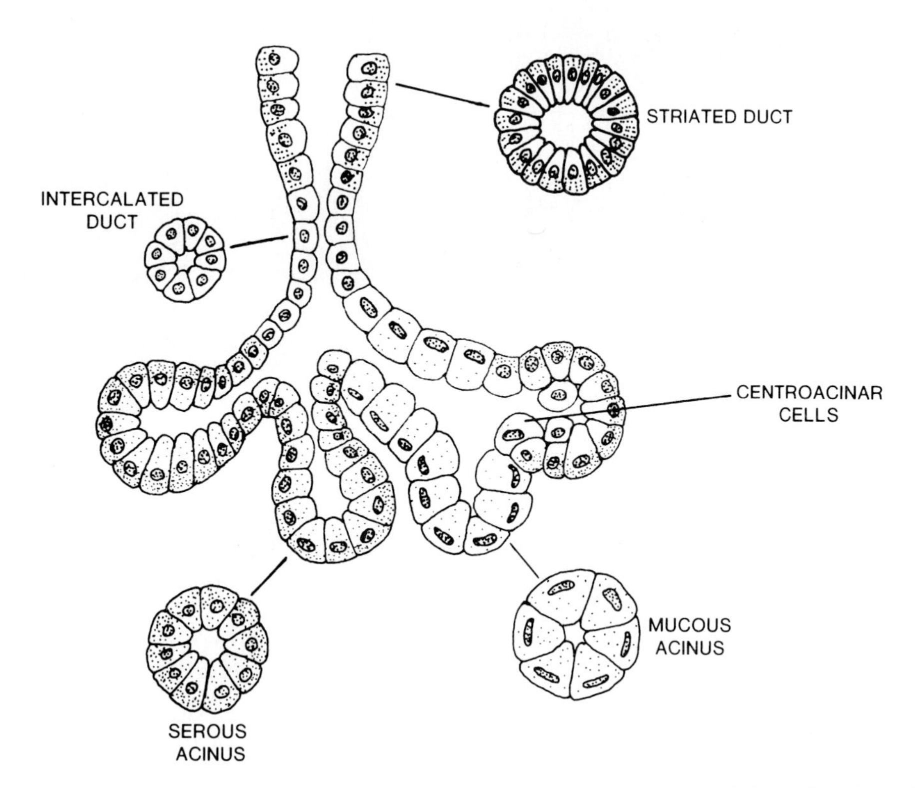

FIGURE 5.5. Diagrammatic illustration of the acinar salivary glands of the cat and their ducts. Both serous and mucous cells secrete glycoproteins which, together with water and other materials, provide lubrication for the mouth and esophagus and solvents for the substances that stimulate the taste buds. The duct cells absorb sodium and secrete potassium to modify the ionic concentration of the saliva. The cat serous cells do not produce salivary amylase for the digestion of starch.

parotid duct crosses the masseter muscle with branches of the facial nerve and a vein and opens to the mouth opposite the last premolar tooth (opposite the second molar in humans).

2. **SUBMAXILLARY** (sub-MAX-si-ler-*e*) **GLAND.** Ventral to the larger parotid at the angle of the jaw. The **submaxillary duct** serves both the submaxillary and sublingual glands and opens through a small papilla on either side of the base of the **frenulum** (FREN-*u*-lum).
3. **SUBLINGUAL** (sub-LING-gwal) **GLAND.** Lies in the tissues surrounding the submaxillary beneath and on each side of the tongue (in humans there are from eight to 29 ducts that vary in size and open under the tongue).
4. **INFRAORBITAL** (in-fra-*O*R-bi-tl) **GLAND** (not illustrated). Found within the orbit. Its ducts enter the mouth just caudal to the molar tooth on each side of the head.
5. **MOLAR** (M*O*-lar) **GLAND** (not illustrated). A very small gland located near the rostral edge of the masseter muscle on the outer surface of the maxilla.

Humans lack the infraorbital and molar glands, though there are numerous other **accessory glands** found at the back of the dorsum of the tongue, near the palatine tonsils, and on the soft palate, lips, and cheeks.

Oral Cavity

1. **VESTIBULE** (VES-ti-b*u*l). The area of the mouth between the lips, cheeks, and teeth.
2. **TEETH.** The incisors, canines, premolars, and molars (see figure 3.10).
3. **TONGUE.** (figure 5.6) The frenulum is the thin membrane on the midventral surface of the tongue, holding the tongue to the floor of the mouth. Intrinsic tongue muscles (**lingualis proprius** [ling-GWA-lis PR*O*-pr*e*-us], see musculature, page 52) alter the shape of the body of the tongue. The lingualis proprius is composed of four fiber bundles.

Lingual Papillae of the Cat (figure 5.7)

Name	Shape	Frequency	Position
CIRCUMVALLATE PAPILLAE (ser-kum-VAL-*at*)	Column	two to three pairs (four to six pairs in human	Lateral base of tongue
FUNGIFORM PAPILLAE (FUN-ji-form)	Mushroom	Few (numerous in human)	Sides and tip of tongue
FILIFORM PAPILLAE (FIL-i-form)	Spiny	Many	Cover dorsum of tongue
FOLIATE PAPILLAE (FOL-*i-at*)	Flat	Few	Lateral and caudal to circumvallate papillae

The circumvallate, foliate, and fungiform papillae may have **taste buds** located in furrows or grooves of the papilla. Each taste bud consists of elongated cells arranged around a central lumen (cavity) that opens to the papillary furrow by a **taste pore.**

There are also microscopic **papillae simplices** in humans. These cover the entire tongue and even the larger papillae. Each contains a capillary.

4. **HARD PALATE** (PAL-at) (not illustrated). The roof of the forepart of the oral cavity.
5. **SOFT PALATE** (not illustrated). The caudal continuation of the hard palate tissues bordered by the internal pterygoid processes.
6. **Incisive** (in-S*I*-siv) **duct** (not illustrated). Opens through the rostral palatine foramen (see page 19) and connects the vomeronasal organ with the mouth.

Pharynx (figure 5.7)

The **PHARYNX** (F*A*R-inks) is the common chamber of the respiratory and digestive systems. Embryonically (and phylogenetically), this area is highly vascularized; the epithelium gives rise to several endocrine glands (thyroid, parathyroids, and ultimobranchial or calcitonin glands) and other structures associated with the circulatory system (the thymus, carotid, and aortic bodies).

1. **ISTHMUS FAUCIUM** (FAW-c*e*-um) (not illustrated). The opening of the oral cavity to the pharynx.
2. **NASOPHARYNX** (n*a*-s*o*-F*A*R-inks) (not illustrated). The cavity above the soft palate leading to the nasal cavities.
3. **INTERNAL NARES** (N*A*-r*e*z) or **CHOANAE** (k*o*-*A*-n*e*) (not illustrated). Opening from the nasal chamber to the nasopharynx.
4. **AUDITORY TUBES** (not illustrated). Found in the lateral walls of the nasopharynx, connect the cavities of the middle ear with the pharynx.
5. **GLOTTIS** (GLOT-is). The opening from the pharynx to the larynx just ventral to the esophagus and held open by cartilaginous rings in its wall.
6. **EPIGLOTTIS** (ep-i-GLOT-is). A raised flap between the tongue and glottis.
7. **ESOPHAGEAL OPENING.** Dorsal to glottis.

Alimentary Tract (figures 5.8–5.11)

1. **ESOPHAGUS** (*e*-SOF-a-gus). Extends from the pharynx to the stomach.
2. **STOMACH.** There are three regions, **cardiac** (KAR-d*e*-ak) (entrance of esophagus), **fundic** (FUN-dik) (large, middle saclike portion), and **pyloric** (p*i*-LOR-ik) (constricted caudal portion that opens to the duodenum). The cardiac lining has mucus-secreting cells, but no enzymes are produced here (figure 5.9). The fundic portion

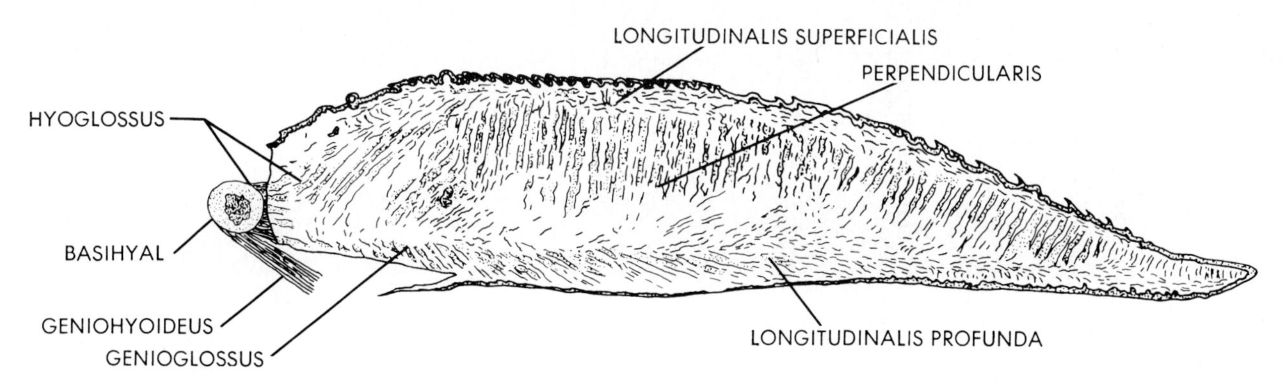

FIGURE 5.6. *M. lingualis proprius* (intrinsic tongue musculature). Transverse fibers are difficult to see in this section. They run approximately at right angles to the perpendicular fibers.

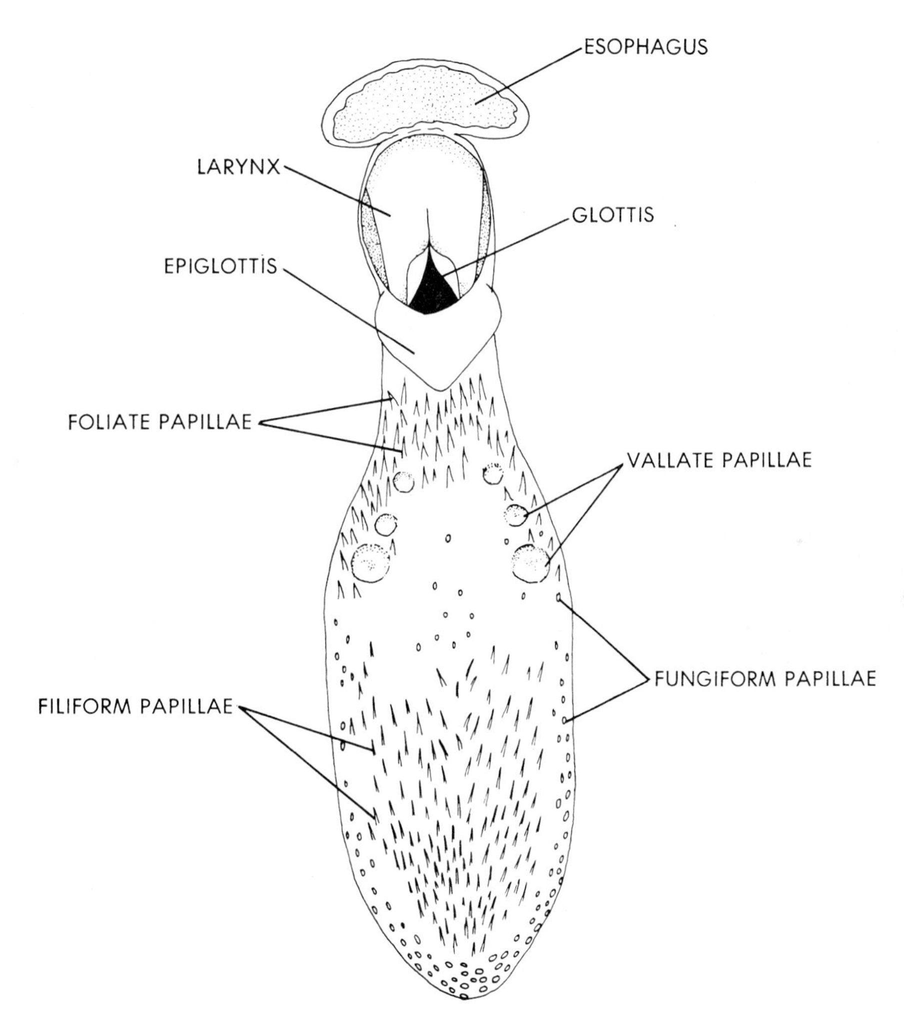

FIGURE 5.7. Dorsal view of the cat tongue and pharynx.

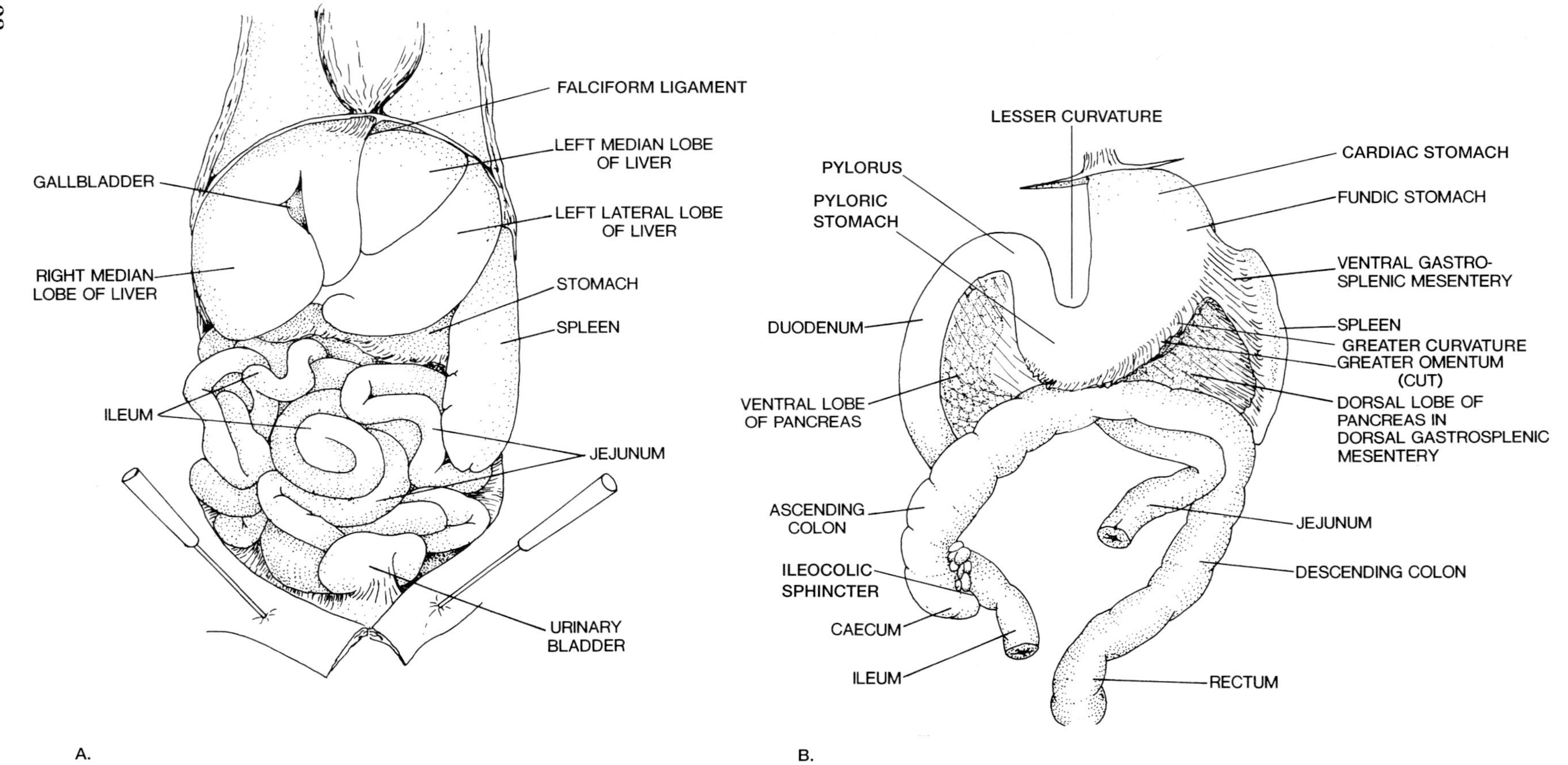

FIGURE 5.8. Ventral views of the abdominal viscera with the greater omentum removed (A), and with the liver and most of the jejunum and ileum removed (B).

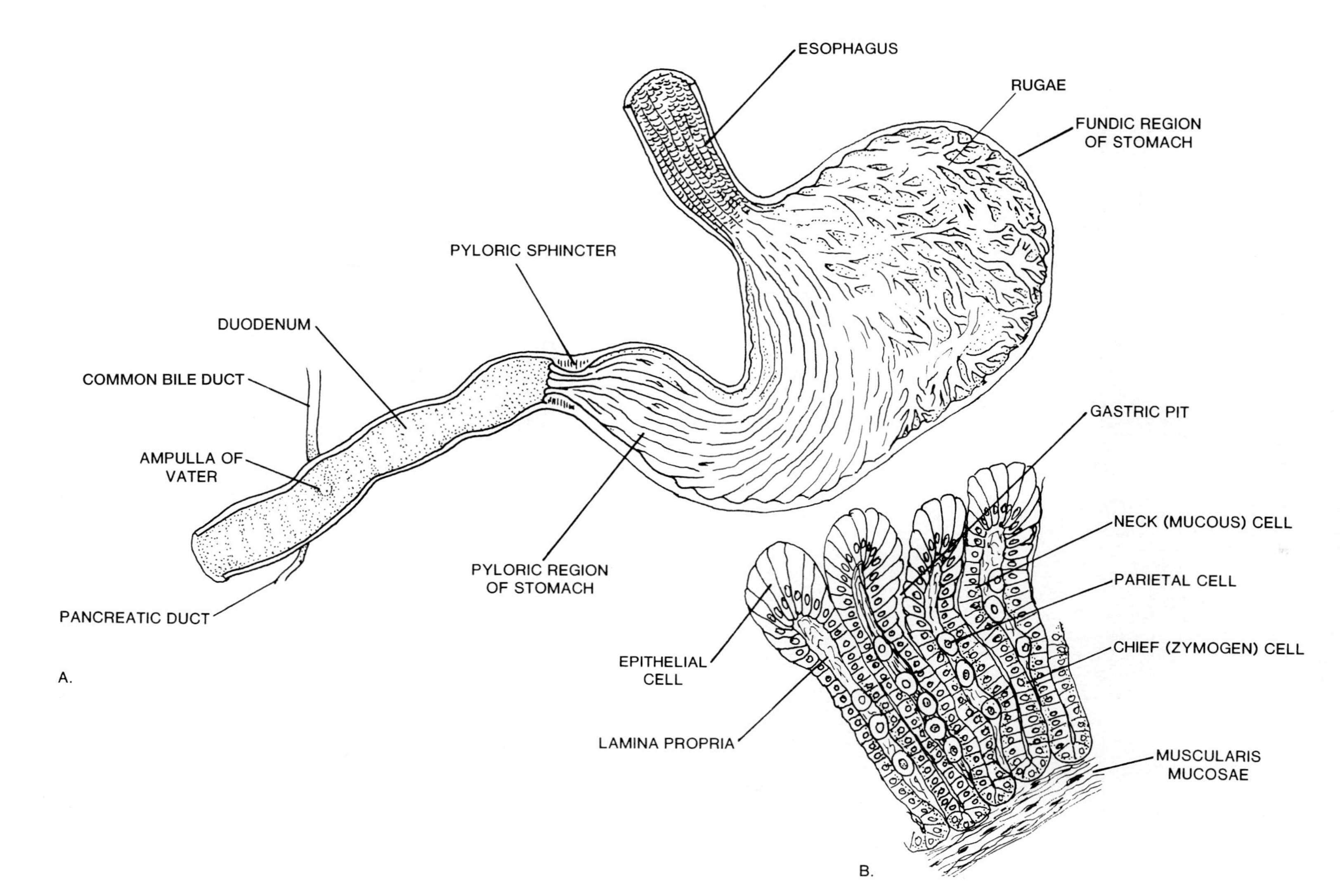

FIGURE 5.9. (A) Internal view of the stomach, esophagus, and duodenum. Part B illustrates the gastric pits of the fundic stomach but does not include the outer layers (submucosa, muscularis externa, and serous) of the stomach wall.

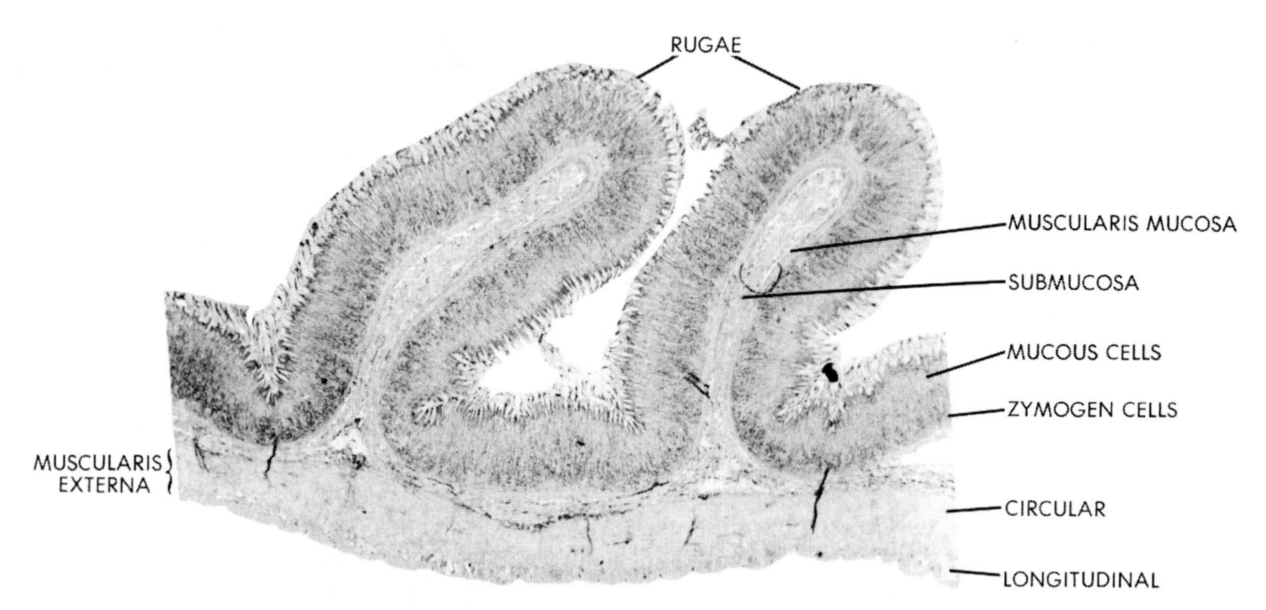

FIGURE 5.10. Photomicrograph of a cross section of the cat fundic stomach. The columnar cells of the mucosa are arranged in deep pits. Cells at the base of the pits secrete gastric enzymes (zymogen) and, thus, stain darker than the more superficial mucus-secreting cells. The mucosa and submucosa of the cat stomach are arranged in longitudinal folds (rugae).

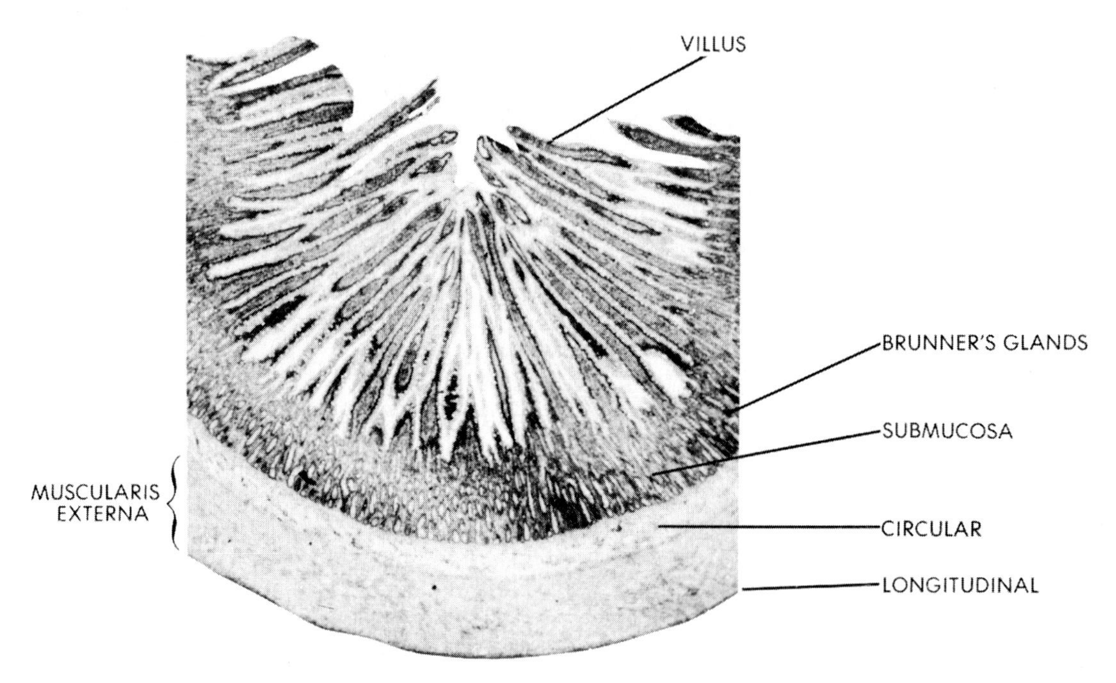

FIGURE 5.11. Photomicrograph of a cross section of the cat duodenum. The duodenum differs from the rest of the small intestine in having large, branched, mucus-secreting Brunner's glands in the submucosa.

secretes all the enzymes and hydrochloric acid produced by the stomach. Pyloric cells secrete mucus. The **greater curvature** of the stomach is the midline convex curvature on the intestinal side, and the **lesser curvature** is the midline concave curvature on the hepatic side.

Slit open the stomach and notice the longitudinal ridges lining the wall; these are called **rugae** [(ROO-je) figure 5.10].

3. **SMALL INTESTINE.** About six times the length of the body of the cat (excluding the tail). In humans, the length is three to four times body length.
   a. **PYLORUS.** The passageway between the stomach and small intestine (duodenum). The muscularis mucosa is composed principally of circular, smooth muscle fibers in this region and is extremely thick, thus forming the **pyloric sphincter** (SFINK-ter).
   b. **DUODENUM.** This first portion of the small intestine receives the ducts of the digestive glands. The epithelium of the duodenum (and all of the small intestine) has numerous tiny projections called **villi** [(VIL-*i*) figure 5.11], which give the lining a soft, velvety appearance. The ventral lobe of the pancreas is adjacent to the duodenum, which will help to distinguish this region from the jejunum.
   c. **JEJUNUM** (je-JOO-num). The second portion of the small intestine (following the duodenum). The jejunum is similar to the duodenum in structure.
   d. **ILEUM** (IL-*e*-um). The remaining portion of the small intestine. It opens to the large intestine by the **ileocolic** (il-*e*-*o*-K*O*L-ik) **sphincter.** The ileum is characterized by lymph nodules **(Peyer's patches)** in the submucosa. Enlargements on the surface of the ileum caused by the presence of Peyer's patches are usually grossly evident.
4. **COLON** (K*O*-lon). Part of the **large intestine,** between the caecum and rectum so called because the lumen of the large intestine has a greater diameter than the lumen of the small intestine. At the entrance to the colon is a short caudally projecting sac, the **CAECUM** (S*E*-kum), and a forward continuation of the colon proper, the **ascending colon.** At the level of the stomach, the colon loops caudally, as the **descending colon,** to the rectum. The epithelium of the colon has a large population of mucus-secreting goblet cells.

> In humans, the colon is disproportionally longer and has more subdivisions. The ascending colon is about 15 cm long, followed by a **transverse colon** of nearly 50 cm. The descending colon (15 cm) connects the transverse colon to the **sigmoid colon,** which is about 40 cm long. This is followed by the **rectum** and **anus.**

5. **RECTUM** (REK-tum). The straight terminal portion of the colon. The epithelium of the rectum has a very dense population of goblet cells that provide mucinous lubrication for the feces.
6. **ANUS** (*A*-nus). The very short termination of the alimentary tract.

Digestive Glands (figures 5.8, 5.12, 5.13)

1. **LIVER** (figure 5.8). The largest gland in the cat, is divided into right and left halves by the **falciform** (FAL-si-f*o*rm) **ligament.** The left half of the liver consists of two lobes, a rostral **left median lobe** and a caudal **left lateral lobe.** The right half has three lobes, but the **right median lobe** is so large that it obscures the other two. The **right lateral lobe** is dorsal to the right median lobe. Together, the right median and right lateral lobes surround the small **caudate,** or **spigelian** (sp*i*-J*E*-l*e*-an) **lobe** (not illustrated).

> The liver of humans consists of two large lobes, the right and left, and two much smaller lobes, the **caudate** and **quadrate.** The right lobe is six times the size of the left and is separated from the left by the falciform ligament, which is attached to the inferior side of the diaphragm, and to the fascia of the rectus abdominis muscle.

The **GALLBLADDER** is in a cleft on the ventral surface of the right median lobe. **Bile canaliculi** (kan-a-LIK-*u*-l*i*) run between individual liver cells and open to **bile ductules** which, in turn, drain into **hepatic** (he-PAT-ik) **ducts** in the various lobes. (These structures, illustrated in figure 5.12, require a microscope for observation.) The hepatic ducts join the **cystic** (SIS-tik) **duct** (from the gallbladder to the common bile duct) (not illustrated), or connect directly to the **common bile duct,** which opens into the duodenum. The common bile duct projects into the duodenum as the **ampulla of Vater** (FAH-ter); its opening is controlled by the **sphincter of Oddi** (*O*D-*e*) (not illustrated), which surrounds the ampulla.

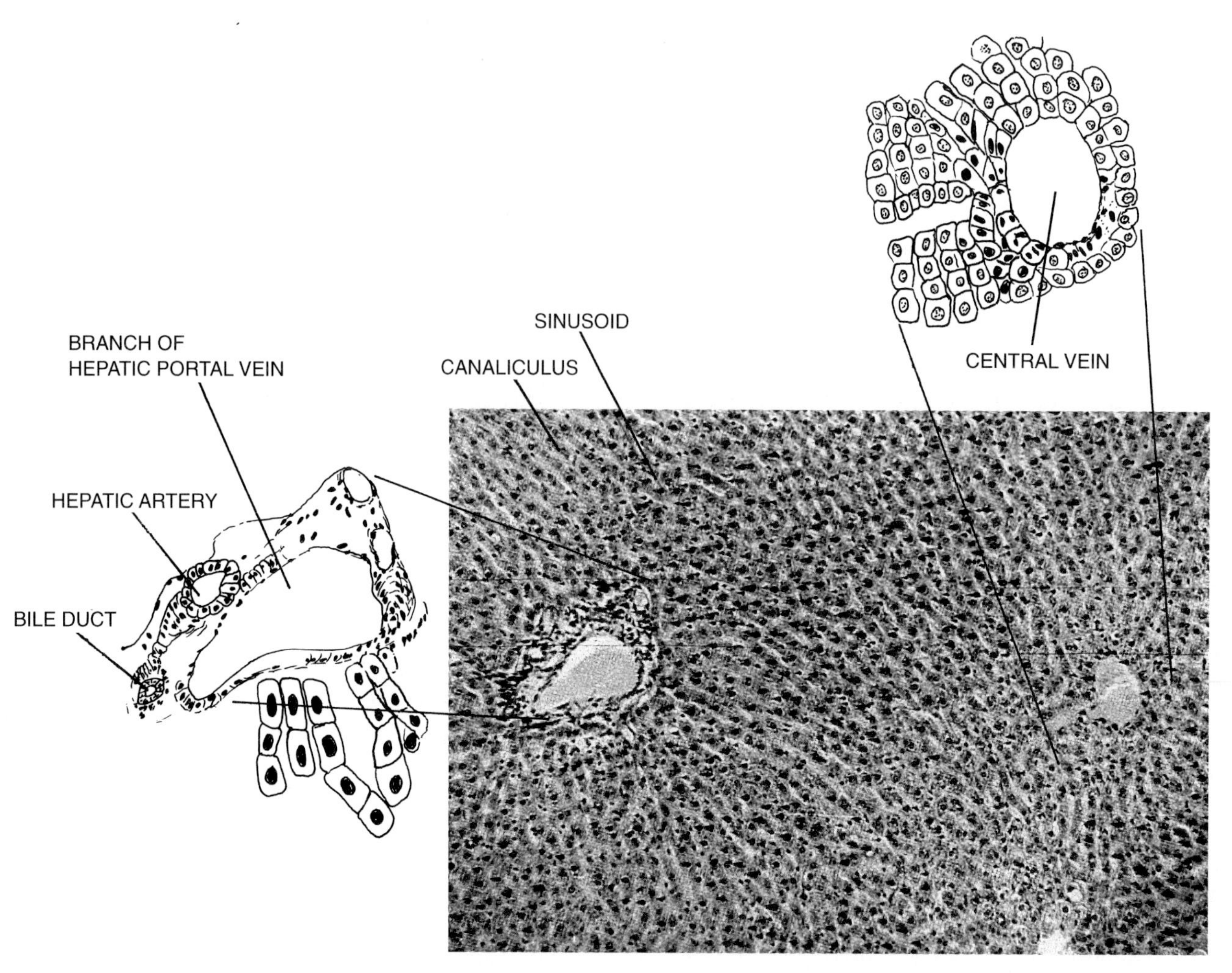

FIGURE 5.12. Photomicrograph and diagrams of the microstructure of the mammalian liver. The details of the functional units, the hepatic portal vein, hepatic artery, bile duct, canaliculi, sinusoid, and central vein are shown. Materials are extracted from the blood in sinusoids by the liver cells. Bile is secreted into the canaliculi, which empty into the bile duct as a waste product. Blood is delivered to the sinusoid from the hepatic portal vein and hepatic artery. From the sinusoid, blood drains into the central vein, to the hepatic vein, and finally to the inferior vena cava (×100).

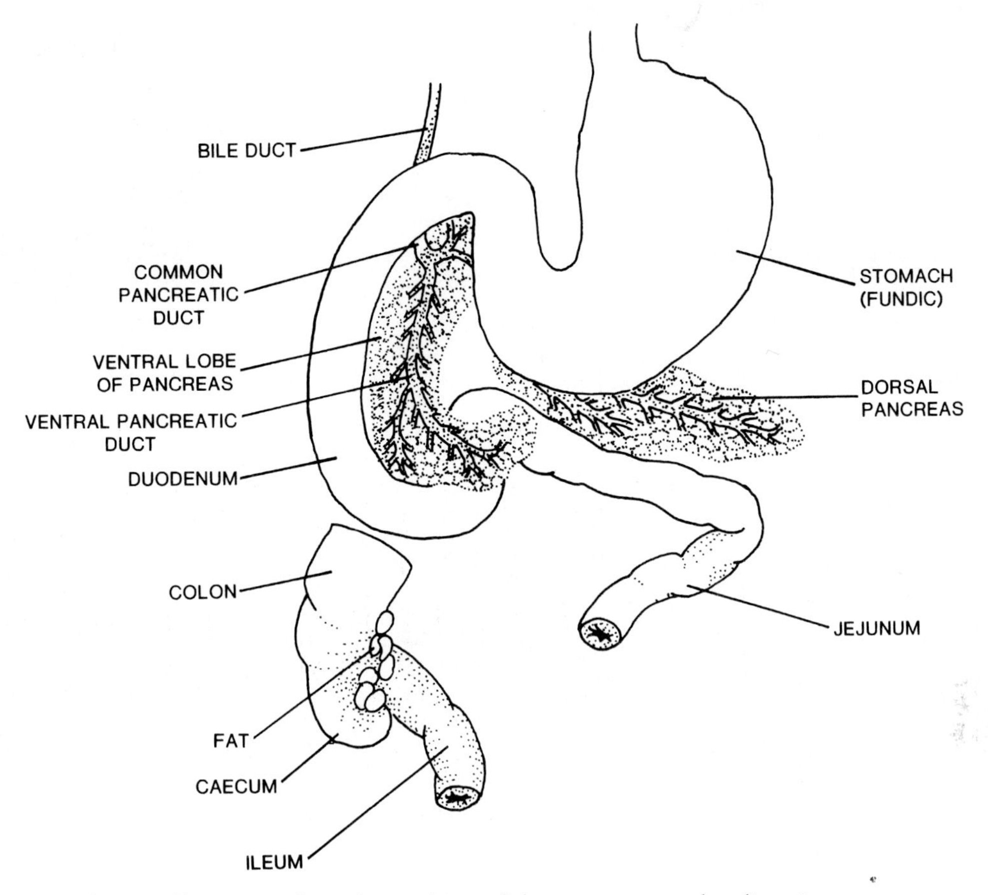

FIGURE 5.13. Ventral view illustrating the relationships of the pancreas to the digestive tract.

2. **PANCREAS** (PAN-kre-as) (figure 5.13). A diffuse, bilobate gland with a **dorsal lobe** in the lienogastric mesentery and a **ventral lobe** in the mesoduodenum. The cells of the pancreas (not illustrated) are arranged in groups similar to those of the salivary glands, so their secretion flows into a central cavity between the secreting cells. This arrangement is termed **acinar** (AS-i-nar). The acinar cavity is drained by ducts that collectively drain into either the common bile duct or an accessory, which drains directly into the duodenum.

   Some of the acinar groups have enlarged cells that obliterate the central cavity and have no drainage duct. These bundles are known as **pancreatic islands** (islets of Langerhans [LANG-er-hanz]); they secrete their products into adjacent blood vessels. The pancreatic islands are endocrine glands, and their secretions are the hormones insulin and glucagon.

> In the human, the **pancreas** has just one lobe occupying the curve of the duodenum. The **pancreatic duct** empties into the duodenum by a common orifice with the bile duct. There may be an accessory pancreatic duct that opens into the duodenum above the main duct. In nearly half of the cases, the main pancreatic duct opens independently of the bile duct.

Suggested Readings

Bosma, J. F. 1956. Myology of the pharynx of cat, dog, and monkey with interpretation of the mechanism of swallowing. *Annals Otol.* 65:981–992.

LeBlance, B., M. Wyers, and M. Lagadic. 1981. The digestive tract: An endocrine organ. *Recl. Med. Vet. Ec. Alfort* 157(9):629–638.

Thomason, J. J., and Russell, A. P. 1986. Mechanical factors in the evolution of the mammalian secondary palate: A theoretical analysis. *Jour. Morphology* 189:199–213.

CHAPTER 6

Respiratory System

The respiratory system functions as a gas exchange mechanism between the external medium (air) and blood. Oxygen diffuses across the lung membranes to the blood, which carries it to the tissues. Carbon dioxide is acquired by the blood at the tissues and is released to the medium by diffusion through the lung membranes. Oxygen is used by the mitochondria of the cells during cellular respiration (metabolism of glucose). Carbon dioxide is produced as a waste product of this same cellular metabolism.

Preparation of the Cat for Observation of the Respiratory System

You may have already opened the body cavity. If not, follow the instructions on page 90 and then make a midventral incision along the length of the larynx and cranial trachea to see the inside of the larynx.

Thoracic Cavity

The thoracic cavity is separated into three parts: the **pericardial cavity,** enclosing the heart; the **right pleural cavity,** with the right lung; and the **left pleural cavity,** with the left lung. The **pleura** is the lining of the thoracic coelomic cavity and the lung. The **parietal pleura** lines the cavity, while the **visceral pleura** covers the surface of the lungs.

Upper Respiratory Tract

1. **NASAL CAVITY.** Divided into separate cavities by a bone and cartilage **nasal septum.** These cavities warm (or cool) the inhaled air to body temperature and filter and humidify it. Air enters the nasal cavities through the **external nares** (N*A*-r*e*z). The mucus is supplied by glands in the nasal epithelium. Air then leaves the nasal cavities through the caudal nares and enters the pharynx.
2. **CAUDAL NARES** (choanae). The openings at the caudal end of the nasal cavities into the pharynx.
3. **PHARYNX.** A passageway common to openings from the nasal cavities **(nasopharynx),** the mouth **(oropharynx),** trachea, and larynx **(laryngopharynx).**
4. **GLOTTIS** (GLOT-is). The opening into the larynx from the pharynx.

Lower Respiratory Tract (figure 3.9)

1. **LARYNX** (L*A*R-inks). The "voice box." The wall of the larynx is held rigid by five cartilages.
   a. **Epiglottis** (ep-i-GLOT-is). A rostral cartilage that projects into the pharynx above the glottis.
   b. **Arytenoid** (ar-e-TE-noyd) **cartilages.** The caudal border of the glottis is supported by small, paired arytenoid cartilages. The **vocal cords (folds)** are attached to the arytenoid and epiglottal cartilages. There are two pairs of vocal cords: 1) the false or cranial pair and 2) the true or caudal pair. Both are folds of the mucous membranes lining the larynx (not illustrated).
   c. **Thyroid** (TH*I*-royd) **cartilage.** The center of the larynx is stiffened by a single thyroid (shield-shaped) cartilage, having paired articulations with the hyoid apparatus.
   d. **Cricoid** (KR*I*-koyd) **cartilage.** Unlike the thyroid cartilage, the more caudal cricoid cartilage forms a complete ring surrounding the larynx.
2. **TRACHEA** [(TR*A*-k*e*-a) figures 6.1, 6.2]. The "windpipe" is held open by a series of U-shaped **tracheal cartilages** in the wall of the trachea. The lining of the trachea is composed of ciliated, mucus-secreting cells that keep the lining moist.

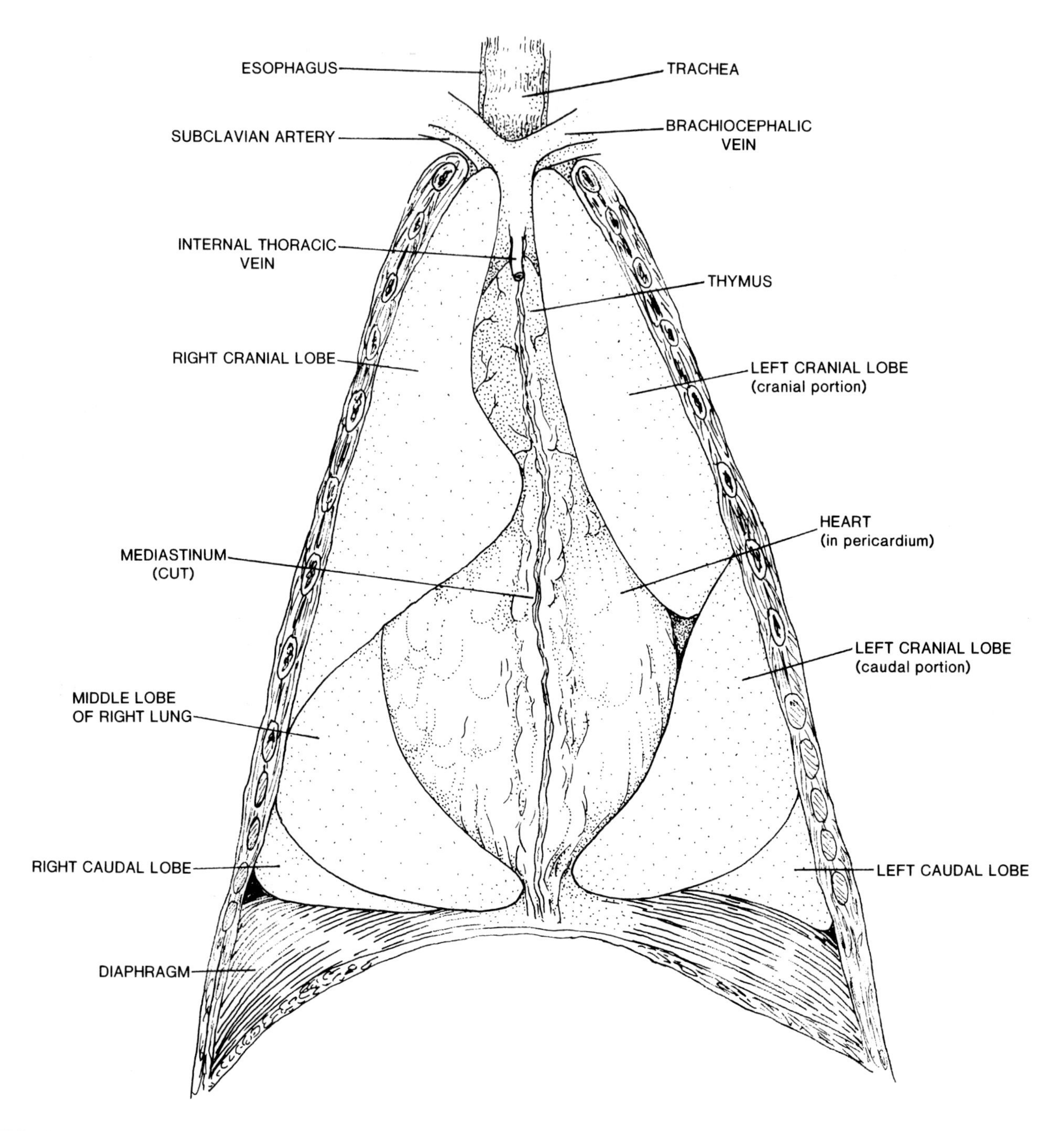

FIGURE 6.1. Ventral view of the thoracic viscera of the cat.

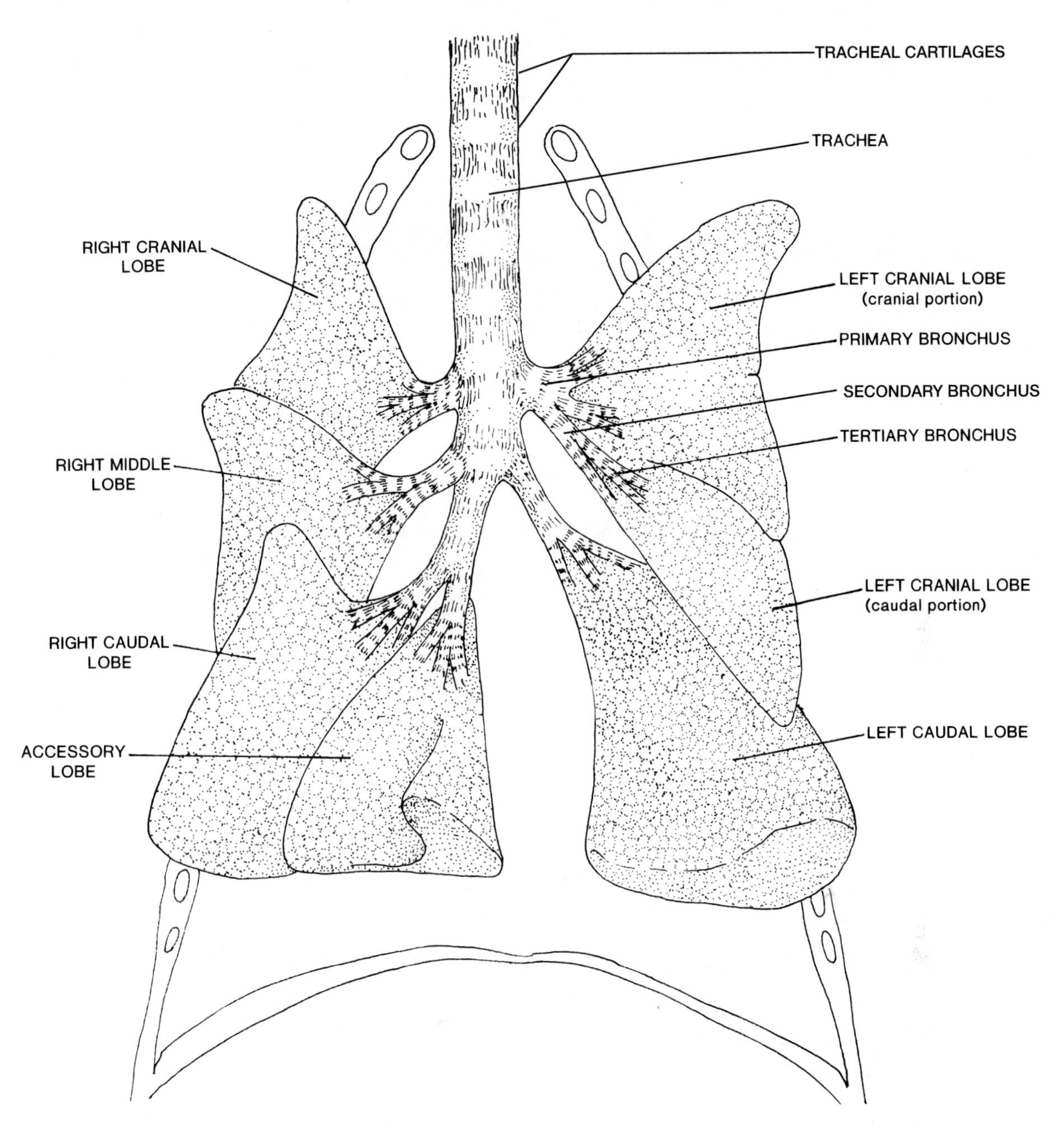

FIGURE 6.2. Diagram of the distribution of bronchi in the cat lungs. The mediastinal viscera have been removed and the lungs have been pulled laterad.

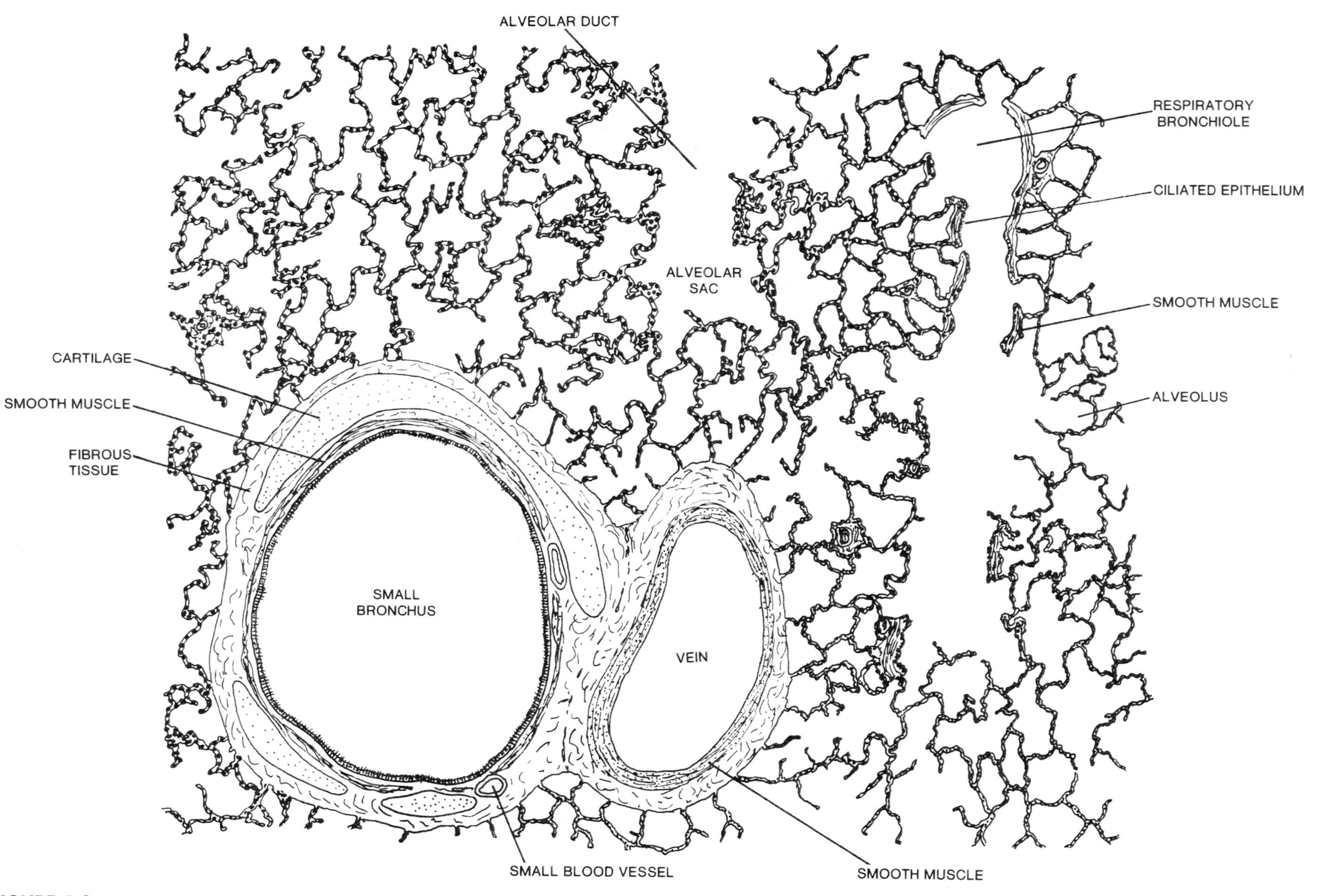

FIGURE 6.3. Microstructure of the mammalian alveoli and their ducts.

3. **BRONCHI** (BRONG-k*i*) (figure 6.2).
   a. **Primary bronchi.** The trachea bifurcates to form the primary bronchi at the level of the heart, just after it enters the thoracic cavity, with one bronchus going to each lung. The bronchial tubes are supported by U-shaped cartilages and lined by ciliated cells.
   b. **Secondary** and **tertiary bronchi.** Within the lungs, the primary bronchi form secondary and tertiary bronchi with cartilage rings.
   c. **Terminal bronchioles** (BRONG-k*e*-*o*l) (not illustrated). The smallest division of the bronchi lack cartilaginous support. Some, called **respiratory bronchioles** (figure 6.3), have alveoli in their walls.
4. **ALVEOLI** (al-V*E*-*o*-l*i*). The respiratory or terminal bronchioles finally branch into **alveolar ducts** that lead to **alveoli** or **alveolar sacs**. Gas exchange takes place through blood capillaries and the alveolar walls. All alveolar ducts and bronchioles have **smooth muscle fibers** in their walls, but the alveoli and alveolar sacs do not.
5. **LOBES.** The lobes of the lungs are named according to the division of the bronchi, rather than by the superficial fissures that appear to divide them. Thus, the right lung has four lobes **(cranial, middle, caudal,** and **accessory),** and the left has only two lobes **(cranial** and **caudal),** but the left cranial lobe is partially divided into cranial and caudal portions by an incomplete fissure.

There is a pattern of lobes in humans; three are on the right **(superior, inferior** and **middle)** and two are on the left **(superior** and **inferior).** Bronchi supplying the lobes are termed **lobar bronchi.** The lobar (secondary) bronchi then divide to form the **tertiary bronchi,** which supply the **bronchopulmonary segments.** Therefore, the tertiary bronchi are also called **segmental bronchi.**

Suggested Readings

Bosma, J. F. 1956. Myology of the pharynx of cat, dog, and monkey with interpretation of the mechanism of swallowing. *Ann. Otol.* 65:981–92.

Mortola, J. P., and J. T. Fisher. 1980. Comparative morphology of the trachea in newborn mammals. *Respir. Physiol.* 39(3):297–302.

Moss, M. L. 1969. Phylogeny and comparative anatomy of oral ectodermal-ectomesenchymal inductive interactions. *J. Dent. Res.* 48:732–37.

Niewenhuis, R. 1966. Comparative histology of the trachea of the dog and cat. *Anat. Rec.* 154:454.

Sobin, S. S., H. M. Tremor, and Y. C. Fung. 1970. Morphometric basis of the sheet-flow concept of the pulmonary alveolar microcirculation in the cat. *Circ. Res.* 26:397–414.

Sobin, S. S., Y. C. Fung, H. M. Tremor, and T. H. Rosenquist. 1972. Elasticity of the pulmonary alveolar microvascular sheet in the cat. *Circ. Res.* 30:440–50.

Thomason, J. J., and A. P. Russell. 1986. Mechanical factors in the evolution of the mammalian secondary palate: a theoretical analysis. *J. Morphol.* 189:199–213.

CHAPTER 7

Circulatory System

The circulatory system, consisting of the heart, blood vessels, and blood, functions to transport nutrients, wastes, hormones, gases, and cells within the body. This transportation system has a pump (the heart) and a pipeline (the arteries and veins) that convey a fluid, the blood. Blood also transports heat; it is warmed by the heat generated during muscle contraction and cooled by contact with low environmental temperature.

Regardless of whether they transport oxygenated or carboxylated blood (figure 7.1), the vessels carrying blood from the heart are called **ARTERIES** (AR-ter-*ez*) and those carrying blood to the heart are **VEINS** (v*a*nz). The flow of blood in arteries is generated by pressure originating in the heart. Land vertebrates have valves in their veins that compartmentalize the vessel and direct flow toward the heart (figure 7.2). The contraction of skeletal muscles adjacent to the veins of tetrapods compresses the veins and moves the blood to the next compartment toward the heart. Arteries have thicker walls than veins (to withstand the greater pressure).

Doubly injected specimens in which the arteries are filled with red latex and the veins with blue latex should be used. Notice that the color scheme is reversed in the pulmonary circulation. The hepatic portal system contains dried blood unless the specimens are triply injected; then the hepatic portal system will be filled with yellow latex.

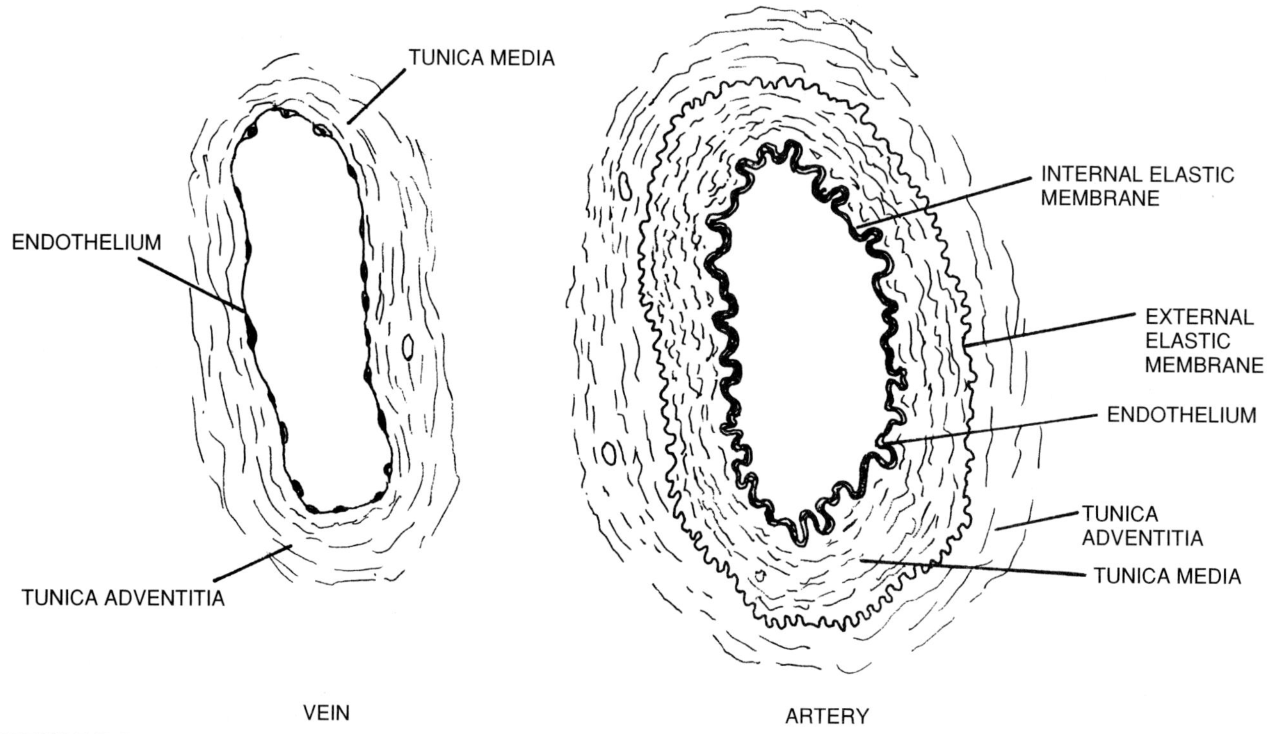

FIGURE 7.1. Microstructure of a median sized artery and vein.

Preparation for Observation of the Circulatory System

Carefully clean the loose connective tissues away from the vessels. Do not use a scalpel; forceps and a blunt probe will be adequate for most of this dissection. Hemostats are useful in dissecting the heart. To remove fat from vessels, study the figures of the vessels carefully; then hold a pair of forceps gently but firmly over the exposed portion of the vessel to be cleaned and pull the forceps over the vessel, thus scraping the fat from the blood vessel. After chunks of fat are removed, wipe the vessel and adjacent areas with a paper towel to remove the grease released when you broke the adipose cells with your forceps. BE EXTREMELY CAREFUL IN THIS DISSECTION, AS VEINS ARE EASILY TORN AND DESTROYED.

The Heart

External Anatomy of the Cat Heart (figure 7.3)

Carefully separate the membranous walls of the mediastinal septum and observe the pericardial membranes within the mediastinum. As with the peritoneum and pleura, the **PERICARDIA** also consist of visceral and parietal portions. *Cut open the parietal pericardium.*

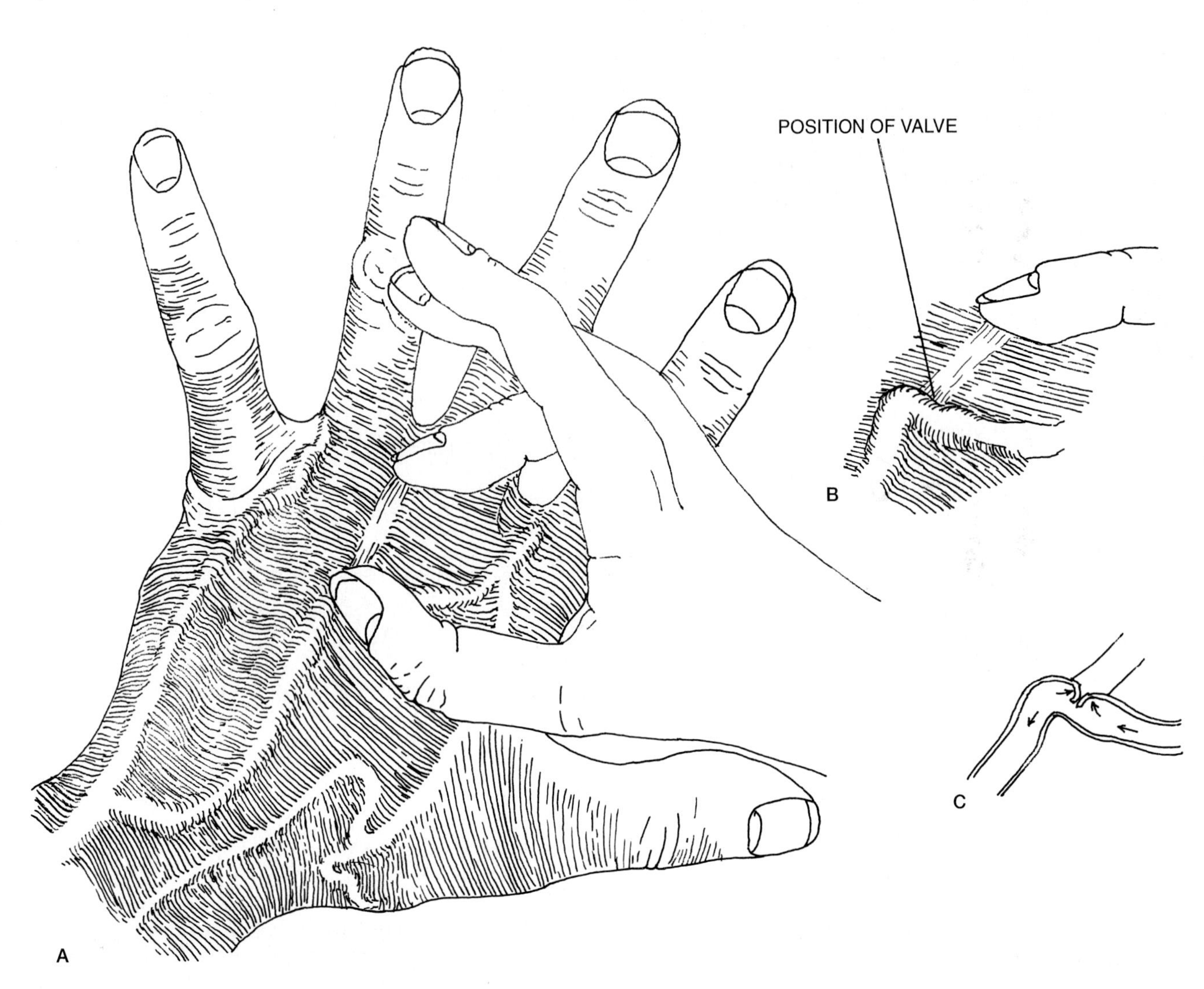

FIGURE 7.2. Venous valves. The presence of valves may be demonstrated by blocking the blood flow to a superficial vein in your hand and pressing over the vein, toward the heart, until you reach a valve (usually an inch or two). (A) The collapsed vein and the position of the valve. (B) Detail of the area of the collapsed vein. (C) A diagrammatic longitudinal section of the valve.

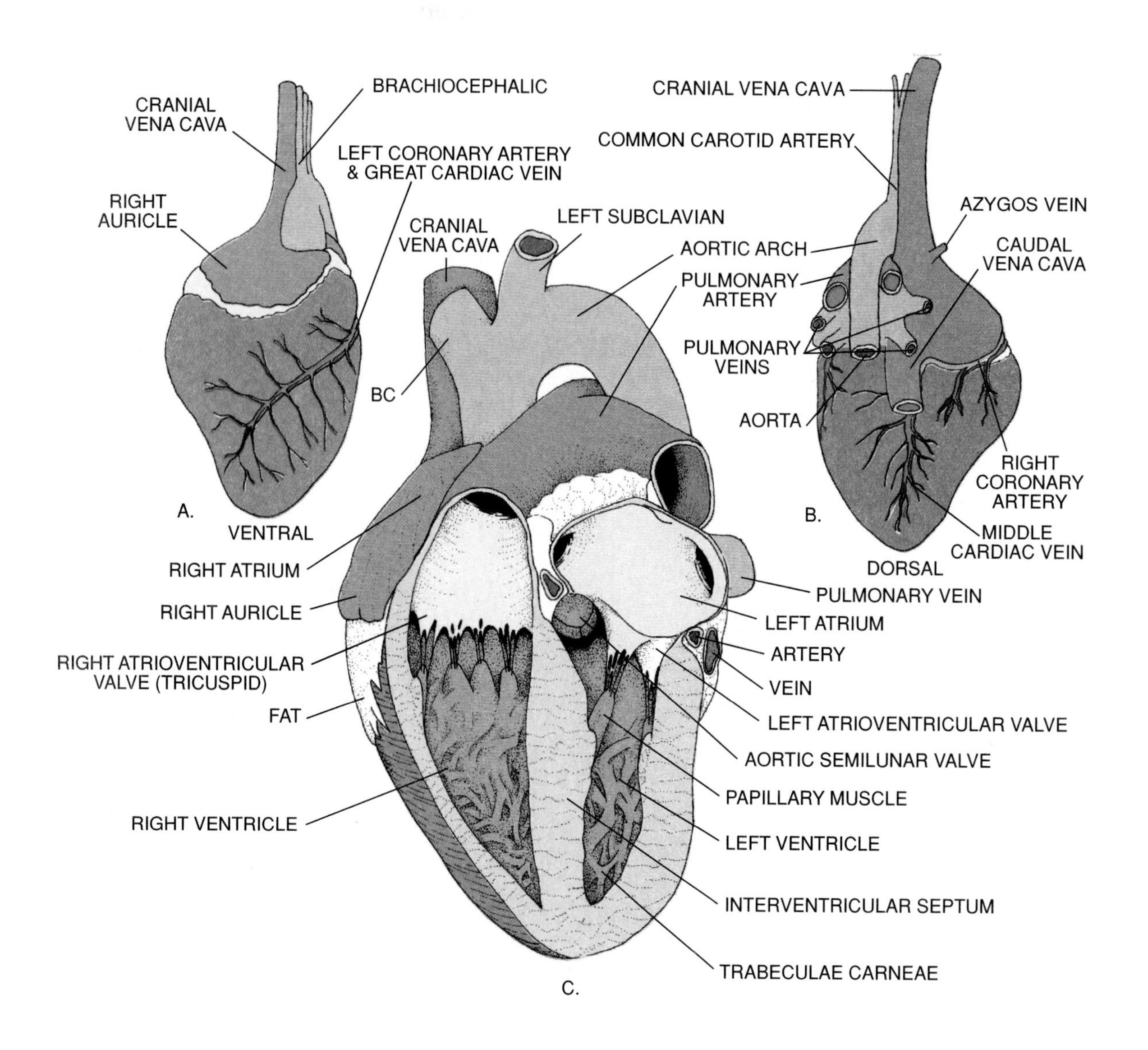

FIGURE 7.3. The heart of the cat. (A) Ventral view. (B) Dorsal view. (C) Left view with most of the wall dissected away to show internal structures.

Observe the **PERICARDIAL SPACE** beneath the membrane. The visceral pericardium adheres to the heart. Find and study the following:

1. **ATRIA** (*A*-tre-a). Two small chambers on either side of the cranial (superior in the human) portion of the heart. The **RIGHT ATRIUM** receives blood from the cranial and caudal vena cava (superior and inferior vena cava in the human) and from the **coronary sinus,** which receives blood from the muscle of the heart itself. The **LEFT ATRIUM** drains blood from the lungs via the pulmonary veins.

Spend a few minutes to become familiar with the differences in anatomical directions for cats and humans. **Cranial** in the cat corresponds to **superior** in the human. **Caudal** in the cat is **inferior** in the human. **Dorsal** and **ventral** in the cat are **posterior** and **anterior,** respectively, in the human. In this text, cat directional terms will be used. When you are concentrating on the human anatomy, be sure to make the appropriate conversions (figure 1.1).

2. **AURICLES** (AW-ri-kls). Small muscular flaps extending caudad from each of the atria.
3. **VENTRICLES** (VEN-tri-kls). The muscular caudal portion of the heart, consisting of two ventricles separated internally by an **INTERVENTRICULAR SEPTUM** (SEP-tum). Note that the **LEFT VENTRICLE** is larger and more muscular than the **RIGHT VENTRICLE.** The right ventricle does not reach the **APEX** (*A*-peks) of the heart. The ventricular chambers are separated from the atria, pulmonary artery, and aorta by valves.
4. **AORTA** (*a*-*O*R-ta). In all mammals, the single aorta originates from the left ventricle and passes cranial to the heart. Dorsal to the heart, the aorta arches to the left and proceeds caudally along the dorsal midline of the thoracic and abdominal cavities. The aorta distributes blood to all the systemic parts of the body.
5. **PULMONARY** (PUL-m*o*-ner-*e*) **ARCH.** A large artery comparable to the aorta but originating in the right ventricle. The arch divides just cranial to the heart to form the **RIGHT** and **LEFT PULMONARY ARTERIES,** which lead to the lungs.
6. **VENA CAVA** (V*E*-na K*A*-va). The **CRANIAL** and **CAUDAL VENAE CAVAE** are best seen on the dorsocranial (posterior superior in human) aspect of the heart, where they enter the right atrium. These vessels drain blood from all the body structures and organs except the heart muscle and lungs.
7. **PULMONARY VEINS.** There are four pulmonary veins—two from each lung. These veins drain the pulmonary capillaries of the lungs into the left atrium.

Internal Anatomy of the Cat Heart (figure 7.3)

Sever the aorta, pulmonary artery, pulmonary veins, and cranial and caudal venae cavae near the heart. Remove the heart from the thoracic cavity. With a single-edge razor blade, make a cut through the apex of the ventricles to the left atrium and the pulmonary arch to expose the interior of the heart. Remove the dried blood or latex from the ventricles.

1. **TRABECULAE CARNEAE** (tra-BEK-*u*-le KAR-ne-*e*). Muscular bands on the inner wall of the ventricles.
2. **PAPILLARY** (PAP-i-l*a*r-*e*) **MUSCLES.** Muscle papillae similar to the trabeculae carneae but with tendons (chordae tendineae) attached to them and to the flaps of the atrioventricular valves.
3. **CHORDAE TENDINEAE** (K*O*R-d*e* TEN-di-n*e*). Tendinous cords that attach the papillary muscles to a valve **CUSP** (kusp).
4. **TRICUSPID** (tr*i*-KUS-pid) **VALVE.** The valve located between the right atrium and right ventricle.
5. **BICUSPID** (b*i*-KUS-pid) or **MITRAL** (M*I*-tral) **VALVE.** The valve on the left side between the left atrium and left ventricle (collectively, the tricuspid and mitral valves are termed the **atrioventricular valves**). Blood passes to the pulmonary artery from the right ventricle and into the aorta from the left ventricle.
6. **SEMILUNAR** (sem-*e*-LOO-nar) **VALVES.** The valves between the ventricles and the aortic and pulmonary arches.

Systemic Arteries

The Aortic Arch and Its Branches (figure 7.4)

The **AORTA** is the large vessel exiting the left ventricle.

1. **CORONARY** (K*O*R-o-n*a*-re) **ARTERIES.** *With a pair of forceps, carefully remove the rubber injection mass from the aorta. Take care not to break the latex mass during removal.* The latex is a negative impression of the aorta, its semilunar valve, and the openings to the coronary arteries. The two coronary arteries (right and left) serve the heart muscle and branch from the aorta just cranial to the two ventral cusps of the left semilunar valve. The right coronary artery turns toward the apex on the dorsal surface, and the left coronary artery branches to the ventricular apex on the ventral surface of the ventricles. Both vessels lie in a groove, the **INTERVENTRICULAR SULCUS,** marking the division between the two ventricles.
2. **BRACHIOCEPHALIC** (br*a*k-*e*-*o*-se-FAL-ik) **ARTERY.** A single unpaired vessel arising from the base of the aortic arch (the dorsal aorta as it leaves the heart and arches left and dorsal to the heart). The brachiocephalic gives rise to the right and left common carotid arteries and the right subclavian artery.
   a. **COMMON CAROTID** (ka-ROT-id) **ARTERIES.** Extending craniad on either side of the trachea to their division at the level of the mastoid process into internal and external carotids.

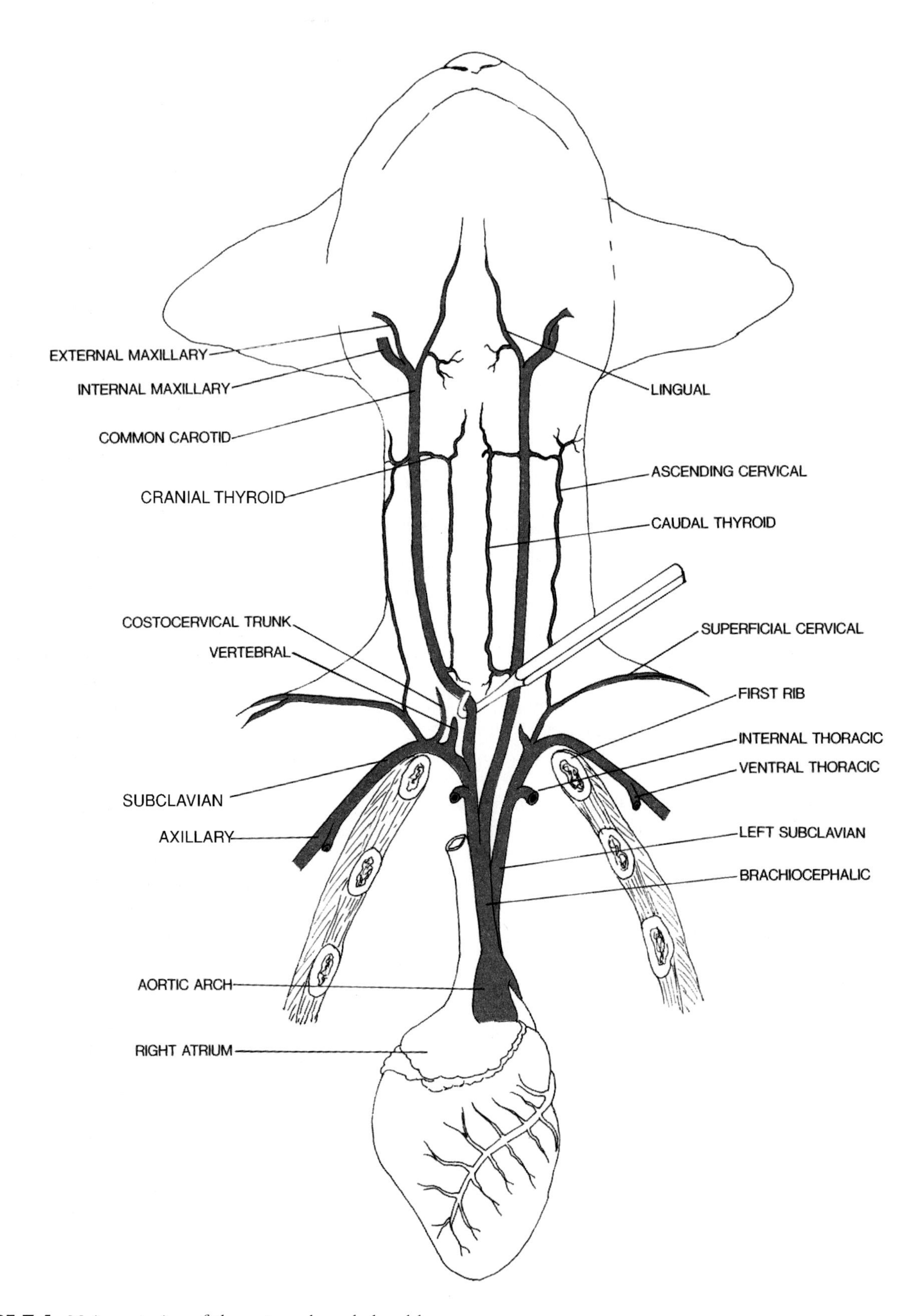

FIGURE 7.4. Major arteries of the cat neck and shoulder.

In the human, the left common carotid artery generally arises directly from the aorta rather than from the brachiocephalic artery as in the cat, although a common variation in the human is for the artery to branch from the brachiocephalic. In about 12 percent of humans, the right common carotid artery arises directly from the aorta. These normal variations are not defects, and create no health problems.

i. **CAUDAL THYROID** (TH*I*-royd) **ARTERIES.** Branch from the common carotids at the base of the neck and pass forward on the lateral walls of the trachea to serve the thyroid gland and larynx.

The **inferior thyroid** of the human arises from the **thyrocervical trunk,** which is, in turn, a branch of the subclavian artery near the **internal thoracic artery.** The inferior thyroid supplies blood to the larynx, trachea, esophagus, bodies of the cervical vertebrae, and infrahyoidal muscles.

ii. **CRANIAL THYROID ARTERIES.** Branch from the common carotids at the level of the thyroid gland.

In humans, the **superior thyroid artery** arises from the **external carotid artery** (see below). It supplies blood to the thyroid gland, the infrahyoidal muscles, the sternocleidomastoid muscle, and the larynx.

iii. **INTERNAL CAROTID ARTERIES.** Very small branches that pass through the rostral end of the jugular foramina, traversing the dorsal surface of the tympanic bullae and leaving the tympanic cavities just medial to the auditory (Eustachian) tubes. The internal carotids then branch to form several small vessels and turn mediad to join the **cerebral** (se-R*E*-bral) **arterial circle** at the base of the brain (not illustrated).

The **internal carotid arteries** are proportionally larger in man than in the cat. They enter the middle cranial fossa of the skull through the **carotid canals** and the **foramen lacerum.** In humans, the internal carotid arteries supply the majority of the blood to the brain, though the vertebral arteries are a source of arterial blood, too.

iv. **EXTERNAL CAROTID ARTERIES.** These vessels give off several small muscular branches and pharyngeal branches, but mainly they continue as large trunks on the ventrolateral borders of the bullae, where they are termed the **internal maxillary** (MAX-si-ler-*e*) **arteries.** The internal maxillaries supply the infraorbital areas, the palate, and the floor of the nasal chamber. The **external maxillary arteries** originate from the external carotids next to the internal maxillaries and supply the masseter muscles and the pulp cavities of the teeth. Also supplied by these vessels are the **external arterial rete plexuses** (R*E*-te PLEKS-us-*ez*) (not illustrated), the epithelium of the nasal cavities, and the **buccal** (BUK-al) **arteries** (not illustrated). Blood in the arterial rete empties into anastomosing vessels that drain into the cerebral arterial circle. The arterial rete is surrounded by a venous sinus with cooled venous blood from the epithelia of the nose, eye, and cheek. Cool venous blood reduces the temperature of the arterial blood before it reaches the brain. **Lingual** (LING-gwal) **arteries** supply the tongue.

The **external carotid arteries** of the human are about the same size as the internal carotids. Branches of the external carotid arise in the following order: **superior thyroid, ascending pharyngeal, lingual, facial, occipital, posterior auricular, superficial temporal,** and **maxillary.** Cooling mechanisms, similar to the cat's, have been described for the human brain, though the rete is formed from the circle of Willis in the human. Venous blood from the face may be used to reduce the temperature of the arterial blood flowing to the brain.

b. **RIGHT SUBCLAVIAN** (sub-KL*A*-v*e*-an) **ARTERY.** This vessel branches from the brachiocephalic artery along with the common carotid arteries and passes out of the thoracic cavity at the first rib.

3. **LEFT SUBCLAVIAN ARTERY.** Branches directly from the aortic arch just distal to the brachiocephalic artery (distal to the left common carotid in the human). Otherwise, the two subclavian arteries (right and left) are similar. Each traverses the axilla as the axillary artery and continues into the arm as the brachial artery.
   a. **VERTEBRAL** (ver-*TE*-bral) **ARTERIES.** Branch from the subclavian arteries at the base of the neck and pass directly to the transverse foramina of the sixth cervical vertebra. The vertebral arteries pass through the transverse foramina of the cervical vertebrae and then pass medially through the intervertebral foramina between the skull and atlas and into the foramen magnum. The right and left vertebral arteries anastomose to form the **basilar** (B*A*S-i-lar) **artery** (not illustrated) on the ventral surface of the medulla. The basilar joins the internal carotids to form the cerebral arterial circle.
   b. **INTERNAL THORACIC** (th*o*-RAS-ik) **ARTERIES.** Branch from the subclavian arteries at the level of the first rib and send branches to the musculature of the ventral chest region.
   c. **SUPERFICIAL CERVICAL** (SER-vi-kl) **ARTERIES.** Arise from the subclavian arteries and serve the dorsal muscles of the neck and chest, trachea, larynx, esophagus, thyroid gland, and cranial vena cava. In humans, this vessel is a branch of the **thyrocervical trunk.**
   d. **COSTOCERVICAL** (kos-t*o*-SER-vi-kl) **TRUNKS.** Branch from the subclavian arteries at the same level as the superficial cervical arteries, but serve the neck and shoulder muscles, lymph nodes, salivary glands, skin, and cutaneous muscles of the neck. The costocervicals originate distal to the thyrocervical trunks in humans.
   e. **AXILLARY** (AK-si-l*ar*-*e*) **ARTERIES.** Continuations of the subclavian arteries through the armpits. They produce branches to the pectoral and shoulder muscles and then proceed into the arms as the **BRACHIAL** (BRA-ke-al) **ARTERIES,** which pass through the pectoralis profundi (pectoralis minor in humans) with the brachial plexuses and accompany the median nerves on the brachium between the biceps and triceps muscles (see figure 7.10). Branches serve the muscles of the brachium. The brachials divide into several branches at the elbows; these serve the muscles of the forearms (**radial** and **ulnar arteries**).

The Aorta Caudal to the Arch (figures 7.5–7.7)

The dorsal aorta caudal to the aortic arch and continuing through the thoracic cavity is the **THORACIC AORTA.** This vessel then becomes the **ABDOMINAL AORTA** in the abdominal cavity.

1. **INTERCOSTAL** (in-ter-KOS-tal) **ARTERIES.** These branch from the thoracic aorta and serve the intercostal musculature. These are the **posterior intercostal arteries** of the human.
2. **PHRENIC** (FREN-ik) **ARTERIES.** The first branches of the abdominal aorta. These vessels serve the diaphragm.

> In humans, the origin of the **phrenic arteries** varies considerably. They may arise from the aorta directly, from the renal arteries, or as a common trunk from the aorta or celiac artery.

3. **CELIAC** (S*E*-l*e*-ak) **ARTERY** (figure 7.6). A large unpaired branch of the abdominal aorta serving the viscera at the cranial end of the abdominal cavity. The branches of the celiac artery are named according to the visceral organ that they serve, but all of the branches of the celiac are interconnected.
   a. **LEFT GASTRIC** (GAS-trik) **ARTERY.** This vessel branches from the celiac to the lesser curvature of the stomach and passes over the dorsal surface of the stomach to join the **right gastric artery.**
   b. **LIENAL** (l*i*-*E*-nal; or splenic [SPLEN-ik]). This vessel branches from the celiac at approximately the same level as the hepatic and left gastric arteries. The lienal branches almost immediately into the **right** and **left lienal arteries.** Both lienal arteries then branch into additional vessels. The left lienal has branches to the pancreas, but all other branches go to the spleen.

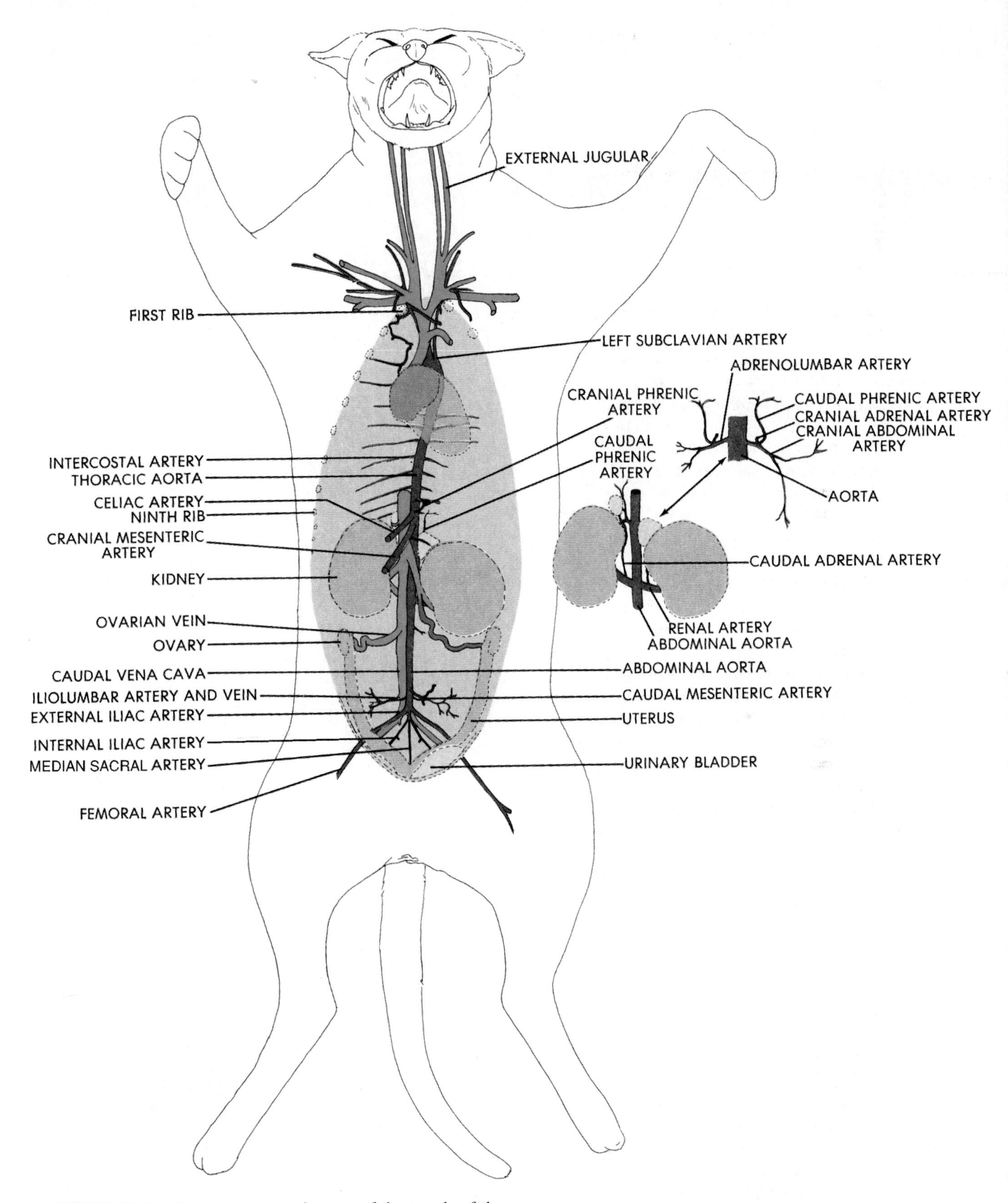

FIGURE 7.5. Major arteries and veins of the trunk of the cat.

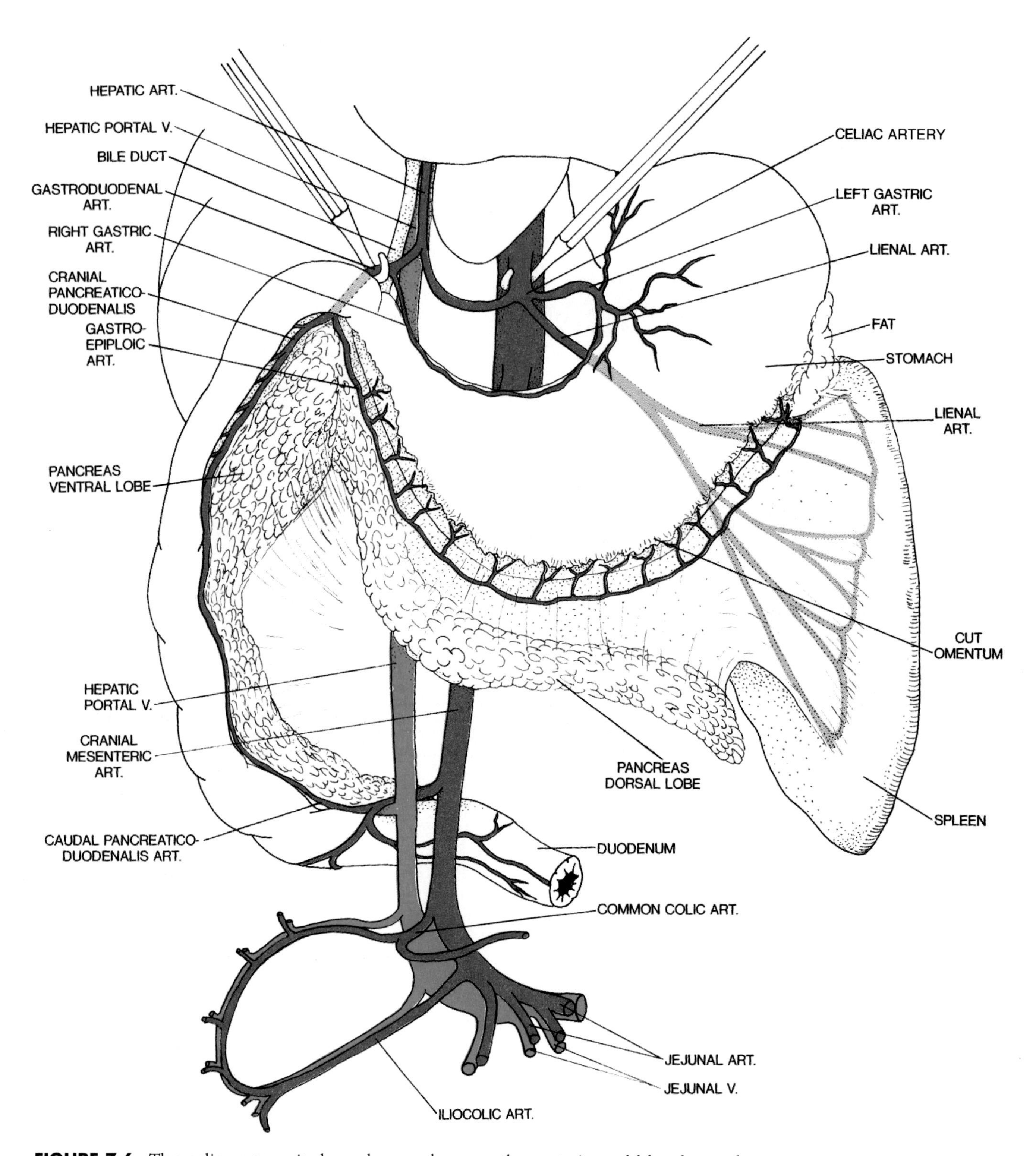

FIGURE 7.6. The celiac artery, its branches, and some other cat visceral blood vessels.

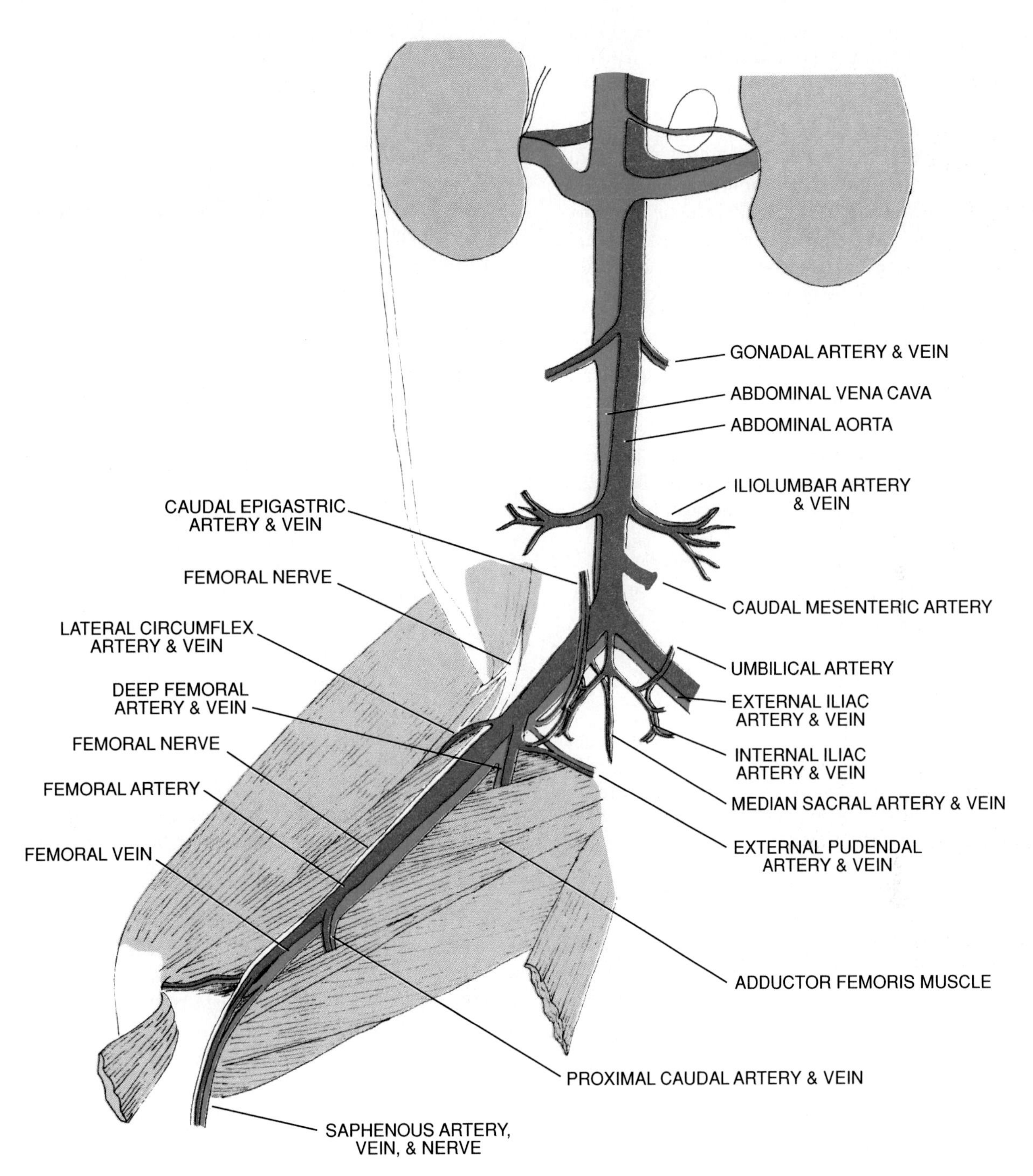

FIGURE 7.7. Details of the arteries, veins, and nerves of the iliac and inguinal regions of the cat.

The lineal artery is the largest branch of the celiac trunk in the human. It branches to form the **pancreatic, left gastroepiploic, short gastric,** and **splenic arteries.** These serve the pancreas, greater curvature of the stomach, and spleen.

c. **HEPATIC** (he-PAT-ik) **ARTERY.** This third branch of the celiac trunk continues into the liver. It arises at the same level as the left gastric and lienal in humans where it is called the **common hepatic artery.** Its branches form anastomoses with branches of the hepatic portal vein.
d. **GASTRODUODENAL** (gas-tr*o*-dyoo-*AH*-de-nal) **ARTERY.** This branch of the celiac artery gives rise to the **right gastric artery** to the lesser curvature of the stomach, the **gastroepiploic** (gas-tr*o*-ep-i-PL*O*-ik) **artery** to the greater curvature of the stomach, and the **cranial pancreaticoduodenal artery** to the duodenum and ventral pancreas. The right gastric artery forms an anastomosis with the left gastric artery and the gastroepiploic anastomoses with the cranial mesenteric artery via the cranial and caudal portions of the pancreaticoduodenal artery.

In humans, the **gastroduodenal artery** is a branch of the common hepatic artery, though it is known to vary and stem directly from the celiac artery. The gastroduodenal branches to form the gastroepiploic and right pancreaticoduodenal, as in the cat. The right gastric artery is a branch of the common hepatic in the human. The anastomosis with the **superior mesenteric artery** (cranial mesenteric of the cat) is via an additional branch, the **retroduodenal artery.**

4. **CRANIAL MESENTERIC** (mes-en-TER-ik) **ARTERY** (figures 7.5, 7.6). This is the **superior mesenteric artery** of the human. The second unpaired branch of the abdominal aorta, it supplies the mesentery and small intestine. The cranial mesenteric artery branches into numerous anastomosing **intestinal arteries** (ileocolic, jejunal). The anastomosing segments adjacent to the intestine have short branches to the intestinal walls. The other branches of the cranial mesenteric are:
   a. **CAUDAL PANCREATICODUODENAL** (pan-kr*e*-at-i-ko-dyoo-AH-de-nal) **ARTERY.** It runs craniad to anastomose with the celiac group.
   b. **COMMON COLIC** (K*O*L-ik) **ARTERY.** This vessel branches from the cranial mesenteric artery and joins the caudal mesenteric group.

In the human, there is a **right** and a **middle colic artery** leaving the superior mesenteric artery. The right colic joins the middle colic that, in turn, forms anastomoses with the **left colic,** leading to another anastomosis with the **inferior mesenteric artery.**

   c. **ILEOCOLIC** (il-*e*-*o*-K*O*L-ik) **ARTERY.** It joins the caudal mesenteric artery near the ileocolic valve.
5. **CRANIAL ADRENAL** (ah-DR*E*-nal) **ARTERIES** (figure 7.5). Paired branches from the aorta to the adrenal glands. These are the **suprarenal** of humans.
6. **RENAL** (R*E*-nal) **ARTERIES.** Paired branches from the aorta that enter the kidney at its medial indentation, the **hilus** (H*I*-lus). These vessels also form loops by branching and rejoining. The anastomosing vessels [**arcuate** (AR-k*u*-at) **arteries**] join at the border between the cortex and medulla of the kidneys. Arterioles from the anastomosing loops branch into glomeruli in the cortex (see figure 8.1 and page 134).
7. **GONADAL ARTERIES** (figure 7.7). Branching directly from the aorta are the:
   a. **Spermatic** (sperm-MAT-ik) **arteries.** Paired vessels in the male that branch from the aorta just caudal to the renal arteries and extend to the testes.
   b. **Ovarian** (*o*-V*A*-r*e*-an) **arteries.** Paired vessels in the female that branch from the aorta to the ovaries and uteri.
8. **ILIOLUMBAR** (il-*e*-*o*-LUM-bar) **ARTERIES.** Paired vessels branching from the abdominal aorta near the caudal mesenteric vein and supplying the ventral loin musculature.

In humans, the **iliolumbar** is a branch of the internal iliac artery. It supplies blood to the quadratus lumborum, psoas major, iliacus, iliac crest, and other structures in the area.

9. **CAUDAL MESENTERIC ARTERY** (**inferior mesenteric** of humans). A single vessel to the colon and rectum. This vessel branches and forms loops similar to the celiac and cranial mesenteric vessels. A branch of the caudal mesenteric joins the cranial mesenteric at the ileocolic valve.

10. **Deep iliac circumflex** (SER-kum-fleks) **artery** (not illustrated). Branches from the aorta in the cat to the ventral lumbar muscles and pectineus. In humans and other mammals, this is a branch of the external iliac.
11. **EXTERNAL ILIAC** (IL-*e*-ak) **ARTERIES** (figure 7.7). Paired branches of the abdominal aorta that form the deep femorals to the groin and the femorals to the limbs.
    a. **DEEP FEMORAL** (FEM-*or*-al) **ARTERIES.** Branch into **caudal epigastric** (ep-*e*-GAS-trik) **arteries** to the inner caudal abdominal walls, and **external pudendal** (p*u*-DEN-dal) **arteries** to the outer caudal abdominal walls, urinary bladder, and umbilical ligaments. The deep femoral proper serves the deeper thigh muscles (adductor muscles).

The **deep femoral** of humans (profunda femoris) is a branch of the femoral artery, as is the **external pudendal.** The **inferior epigastric** does branch from the external iliac artery in the human and supplies the abdominal musculature **(muscular branches)** and spermatic cord **(pubic branch).**

    b. **FEMORAL ARTERIES.** The major continuations of the external iliac arteries into the thighs. The branches from these vessels in the thighs are:
        i. **Lateral femoral circumflex arteries.** Supply the deep cranial muscles of the thigh (vastus and quadriceps muscles).
        ii. **Proximal caudal femoral arteries.** Branch caudally to the adductor, gracilis, and semimembranosus muscles.
        iii. **Genus descendens arteries.** Branch to the knees and proximal end of the tibia.
        iv. **Saphenous** (sah-F*E*-nus) **arteries.** Serve the distal legs and feet.

The **femoral artery** gives rise to the **superficial epigastric, superficial iliac circumflex, superficial external pudendal, deep external pudendal, muscular profunda femoris,** and **descending genicular arteries** in the human. The first three of these run to the pelvis and abdomen (the superficial external pudendal artery also supplies blood to the penis and scrotum of the male and the labia majora of the female). The remaining three vessels reach the gluteal, thigh, and deep rotator muscles.

12. **INTERNAL ILIAC ARTERIES** (figure 7.7). Paired branches from the terminal end of the aorta, about one centimeter caudal to the branching of the external iliac arteries. From this point the dorsal aorta continues into the tail as the **median sacral** (S*A*-kral) **artery.** Branches of the internal iliacs are:
    a. **Umbilical** (um-BIL-i-kal) **arteries.** Supply the placenta in the embryo and the urinary bladder postnatally.
    b. **Internal pudendal** (pyoo-DEN-dal) **arteries** (not illustrated). The last branches of the internal iliac arteries. Branches of these vessels in the male serve the prostate gland, scrotum, penis, and rectum. In the female, branches serve the vagina, uterus, clitoris, and rectum.

The **internal iliac** (hypogastric) **artery** is the most variable artery in the human body; in general, it supplies blood to the genitalia, pelvic viscera, gluteal region, and pelvic floor. As in the cat, the internal iliac arteries of the human fetus give rise to the umbilical arteries. The iliolumbar artery (see above) is one of the many branches.

Systemic Veins

Veins Cranial to the Heart (figures 7.8–7.11)

1. **CRANIAL VENA CAVA** (figure 7.8). The cranial vena cava (**superior vena cava** of humans) is a single large vessel formed by the union of the right and left brachiocephalic veins. In addition, the unpaired internal thoracic and azygos (AZ-i-gos) veins drain into the cranial vena cava.
    a. **AZYGOS VEIN.** Located on the right side only, this vessel receives the **intercostal veins** (not illustrated) from the intercostal muscles and drains to the right dorsal side of the cranial vena cava.
    b. **INTERNAL THORACIC VEIN.** A single trunk as it enters the cranial vena cava but originating as two vessels, one on each side of the sternum, that receive smaller veins from the ventral intercostal tissues.

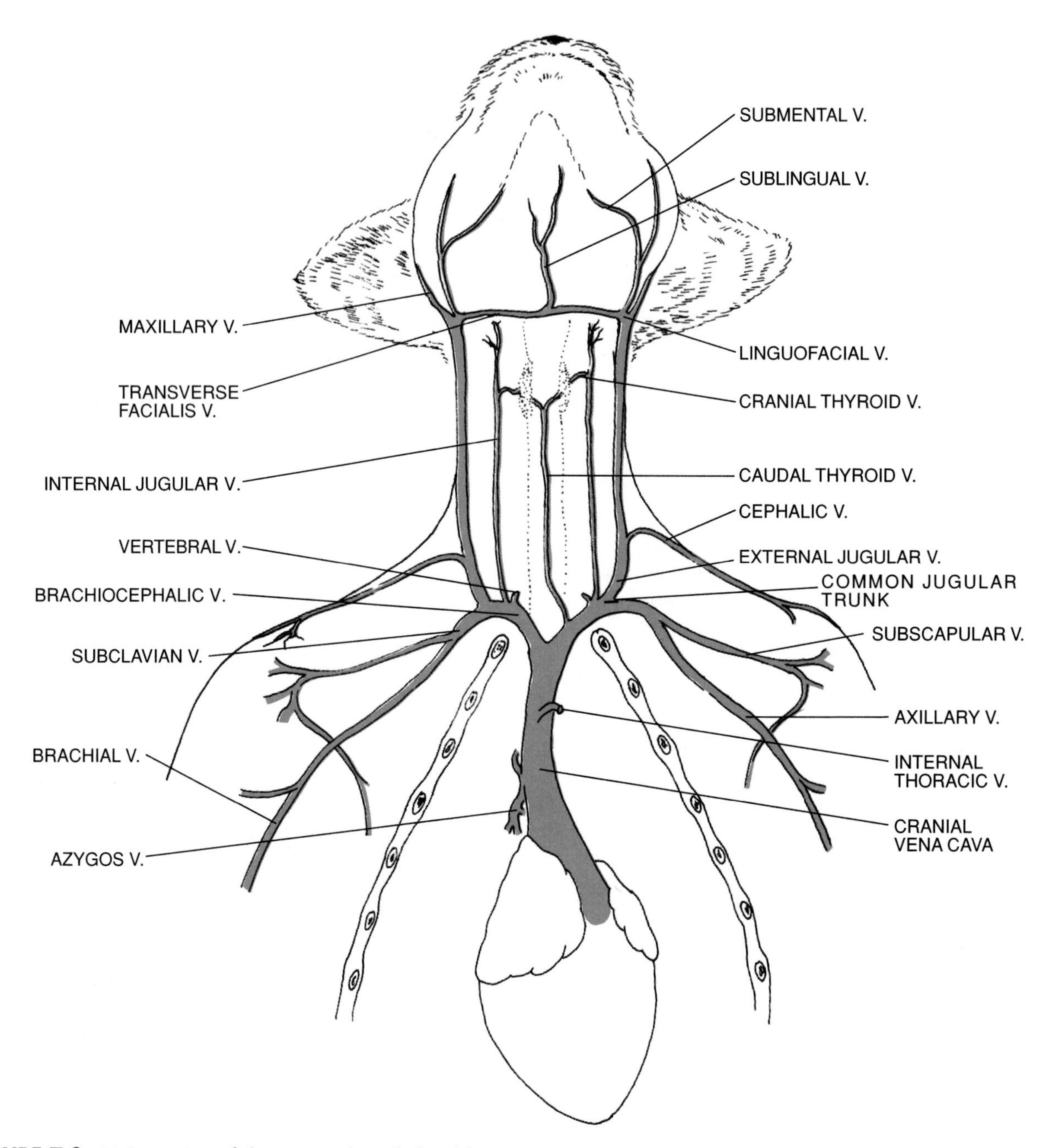

FIGURE 7.8. Major veins of the cat neck and shoulder.

In humans, commonly, the **right internal thoracic vein** is a tributary to the superior vena cava and the **left internal thoracic** enters the left brachiocephalic vein.

c. **Right costocervical vein.** The right side of this asymmetric pair usually enters the cranial vena cava on the dorsal side between the internal thoracic and the brachiocephalic veins (not illustrated).

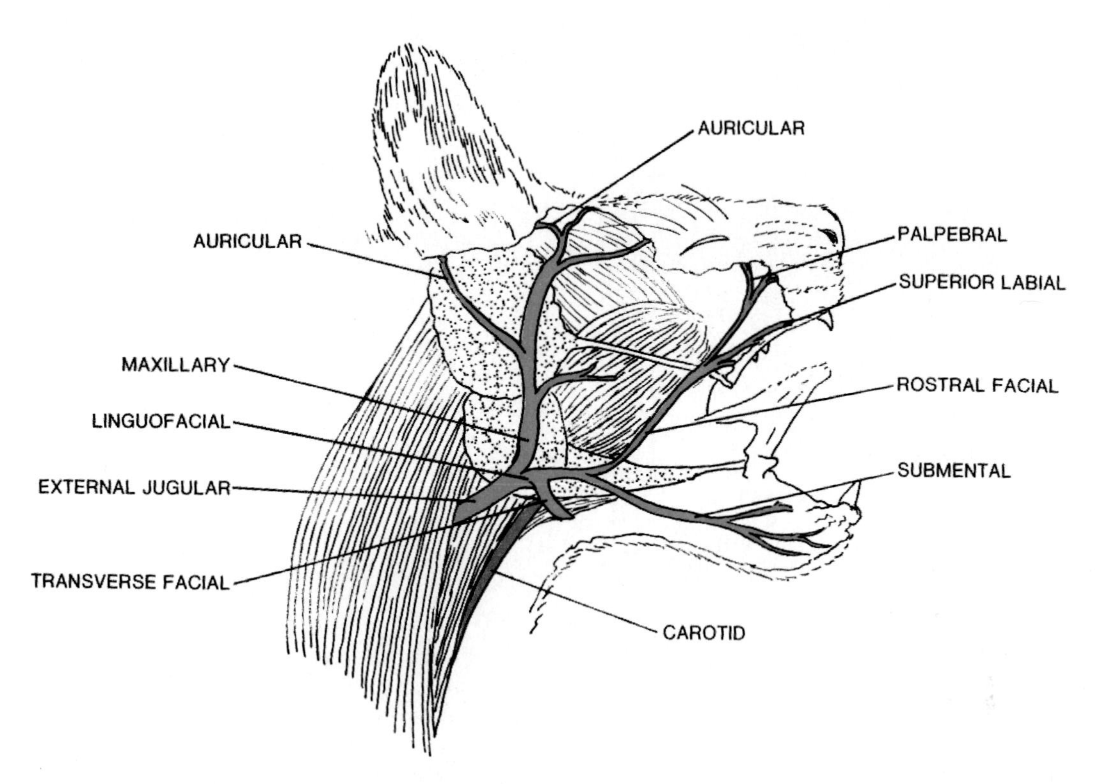

FIGURE 7.9. Superficial veins of the cat head and neck.

> There are no veins termed costocervical in humans, but a number of vessels drain this general area. For example, the **highest intercostal** (a tributary of the azygos vein from the upper intercostal spaces) and the **deep cervical** (a tributary of the vertebral vein) from the deep muscles of the back and neck.

d. **BRACHIOCEPHALIC VEINS.** Paired vessels that originate at the level of the first rib by the union of the common jugular and subclavian veins. They receive the internal jugulars, left costocervical, and vertebral trunks at the base of the neck.

   i. **Left costocervical vein.** Joins the left brachiocephalic vein. This vein receives branches from the back and neck muscles (not illustrated).

> Again, there is no costocervical vein in humans, but the highest intercostal vein of humans is a tributary to the left brachiocephalic vein.

   ii. **Vertebral veins**. Formed by the **intervertebral veins** (that pass through the intervertebral foramina), the **dorsal scapular veins, dorsal intercostal veins,** and the small muscular veins from the dorsal thoracic musculature (only the vertebral vein is illustrated).

   iii. **Caudal thyroid vein.** Formed of small branches from the thyroid gland, the tracheal and laryngeal muscles, and the neck muscles. The tributaries form a single vein on the ventral surface of the trachea; this vessel enters the left brachiocephalic vein.

> There are generally two **inferior thyroid veins** in the human, though they may join to form a common trunk draining into the brachiocephalic vein.

   iv. **Common jugular trunks.** Very short trunks formed by the internal and external jugular veins. Together with the subclavian veins they form the brachiocephalic veins.

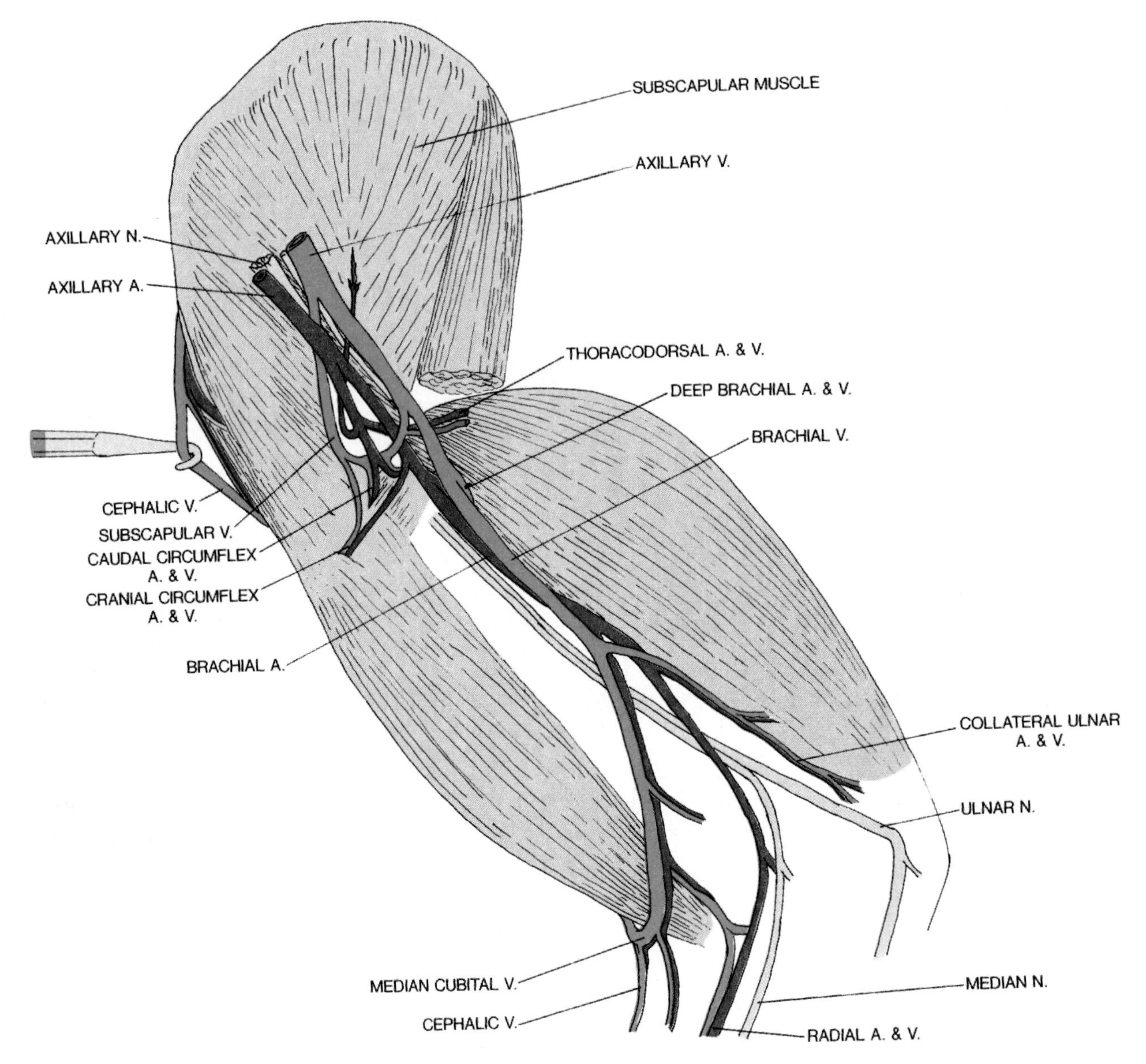

FIGURE 7.10. Medial view of the arteries, veins, and nerves of the axilla and brachium in the cat.

In humans, the external jugular vein unites with the subclavian vein to form the brachiocephalic vein. The internal jugular vein, along with the vertebral vein, is a tributary of the brachiocephalic vein.

a. **INTERNAL JUGULAR** (JUG-*u*-lar) **VEINS.** Drain blood from the sinuses surrounding the external rete plexuses and from more medial structures of the head, especially the brain. Occipital, pharyngeal, and cranial thyroid veins plus several small tributaries enter the internal jugular at the base of the head.

b. **EXTERNAL JUGULAR VEINS** (figure 7.9). Receive **cephalic veins** from the shoulders, **maxillary** (MAX-si-ler-*e*) and **linguofacial** (ling-gw*o*-*FA*-shal) **veins** from the head, and several other small muscular vessels from the base of the head. The cephalic veins lie on the cranial borders of the shoulders adjacent to the deltoid muscle. The cephalics form an anastomosis with the **BRACHIAL** (BR*A*-k*e*-al) **VEINS** via the **medial cubital** (K*U*-bi-tl) **veins** at the cranial surface of the elbows. The cephalics continue over the lateral surface of the forearm musculature, pass over the shoulder to the medial surface of the scapulae (where they are sometimes called the transverse scapular veins), and receive muscular branches before entering the external jugular veins.

> The human pattern is different. The **external jugular vein** receives blood from the superficial muscles and tissues of the upper back, neck, and head, including the face and tissues external to the cranium. Rarely, the **cephalic vein** will join the external jugular, usually terminating at the axillary vein.

v. **SUBCLAVIAN VEINS.** Receive the **AXILLARY VEINS** from the armpits (axilla) and these, in turn, receive the **brachial veins** from the medial surface of the arms. The brachial veins originate at the inner angle of the elbow as a union of the **radial** (R*A*-d*e*-al) and **median cubital veins** (figure 7.10). The brachial veins receive small muscular branches from the biceps muscles, **collateral ulnar veins** from the elbows, **deep brachial veins** from the triceps, and other small vessels. The radial veins receive **interosseus** (in-ter-OS-*e*-us) and **ulnar** (UL-nar) **veins** from the arms (not illustrated).

> In the human pattern, the **axillary vein** receives the **cephalic vein** and **basilic vein,** which drain the superficial arm, forearm, and hand. The axillary also receives the deeper brachial veins at the point of junction with the basilic. The **brachial veins** are doubled, lying on each side of the brachial artery. These receive the **radial** and **ulnar veins** (also deep). There is a complex of vessels at the anterior elbow connecting the basilic and cephalic veins and often an anterior **medial antebrachial vein.** This complex, the **median cubital vein,** is commonly used for venipuncture.

Veins Caudal to the Heart (figures 7.7, 7.11)

1. **CAUDAL VENA CAVA.** A single vessel extending from its origin at the caudal end of the abdominal cavity to its entrance into the right atrium with the cranial vena cava. Smaller veins enter the caudal vena cava in the abdominal cavity.
   a. **HEPATIC VEINS.** Enter the vena cava as it passes through the liver. At least one large hepatic vein drains each of the lobes of the liver. Hepatic veins are formed by the union of central veins in the liver.
   b. **Cranial phrenic veins** (not illustrated). Enter the vena cava from the diaphragm.
   c. **Adrenolumbar** (ad-r*e*-n*o*-LUM-bar) **veins.** Paired vessels draining the adrenal gland and receiving the **caudal phrenic veins** (not illustrated).

> In humans, the **suprarenal veins** are asymmetrical. The right enters the inferior vena cava and the left terminates either in the left renal or left inferior phrenic veins. The **inferior phrenic veins** are also asymmetrical. The right enters the inferior vena cava and the left, often doubled, ends in the left renal or suprarenal and inferior vena cava.

   d. **RENAL VEINS.** Paired vessels draining the kidneys. The left renal vein also receives the **LEFT GONADAL (SPERMATIC** or **OVARIAN) VEIN** from the gonad. The right renal vein enters the vena cava somewhat cranial to the left.

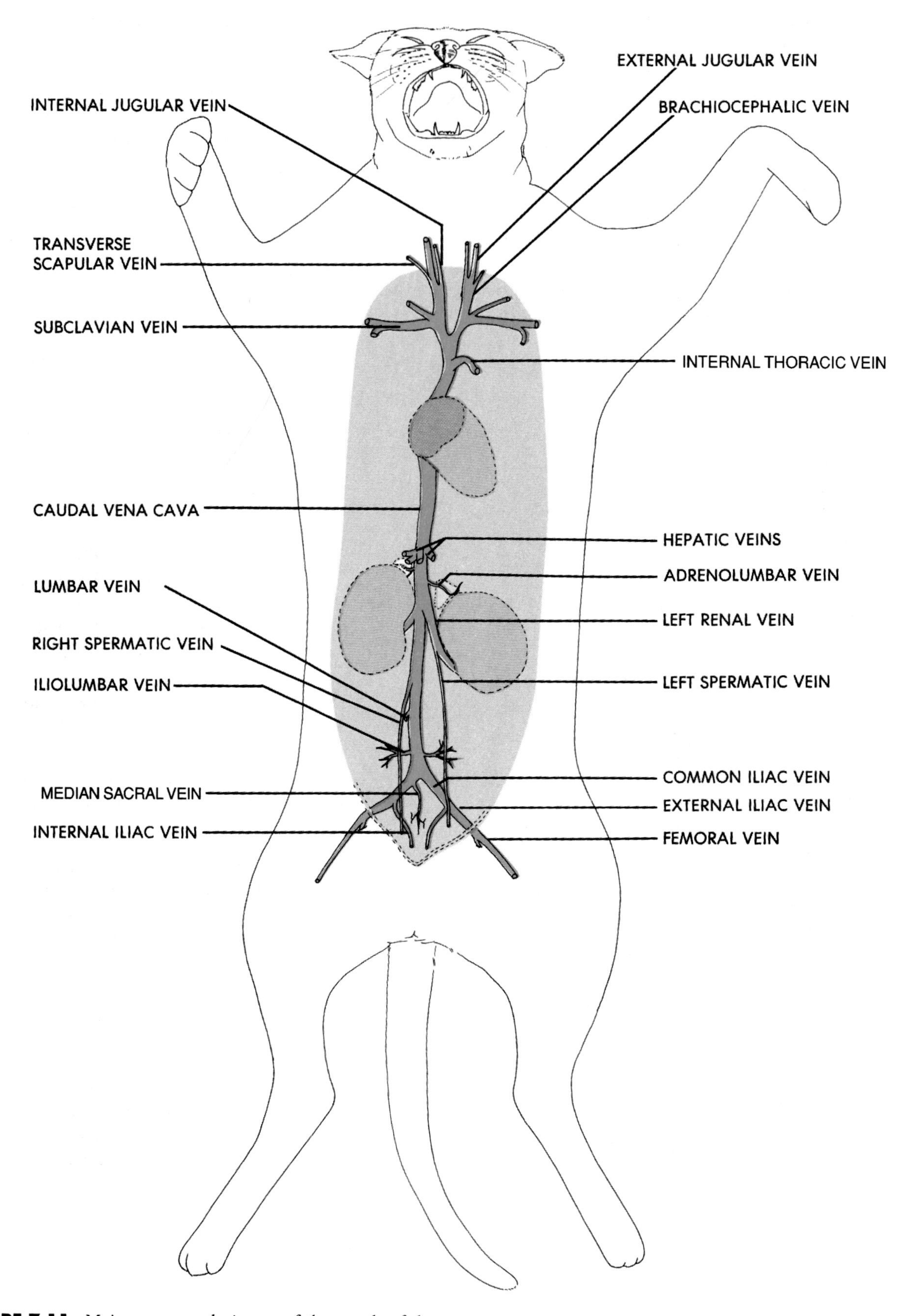

FIGURE 7.11. Major venous drainage of the trunk of the cat.

e. **RIGHT GONADAL (SPERMATIC** or **OVARIAN) VEIN.** Enters the caudal vena cava about two centimeters caudal to the right renal vein. Asymmetry between the left and right gonadal veins is produced during embryogenesis in both cats and humans. Most of the inferior vena cava and azygous vein of the adult is formed from the right embryonic posterior cardinal vein. The left ovarian and spermatic veins are remnants of the left posterior cardinal vein.

f. **LUMBAR VEINS.** Drain blood from the dorsal subvertebral muscles to the vena cava. These vessels are usually unpaired and arise at several intervals on the dorsal side of the vena cava.

g. **Iliolumbar** (il-*e*-*o*-LUM-bar) **veins.** Right and left iliolumbars receive blood from the musculature of the dorsal lumbar area.

h. **COMMON ILIAC VEINS**. Join to form the origin of the caudal vena cava. These veins, in turn, are formed by the union of the internal and external iliac veins. The right common iliac vein also receives the **median sacral** (S*A*-kral) **vein** from the tail. In humans, the **middle sacral** terminates in the left common iliac vein.

   i. **INTERNAL ILIAC VEINS.** Smaller than the external iliacs, these paired vessels receive additional vessels from the perineum. The veins drain the vas deferens, prostate, rectum, and penis of the male; the vagina, uterus, clitoris, and rectum of the female; and the obturator muscles, gluteal muscles, and dorsal tail of both sexes. The following are not illustrated.

      (1) **Obturator veins.** From the obturator muscles.
      (2) **Prostatic vein.** Serves the male vas deferens, prostate gland, and middle rectum.
      (3) **Vaginalis vein.** Serves the female vagina, uterus, and middle rectum.
      (4) **Cranial gluteal vein.** From the gluteal muscles.
      (5) **Lateral caudal vein.** From the dorsal tail. This is a counterpart of the medial sacral vein.
      (6) **Caudal gluteal vein.** From the gluteal muscles.
      (7) **Internal pudendal vein.** From the penis in the male, the clitoris in the female.

   ii. **EXTERNAL ILIAC VEINS**. Receive **caudal abdominal veins** (not illustrated) from the lateral abdominal muscles, a **pudendoepigastric vein** (not illustrated) from the caudal epigastric region, and the **deep femoral veins** from the pelvis. This vessel continues into the thigh as the **femoral vein.** The femoral veins pass through the body wall (on each side) from the medial surface of the thigh and open to the external iliacs. Each femoral vein receives the following vessels of the thigh:

      (1) **Lateral femoral circumflex veins.** Drain the gluteal and other hip muscles.
      (2) **Proximal caudal femoral veins.** Drain deep thigh muscles.
      (3) **Genus descendens** (d*e*-SEN-denz) **veins.** From the fascia of the knees.
      (4) **MEDIAL SAPHENOUS** (sah-F*e*-nus) **VEINS.** From the legs and feet.
      (5) **POPLITEAL VEINS.** The continuation of the femoral vein below the knee. It receives a large **cranial tibial vein** and a small **caudal tibial vein.**

The **femoral veins** of humans receive the **inferior epigastric vein, deep iliac circumflex vein, pubic vein,** and **superficial** and **deep veins.** As is the case in the superior appendage, the veins of the lower limb may be grouped into superficial (those just under the skin) and deep (those that run along with the arteries). The superficial veins are the **great saphenous vein** (a tributary to the femoral vein), the **small saphenous vein** (a tributary to the popliteal vein, which is an extension of the femoral vein), the **dorsal** and **plantar venous arches,** and the **digital veins.** The deep veins are the **deep venous arches,** the **posterior tibial vein,** the **anterior tibial vein,** the **popliteal vein** (receives the two tibial veins and the small saphenous vein), the **femoral vein,** and the **deep femoral vein** of the thigh.

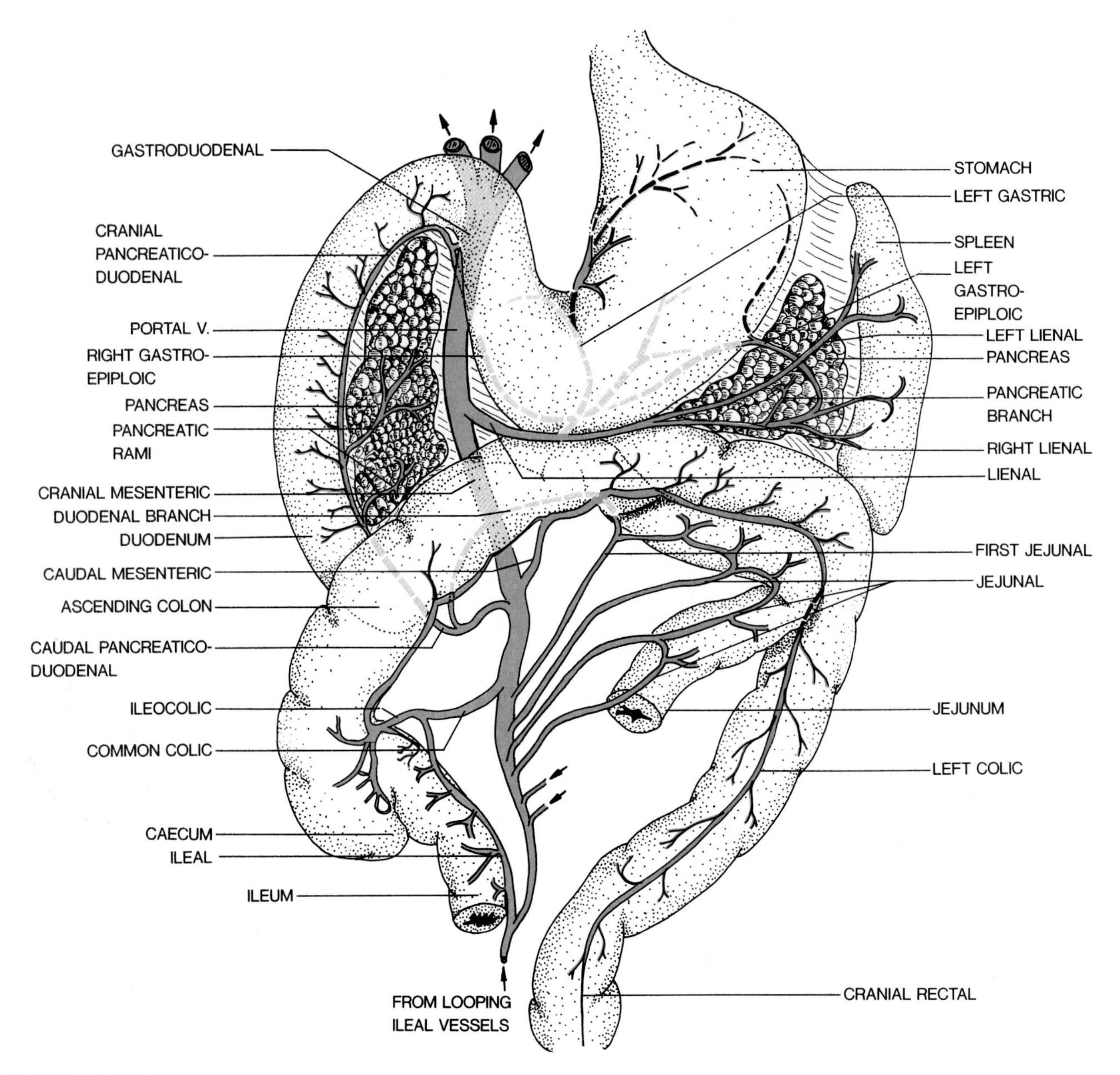

FIGURE 7.12. The hepatic portal system of the cat.

Hepatic Portal System (figure 7.12)

The **HEPATIC PORTAL** (P*O*R-tal) **VEIN** receives vessels from the various abdominal visceral organs that are suspended by mesentery, and empties to **sinusoids** in the liver. Sinusoids then drain to the **central veins,** hepatic veins, and finally, the inferior vena cava. The portal vessel is formed by four major veins, a gastroduodenal, a lienal (gastrosplenic), and cranial and caudal mesenteric.

Tributaries to these vessels are abundant and complex, and anastomose frequently. Only the major vessels are listed. Locate the following visceral vessels:

1. **GASTRODUODENAL** (gas-tr*o*-dyoo-AH-de-nal) **VEIN.** Formed by the
   a. **Cranial pancreaticoduodenal** (pan-kr*e*-at-i-k*o*-dyoo-AH-de-nal) **vein.** Drains the duodenum and ventral lobe of the pancreas.
   b. **Right gastroepiploic** (gas-tro-ep-i-PL*O*-ik) **vein.** From the cranial greater omentum and greater curvature of the stomach.

2. **LIENAL** (li-*E*-nal) or **GASTROSPLENIC VEIN.** Formed by the
   a. **Left gastric vein.** Drains the lesser curvature of the stomach.
   b. **Right and left lienal veins.** Found leaving the spleen to join just distal of the left gastric vein.
   c. **Left gastroepiploic vein.** From the greater omentum and greater curvature of the stomach.
   d. **Pancreatic branches.** A number of small tributaries from the dorsal pancreas entering the right lienal and the lienal proper.
3. **CRANIAL MESENTERIC** (mes-en-TER-ik) **VEIN.** Formed by numerous tributaries from the small intestine (ileum and jejunum). These vessels form connections near their origin to make anastomosing loops. The caudal pancreaticoduodenal vein and the caudal mesenteric vein enter the cranial mesenteric vein very near each other at the caudal end of the ventral pancreas. The first jejunal and ileocolic veins enter the cranial mesenteric vein near to each other and close to the duodenal-jejunal junction.
   a. **Caudal pancreaticoduodenal vein.** This vessel has two main tributaries: one serves the cranial duodenum and ventral mesentery; the other serves the caudal duodenum. The cranial tributary forms an anastomosis with the cranial pancreaticoduodenal vein. The caudal tributary joins with the first jejunal loop of the cranial mesenteric vein.
   b. **CAUDAL MESENTERIC VEIN.** Originates by several tributaries, including the **left colic** (K*O*L-ik) **vein,** the **cranial rectal** (REK-tal) **vein,** and the **sigmoid** (SIG-moyd) **vein** (not illustrated) from the large intestine.
   c. **Ileal** (IL-*e*-al) **vein.** Forms looping anastomoses to the ileum of the small intestine and drains to the cranial mesenteric vein.
   d. **Jejunal** (je-JOO-nal) **vein.** The first looping anastomosis of this vessel joins with the duodenal branch of the caudal pancreaticoduodenal vein. The remainder of the loops lie along the distal jejunum and enter the cranial mesenteric artery.
   e. **Ileocolic vein.** Joins the cranial mesenteric vein (via the common colic vein) near the first jejunal vein. Drains the transverse and ascending colon, the cecum, and portions of the ileum.

The **hepatic portal system** of humans consists of the following tributaries to the **portal vein:**

(1) **Lineal,** or **splenic, vein.** From the stomach, greater omentum, and pancreas. In humans, the lineal also receives the **inferior mesenteric vein** from the descending colon, sigmoid colon, and rectum.
(2) **Superior mesenteric vein.** From the small intestine, ascending colon, and transverse colon.
(3) **Coronary vein.** This vessel drains the stomach and esophagus.
(4) **Pyloric vein.** From the pyloric portion of the stomach.
(5) **Cystic veins.** These small vessels serve the gallbladder and ducts.

Lymphatics

The **LYMPHATIC** (lim-FAT-ik) **SYSTEM,** well-developed in mammals, functions as a path by which liquids and leukocytes enter general circulation from the tissues. Lymphatics consist of vessels paralleling the veins, and are comparable to capillaries and veins of the blood vessels. They differ from vessels of a true circulatory system in that the fluid (**lymph** [limf]) is transported only toward the heart; that is, the lymphatic capillaries arise blindly in the tissues. Lymphatic capillaries tend to be larger in diameter and more irregular than blood capillaries. Also, the larger lymphatic vessels have thinner walls than veins do.

Other structures belonging to the lymphatic system include the **THYMUS** (TH*I*-mus) **GLAND,** the **SPLEEN** (spl*e*n) (an elongated reddish body situated in the dorsal mesentery near the stomach), and the **TONSILS** (TON-sils), located in the oral cavity and pharynx. Generally, **LYMPH NODES** are numerous and found in subcutaneous tissues and body cavities. Lymphatic organs are also important as **hemopoietic** (h*e*-m*o*-poy-ET-ik) **tissues** (blood-forming). The spleen is a fetal blood-forming tissue, but not so in adults. Bone marrow forms most of the blood solids in adults.

Because the lymphatic vessels have thin walls, they were not injected in your specimen. Lymph nodes are large, and many of the other lymph organs may be located as well.

Lymphatic Vessels (figure 7.13)

1. **THORACIC DUCT.** This large lymph vessel drains the caudal body regions, including the intestines, and joins the vessels from the left forelimb and head. The duct then enters the left

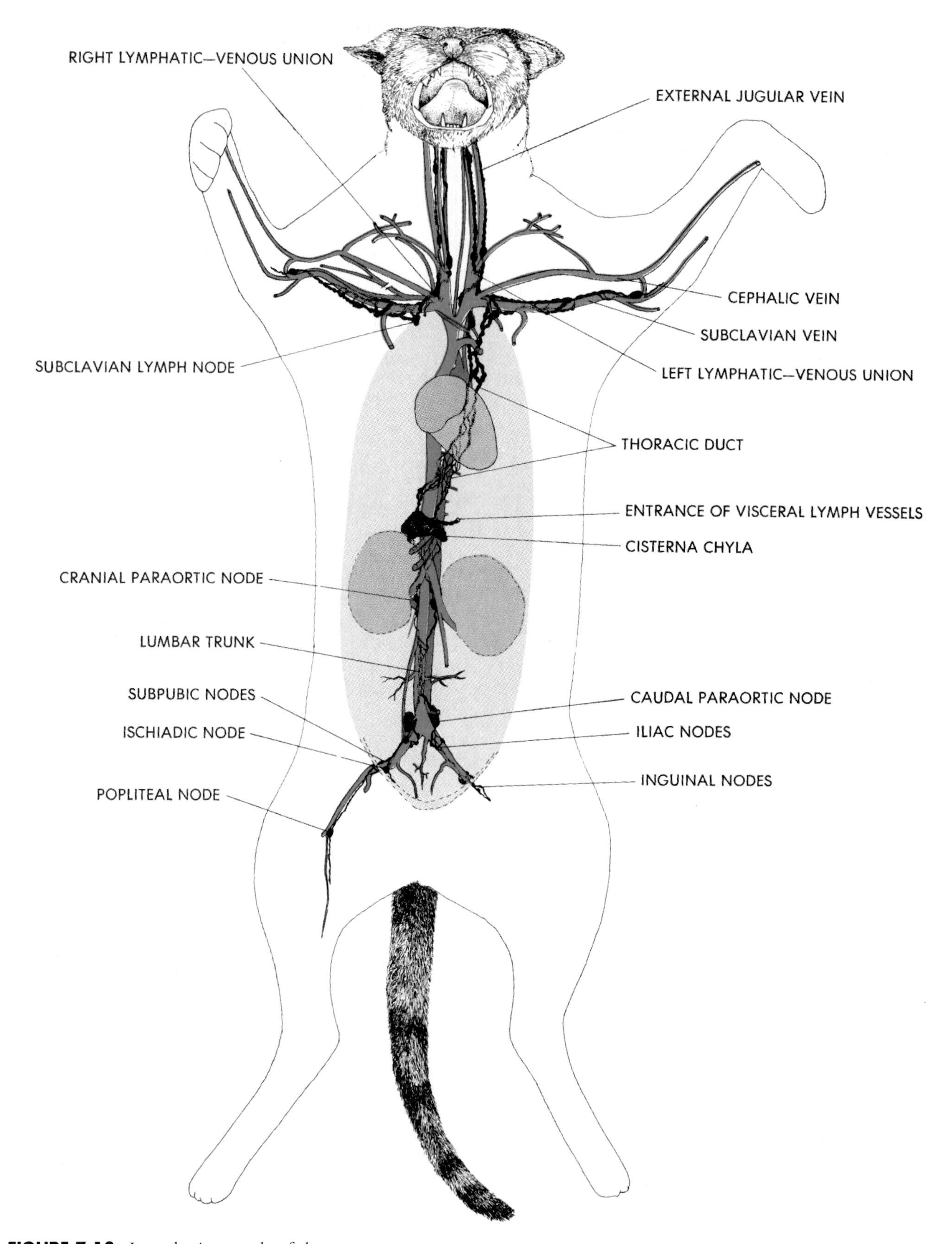

FIGURE 7.13. Lymphatic vessels of the cat.

subclavian vein. The thoracic duct is easiest to locate near the dorsal aorta as it passes through the thoracic cavity. It may also be possible to locate the **CISTERNA CHYLI** (sis-TER-na K*I*-l*i*), just caudal to the diaphragm, as it enters the thoracic duct. The cisterna collects lymphatic vessels from the intestines.

2. **Right lymphatic duct.** The drainage for lymph from the right forelimb and the right sides of the thorax, neck, and head. The duct enters the right subclavian vein.
3. **LYMPH NODES.** You may locate intestinal nodes easily by looking for pinkish swellings in the mesentery. The **lacteals** (LAK-t*e*-als) (not illustrated) are lymph capillaries of the intestine and drain to the intestinal lymph nodes. They are responsible for collecting the products of fat digestion.
4. **SPLEEN** (figure 5.8A, B). This is a flat, elongated organ suspended in the great omentum to the left of the greater curvature of the stomach. The spleen consists of a tissue mass termed the **red pulp,** with nodules of **white pulp** scattered throughout that serve as germinal centers for the production of lymph cells. **Trabeculae** (tra-BEK-*u*-l*e*) of elastic connective tissue containing blood vessels and nerves extend into the red pulp from the outer capsule. *Although details may not be evident, the capsule, trabeculae, and red and white pulp may be seen by preparing a slice of the spleen with a razor blade and observing the slice under low power with a dissecting microscope.*
5. **THYMUS** (figure 6.1). An irregular mass of tissue at the cranial end of the heart and in the ventral mediastinum. In the young, this structure may reach the larynx; later it is a much smaller structure at the bifurcation of the bronchi.

Suggested Readings

Ayer, A. A., and Y. G. Rao. 1958. The coronary arterial patterns in some common laboratory animals: rabbit, dog, and cat. *J. Anat. Soc. India.* 7:5–8.

Baker, M. A. 1979. A brain cooling system in mammals. *Sci. Am.* 240(5):130–39.

Biscoe, T. J., and A. Bucknell. 1963. The arterial blood supply to the cat diaphragm with a note on the venous drainage. *Quart. J. Exp. Physiol.* 48(1):27–33.

Bradshaw, P. 1958. Arteries of the spinal cord in the cat. *Jour. Neurol. Neurosurg. Psychiatry.* 21(4):284–89.

Cabanac, M. 1986. Keeping a cool head. *News Physiol. Sci.* 1:41–4.

Cabanac, M., and M. Caputa. 1979. Natural selective cooling of the human brain; evidence of its occurrence and magnitude. *J. Physiol.* 286:255–64.

Cabanac, M., and H. Brinnel. 1985. Blood flow in the emissary veins of the human head during hyperthermia. *Eur. J. Appl. Physiol.* 54:172–6.

Davis, D. D. 1941. The arteries of the forearm in carnivores. *Field Museum. Nat. Hist. Zool.* 27:137–227.

Davis, D. D., and H. E. Story. 1943. The carotid circulation in the domestic cat. *Field Museum. Nat. Hist. Zool.* 28:1–47.

Ghoshal, N. G. 1972. The arteries of the pelvic limb of the cat (*Felis domesticus*). *Zentralbl. Veterinarmed. [A].* 19:78–85.

Hadziselimovic, H., D. Secerov, and E. Gmaz-Nikulin. 1974. Comparative anatomical investigation on the coronary arteries in wild and domestic animals. *Acta Anat.* 90:16–35.

Huntington, G. C., and C. F. W. McClure. 1920. The development of the veins in the domestic cat. *Anat. Rec.* 20:1–31.

Kampmeir, O. F. 1969. *Evolution and Comparative Morphology of the Lymphatic System.* Springfield, Ill.: Charles C. Thomas, Publishers.

Keller, P., and U. Freudiger. 1983. *Atlas of Hematology of the Dog and Cat.* New York: Paul Parey Scientific Publishers.

CHAPTER 8

Excretory System

The excretory and reproductive systems are closely related, especially in the lower vertebrates, but also in more advanced forms of vertebrates. Excretory tubules of primitive (and embryonic) male vertebrates are also used as passageways for sperm. Pronephric tubules and archinephric ducts transport sperm to the cloaca in most vertebrate males and to the urethra in male mammals. Consequently, some textbooks discuss the excretory and reproductive systems as a single **UROGENITAL SYSTEM.** The metabolic products of protein digestion are toxic and must have a special means of elimination. Some ammonia is excreted directly, but is toxic in small concentrations. Therefore, its conversion to a less toxic compound is necessary to allow the animal to concentrate nitrogenous wastes before excretion. Most mammals excrete urea as the major waste product of protein metabolism, but some also excrete small amounts of other products, including uric acid (human), allantoin (dog), or creatinine (cat).

The **metanephric** kidney, found only in birds and mammals, has tubules with thin loops of Henle. These tubules arise bilaterally caudal to the mesonephros that forms the ureters, collecting ducts, and renal pelvis. The nephrons of mammals (and a portion of those in birds) are metanephric in origin. The typical mammalian kidney is compact and bean-shaped.

1. **KIDNEYS** are embedded in fat on the dorsal body wall. The kidneys are not suspended by a mesentery like the other abdominal organs, but instead are covered by peritoneum only on the portion adjacent to the abdominal cavity. Because of this fact, the kidneys are said to be **retroperitoneal.** *Remove the fat around the kidneys, but be careful not to destroy the adrenal gland and blood vessels that are also embedded in fat in this region. When the left kidney is freed from the parietal peritoneum and fat, lift the kidney up and slice it in half horizontally. Include a portion of the ureter with each half. A single-edge razor blade will be convenient for making this section. Remove and examine half of the kidney with a dissecting microscope* and locate the following (figure 8.1):
   a. **HILUS** (H*I*-lus). The medial concave opening to the kidney, permitting access by the renal blood vessels and ureters.
   b. **URETERS.** Paired ducts extending from the kidneys to the urinary bladder.
   c. **RENAL PELVIS.** The expanded portion of the ureter.
   d. **RENAL PAPILLA.** The apex of the medulla extending into the renal pelvis. The cat has a single papillum in each kidney, but some other mammals, including humans, have several.
   e. **RENAL CORTEX** (K*O*R-teks). The outer portion of the kidney, distinguishable from the medulla by color. The glomeruli and convoluted tubules are located here.
   f. **RENAL MEDULLA** (me-DUL-a). The inner body of the kidney, darker in color than the cortex, with loops of Henle and collecting tubules.
2. **URINARY BLADDER** (figure 9.1). A reservoir for urine from the ureter.
3. **URETHRA** (*u*-R*E*-thra). A duct extending from the neck of the bladder to the exterior (figure 9.1). In the male, the urethra extends through the penis and receives the male sex gland products. In the female, this duct opens to the vestibule inside the urogenital orifice.

Histology of the Mammalian Kidney
(figure 8.1)

The basic functional unit of the kidney is the **NEPHRON** (NEF-ron), consisting of a **renal capsule (Bowman's capsule),** a **proximal convoluted** (KON-vo-l*u*-ted) **tubule,** a constricted central loop **(Henle's** [HEN-l*ez*] **loop),** and a **distal convoluted tubule.** The distal convoluted portion of each tubule

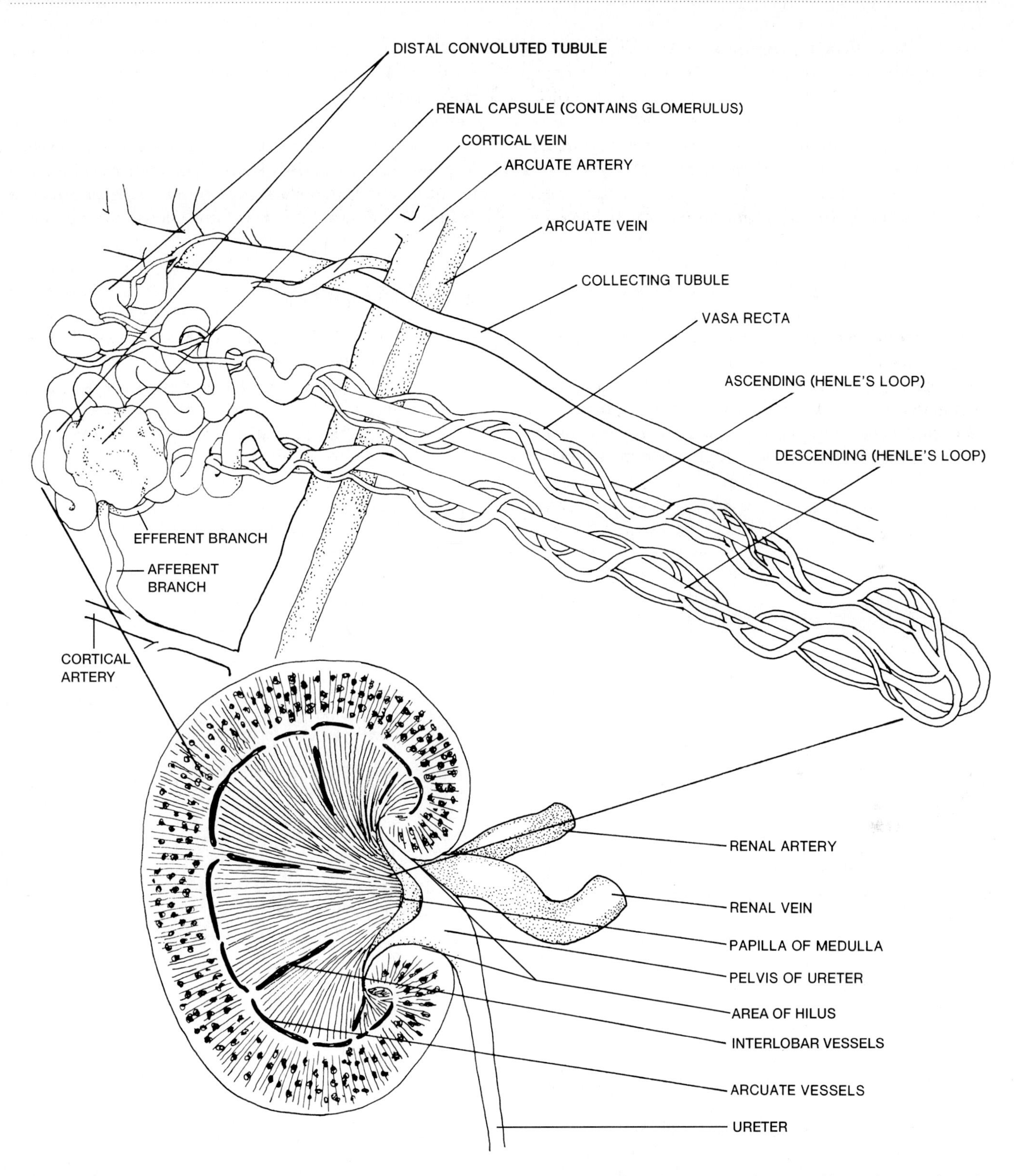

FIGURE 8.1. Microstructure of the mammalian kidney and a diagram of a kidney unit. The capsule and convoluted tubules are in the cortex. Henle's loop and collecting tubules are in the medulla.

opens to a **collecting tubule** that, in turn, opens to the renal pelvis. The loops of Henle and collecting tubules are located in the renal medulla, but the other parts are in the cortex. Nephrons are frequently classified as two types in mammals: (1) **cortical nephrons,** which have the glomeruli in the outer cortex and short loops of Henle mostly contained in the cortex and (2) **juxtamedullary nephrons,** which have the glomeruli near the medulla and where the loops of Henle penetrate deeply into the medulla.

The **renal artery** enters the renal sinus with the ureter and renal vein and divides immediately into **interlobar** (in-ter-LŌ-bar) **arteries.** It divides again at the border between cortex and medulla into **arcuate** (AR-k*u*-at) **arteries** that eventually give rise to numerous **cortical arteries** with **afferent branches** going to a **glomerulus** situated within each renal capsule. The blood is filtered through the glomerulus to the interior of the capsule (which opens to the proximal convoluted tubules), and water is reabsorbed through the walls of the tubules. **Efferent glomerular branches** leave each glomerulus and form a capillary network around the convoluted and loop tubules before joining the **cortical veins.** The cortical veins then join the **arcuate veins** that open to **interlobular veins.** Those then join at the renal sinus into the **renal vein.**

Suggested Readings

Fitzgerald, T. C. 1940. The renal circulation of domestic animals. *Am. J. Vet. Res.* 1:89–95.

Yadave, R. P., and M. L. Calhoun. 1958. Comparative histology of the kidney of domestic animals. *Am. J. Vet. Res.* 19:958–68.

CHAPTER 9

Reproductive System

The basic units of reproduction are the **ova,** from the female of the species, and **sperm,** from the male. The reproductive tracts of both sexes have evolved to ensure the fertilization of the ovum by the sperm, but the male tract in particular is designed to deliver the mobile sperm to the nonmobile ovum. **Internal fertilization** is accomplished by a male intromittent organ that is inserted into the female reproductive tract, where it deposits the sperm.

Male Reproductive System (figure 9.1A)

Male reproductive organs develop earlier than those of the female. The substance of the testes is composed of the lobules and vasa recta. These tubules open to the vasa efferentia; the vasa efferentia, in turn, open to the epididymis. Renal portal vessels of lower vertebrates are homologous to the vessels draining the rete tubules (centripetal veins) and the scrotal wall (pampiniform plexus) of the male mammal.

1. **SCROTUM** (SKR*O*-tum) (not illustrated). A sack of skin, muscle, and connective tissue on the exterior of the body, just ventral to the anus and containing the testes. *Cut open the scrotum and note the following layers from the exterior to the inner wall:* (1) **skin,** (2) **cremaster** (kr*e*-MAS-ter) **muscle** and **fascia,** and (3) **tunica vaginalis** (VAJ-i-nal-is), a tough connective-tissue layer.
2. **TESTES** (TES-t*e*z). Paired reproductive glands within the tunica vaginalis. The substance of the testes is composed of tubular loops called **testes lobules** (not illustrated) and a central tubular plexus, the **vasa recta** (V*A*S-a REK-ta) (not illustrated). The vasa recta open to four to six **vasa efferentia** (not illustrated) that open to the epididymis.
3. **EPIDIDYMIS** (ep-i-DID-i-mis)**.** An intricately coiled tube on the surface of the testis. The tubule originates on the cranial surface of the testis, where it is referred to as the **caput** (KAP-ut) **epididymis.** The coils continue on the lateral surface of the testis as the **corpus epididymis** and on the caudal surface as the **cauda epididymis.**
4. **VAS DEFERENS** (vas-DEF-er-enz). The continuation of the epididymis from the region of the cauda epididymis to the urethra. The vas deferens has a thick muscular wall that is modified at its entrance to the urethra as the **EJACULATORY** (eh-JAK-*u*-lah-t*o*-r*e*) **DUCT.** The urethra carries the sperm from the vas deferens to the female vagina for fertilization of the ovum. All of the following parts of the male tract either open to or support the urethra. The vas deferens is formed from the archinephric duct of ancestral vertebrates, where it serves to drain excretory wastes as well as sperm, in many cases.
5. **PROSTATE** (PROS-t*a*t) **GLAND.** A small lobulated gland, covered by muscles and surrounding the urethra near the entrance of the vas deferens.
6. **PENIS** ([P*E*-nis] figure 9.1B). Consists of three cavernous bodies (two **corpora cavernosa** [K*O*R-p*o*-ra kav-er-N*O*-sa] **penis** (not illustrated) and one **corpus cavernosum** [kav-er-N*O*-sum] **urethra** (not illustrated)), a **glans** ([glanz] the enlarged distal end of the corpus cavernosum urethra), and a **baculum** (BAK-*u*-lum) or **penis bone** (not illustrated). A baculum is not present in the human. *Transect the penis and note the cavernous bodies.*
7. **BULBOURETHRAL** (bul-b*o*-*u*-R*E*-thral) **GLANDS.** Situated on either side of the penis at the attachments of the **bulbocavernosus** and **ischiocavernosus muscles** to the penis.
8. **PREPUCE** (PR*E*-p*u*s). The fold of skin surrounding and enclosing the glans penis. **Preputial** (pr*e*-PYOO-shal) **glands** (not illustrated) open into the area between the prepuce and the glans penis.

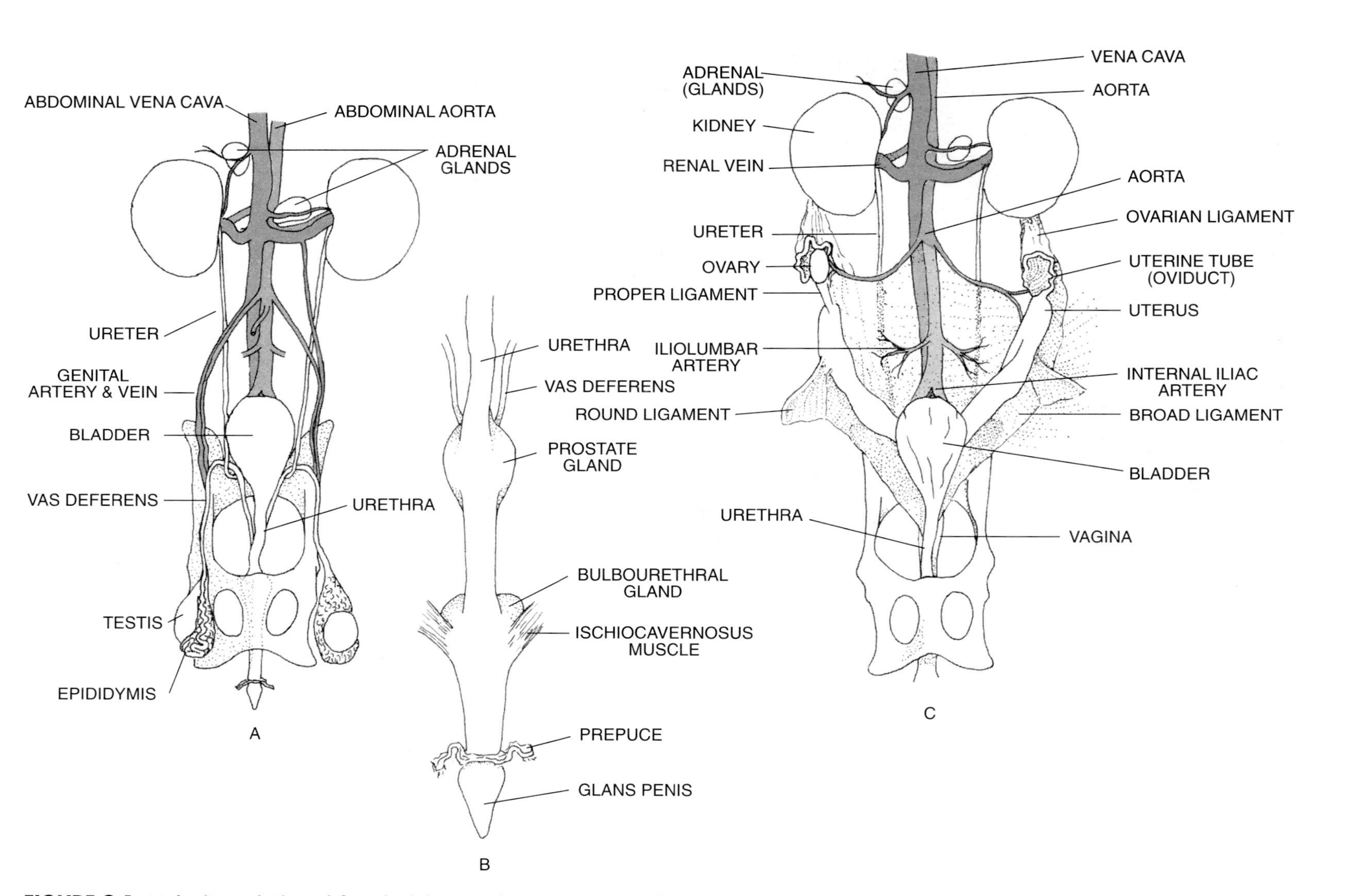

FIGURE 9.1. Male (A and B) and female (C) reproductive systems of the cat. The pelvic girdle is included in parts A and C to orient the reader. Part B is an enlargement of the penis and associated structures in the male.

Female Reproductive System (figure 9.1C)

Embryonically, the ovary develops from tissues peripheral to the primordial testes. The embryonic testes undergoes degeneration as the ovary develops. The oviduct (Fallopian tube) develops as an enfolding of the coelomic lining distinct from the excretory duct.

1. **OVARIES** (*o*-vah-r*e*z). Appear as a mass of **follicles** (FOL-lih-kls) just caudal to the kidneys and usually buried in fat.
2. **OSTIUM** (OS-t*e*-um). The funnel-shaped opening of the oviduct. **Fimbriae** (FIM-br*e*-*e*) (not illustrated) extend as fingerlike projections from the edge of the ostium. Fertilization of the ova usually occurs in the ostium.
3. **UTERINE** (*U*-ter-in) **TUBES** (oviduct or Fallopian tubes). Small, short, highly coiled tubes leading from the ostium to the uteri on each side.
4. **UTERUS** (*U*-ter-us). Divided into two horns, or **cornu uteri** (K*O*R-n*u* *U*-ter-*i*), that fuse caudally into a single body called the **corpus uteri** (not illustrated). The corpus uteri opens to the vagina through a neck, the **cervix** (SER-viks) (not illustrated), that projects into the vagina.

The uterus is *simplex* in the human and is not divided to form two horns. This morphology is typical of animals giving birth to one young at a time.

5. **VAGINA** (va-J*I*-na). Leads from the cervix to the vestibule.
6. **VESTIBULE** (VES-ti-b*u*l) (not illustrated). The external opening for both the reproductive and excretory systems.
7. **Clitoris** (KLI-t*o*-ris) (not illustrated). The female homologue of the penis.

The Pregnant Uterus (not illustrated)

If you have a pregnant female, examine the uterus and carefully cut open the uterine wall and expose the embryos in place. The **placenta** (pla-SEN-ta) is a ring of vascular tissue surrounding the midportion of the embryo with the transparent chorion and amniotic membranes over the cranial and caudal regions of the embryo. This is a zonary type of placenta. The embryo is attached to the placenta by an **umbilical** (um-BIL-i-kal) **cord.** *Carefully remove the embryo and placenta from the uterus.* The membrane surrounding the embryo when the chorion is removed is the **amnion** (AM-n*e*-on). The **chorion** (K*O*-r*e*-on) is a part of the placenta, but the portion of the chorion over the embryo adheres to the uterine wall and is torn loose from the placenta as you remove the embryo from the uterus.

Suggested Readings

Boyd, J. S. 1971. The radiographic identification of the various stages of pregnancy in the domestic cat. *J. Small Anim. Pract.* 12(9):501–6.

Longley, W. H. 1911. The maturation of the egg and ovulation in the domestic cat. *Amer. J. Anat.* 12(2):139–72.

Scott, M. G., and P. P. Scott. 1957. Postnatal development of the testis and epididymis in the cat. *J. Physiol.* 136:40–1.

CHAPTER 10

Nervous System

The nervous system is composed of conducting cells called **neurons** (N*U*-rons) and numerous supporting cells, including **glial (GL*E*-al)** and **Schwann (Shwon) cells.** The function of the neuron is to conduct information (impulses), both within the central nervous system (the brain and spinal cord) and between the central nervous system and various parts of the body. The nervous system is divided into two major anatomical regions. The **PERIPHERAL NERVOUS SYSTEM (PNS)** is made up of peripheral nerves and receptors. These include the cranial and spinal nerves. The **CENTRAL NERVOUS SYSTEM (CNS)** consists of the brain and spinal cord.

There are six functional types of neurons in the nervous system:

1. **Somatic** (s*o*-MAT-ik) **sensory**
2. **Somatic motor**
3. **Visceral** (VIS-er-al) **sensory**
4. **Visceral motor, parasympathetic** (p*a*r-a-sim-pa-THET-ik)
5. **Visceral motor, sympathetic** (sim-pa-THET-ik)
6. **Association**

The **AUTONOMIC** (aw-t*o*-NOM-ik) **NERVOUS SYSTEM (ANS)** consists of visceral motor (efferent) neurons that regulate smooth muscles of the viscera, the blood vessels, and many of the glands that are not under conscious control. These "involuntary" regulatory neurons usually originate in the CNS and pass (via two serial neurons) to a visceral effector. The two divisions of the ANS are often antagonistic, so that one division is excitatory and the other acts as an inhibitor.

Somatic motor (efferent) neurons have one neuron in the pathway from the CNS to the peripheral effector (skeletal muscle). In somatic motor pathways, there is only one type of neuron, and this single neuronal type is always excitatory to the effector.

Sensory neurons (afferent [AF-er-ent] pathways) are either visceral (from the viscera) or somatic (from the skin and special sense organs). These pathways carry information from peripheral sites to the CNS. The nervous system receives external stimuli (information) through both general and special sense organs, integrates this information in the brain or spinal cord (association neurons), and then activates muscles or glands with impulses carried by motor nerves (somatic or visceral efferents). The general and special sense organs (receptors) monitor changes in the external and internal environments. The two general types are **EXTEROCEPTORS** (eks-ter-*o*-SEP-t*o*rs) and **PROPRIOCEPTORS** (pr*o*-pr*e*-*o*-SEP-t*o*rs). The exteroceptors are superficial and sensitive to touch, pressure, pain, and temperature. The proprioceptors are found deep in muscles, tendons, and joints, and monitor muscle and skeletal position to provide the animal with information about the body relative to itself and to the external environment. The general **VISCEROCEPTORS** are located in the viscera and provide information on both conscious (pain) and subliminal (visceral reflexes) visceral activity.

The **SPECIAL SENSE ORGANS** are in association with the head in cephalized animals such as mammals. Senses are vision, audition (hearing), stasis (balance), gustation (taste), and olfaction (smell). Information is relayed to the brain via the cranial nerves.

The CNS of mammals consists of the brain and spinal cord. The brain of mammals is formed embryonically from three primary lobes: **PROSENCEPHALON** (proz-en-SEF-a-lon), **MESENCEPHALON** (mez-en-SEF-a-lon), and **RHOMBENCEPHALON** (rom-ben-SEF-a-lon). During development, these lobes become separated into five subdivisions, and each of the five subdivisions gives rise to the fully developed anatomical structures seen on the adult brain. The major categories, their subdivisions, and the major adult anatomical features for these divisions are as follows:

I. Prosencephalon
  A. Telencephalon
    1. Rhinencephalon—olfactory bulbs and tracts, amygdaloid body, corpus striatum
    2. Cerebral Hemispheres—neopallium
  B. Diencephalon—hypothalamus, thalamus, epithalamus
II. Mesencephalon
  A. Mesencephalon—colliculi and crus cerebri
III. Rhombencephalon
  A. Metencephalon—cerebellum and pons
  B. Myelencephalon—medulla oblongata

A dorsal, hollow nerve cord is a primary chordate characteristic. That is, all members of the phylum chordata have a central cavity in their spinal cord; vertebrates also have cavities in their brain (ventricles) that are continuous with the central cavity of the spinal cord. The central cavities or ventricles are filled with **cerebrospinal** (ser-e-br*o*-SP*I*-nal) **fluid,** which also bathes the outside of the brain and cord.

The spinal cord resides within the neural canal of the vertebral column. It is arranged as **WHITE MATTER,** made up of **fiber tracts,** and a butterfly-shaped **GRAY MATTER,** deep to the white matter and surrounding a **CENTRAL CANAL.** The gray matter consists of cell bodies of association neurons, nonmyelinated motor neurons, and synapses of sensory neurons. Spinal reflexes of all types are mediated here. At each segment, the spinal cord gives off a pair of **ROOTS,** dorsal (sensory) and ventral (motor). These roots join to form the peripheral **SPINAL NERVE** on each side.

The nervous system is a communications system closely allied to the endocrine system, both of which communicate by chemical messengers. These messengers in the nervous system are termed **neurotransmitters** (n*u*-r*o*-TRANS-mit-ers). Neurotransmitters bridge the gaps (**synapses** [si-NAP-s*e*z]) between neurons in a pathway and between neurons and the innervated organs.

Preparation for Observation of the Nervous System

Note: Your instructor may choose for you to dissect the sheep brain instead of the cat brain. If so, turn to appendix 2 and follow the instructions there. If you are to dissect the cat brain, the cranium must be opened now.

Remove the temporalis muscles and muscles of the occipital region. Use bone cutters or triangular wire cutters and cut away the bone of the lambdoidal ridge. This is the thickest bone area roofing the cranium. Be very careful not to cut into the brain. With heavy forceps, chip off the remaining roof of the cranium. Remember that the tentorium is the bony shelf on the inner surface of the parietal bone that is really ossified dura mater. This shelf will need to be removed from its position between the cerebrum and cerebellum. Be sure that you cut (with bone cutters) the tentorium loose from the rest of the parietal bone before lifting it out of its natural position. Leave the eyes and ears intact for further study. First, observe the features of the dorsal brain, and then CUT THE CRANIAL NERVES, LEAVING AS MUCH OF THE NERVE ATTACHED TO THE BRAIN AS POSSIBLE. BE VERY CAREFUL TO LOOSEN THE PITUITARY GLAND FROM ITS SOCKET (sella turcica) IN THE FLOOR OF THE CRANIUM. Cut the spinal cord caudal to the medulla and remove the brain. Place the brain in a receptacle with enough water so the brain is half submerged. Keep the brain moist at all times.

The Brain

Dorsal and Lateral Views (figures 10.1, 10.2)

1. **MENINGES** (me-NIN-j*e*z) (not illustrated). Membranes covering the brain. Mammals have three layers of membranes. The **DURA MATER** (DU-rah-MA-ter) is the outermost of the three and is folded between the cerebral hemispheres as the **FALX CEREBRI** (falks SER-e-br*i*) and between the cerebrum and cerebellum as the **TENTORIUM CEREBELLI** (ten-T*O*-r*e*-um ser-e-BEL-l*i*). In the cat, the tentorium cerebelli is ossified; this bony shelf is fused to the inner surface of the parietal bones (see above for removal).

   The innermost meningeal layer is the thin **PIA** (P*I*-a) **MATER** that adheres to the surface of the brain. Between the dura mater and pia mater is a network of thin, delicate fibers called the **ARACHNOID** (a-RAK-noyd) **LAYER.** The spaces between the arachnoid fibers are the **subdural** (sub-D*U*-ral) and **subarachnoid spaces.**
2. **OLFACTORY** (ol-FAK-t*o*-r*e*) **BULBS.** Found at the extreme rostral end of the brain. The olfactory nerves terminate here after passing through the olfactory foramina of the fenestrated ethmoid plate from the nasal epithelium. The **olfactory tracts** (fiber tracts) extend from the olfactory bulbs to the **hippocampus** (hip-*o*-KAM-pus) (not illustrated).

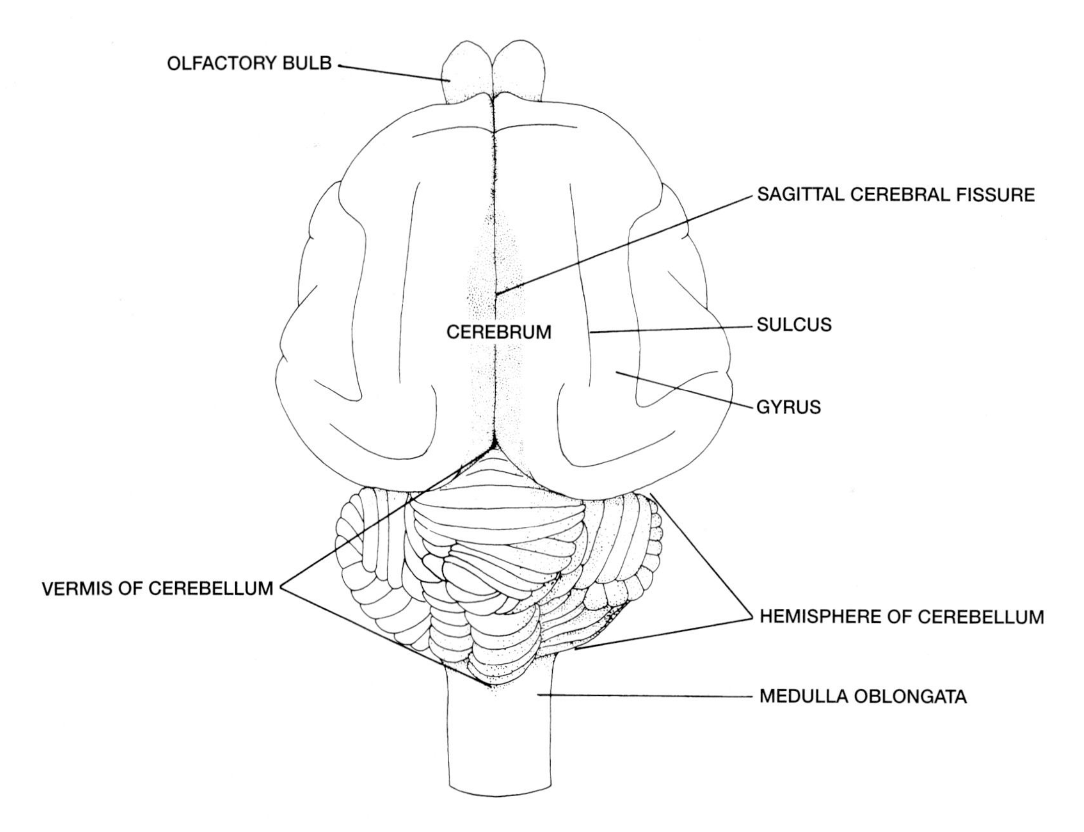

FIGURE 10.1. Dorsal view of the cat brain.

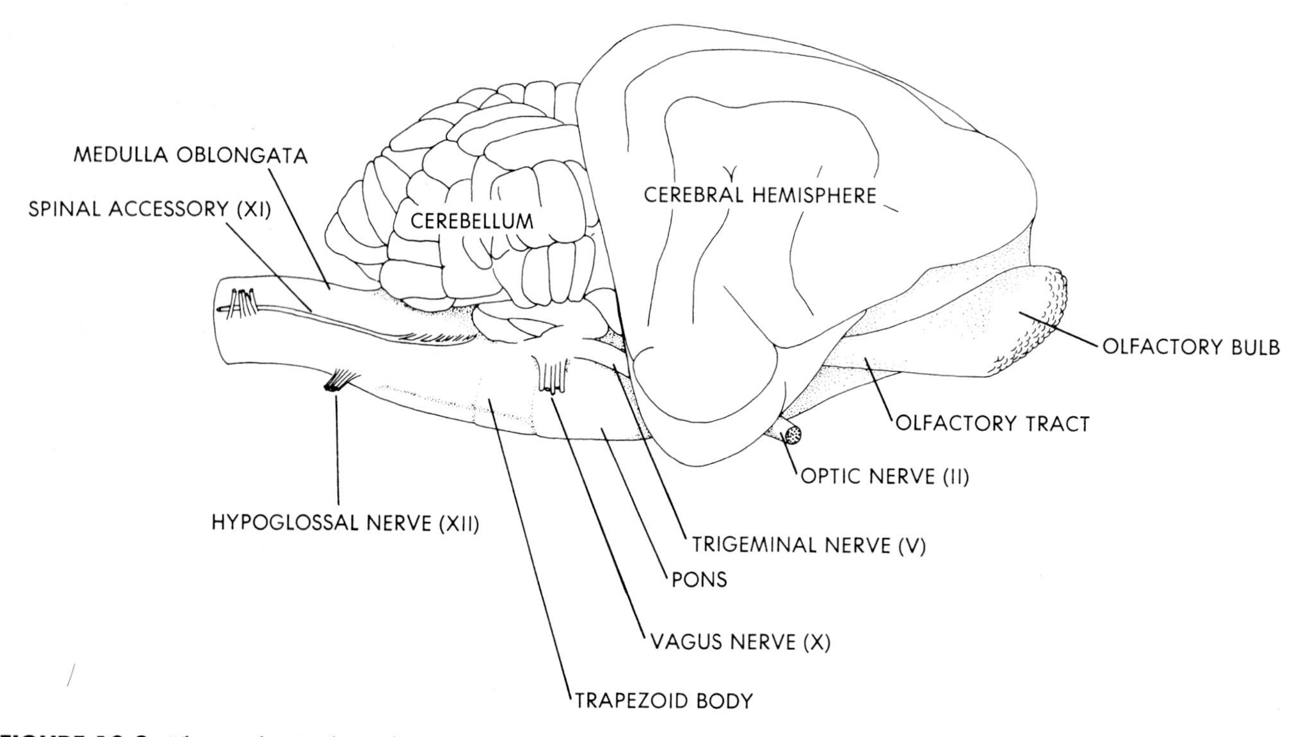

FIGURE 10.2. The cat brain, lateral view.

3. **CEREBRUM** (se-R*E*-brum). Consists of two halves, or **HEMISPHERES,** divided by a deep longitudinal groove, the **sagittal cerebral fissure.** The surface layer of the cerebrum is a layer of gray matter (the cerebral cortex) that is folded, thus increasing the amount of cortical tissue. The grooves of the folds are called **sulci** (SUL-s*i*) and the raised ridges are **gyri** (J*I*-r*i*).

> The cerebrum in humans is proportionately larger than that of cats. The cortex is highly folded resulting in much more gray matter.

4. **CEREBELLUM** (ser-e-BEL-um). The highly convoluted portion of the brain caudal to the cerebrum. It consists of two **lateral hemispheres** and a median **vermis** (VER-mis).
   The cerebellum is principally a motor coordination center. The hemispheres are primarily responsible for coordinating movements of the limbs and digits. The vermis has areas for coordinating movements of the spinal column, shoulders, and hips.
5. **MEDULLA OBLONGATA** (ob-long-GA-ta). The most caudal part of the brain, it constricts imperceptibly into the spinal cord. The functional centers in the medulla regulate respiration, heart rate, and blood pressure, but also play important roles in mediating sensory impulses, endocrine secretions, and general awareness.

Ventral View (figure 10.3)

1. **OLFACTORY TRACTS.** Broad bands of nerve fibers extending caudally and ventrally from the olfactory bulb to merge imperceptibly into the ventral caudal portion of the cerebrum.
2. **OPTIC CHIASMA** (k*i*-AS-ma). Crossing of the optic nerves between the caudal ends of the olfactory tracts.
3. **OPTIC TRACTS.** Continuation of the optic nerves from the chiasma to the rostral colliculi.
4. **Tuber cinereum** (T*U*-ber si-NE-r*e*-um). A small, round elevation just caudal to the optic chiasma. The infundibulum extends from this structure to the pituitary (see sagittal section, below).
5. **Cerebral peduncle** (pe-DUNG-kl). A fiber tract on each side of the pituitary.
6. **PONS** (ponz). Fiber tracts on the ventral, rostral end of the medulla, connecting the hemispheres of the cerebellum.
7. **CRANIAL NERVES.** The twelve pairs of cranial nerves are described in the chart (page 142).

> In humans, all of the cranial nerves are mixed (sensory and motor) except the olfactory, optic, and auditory, which are only sensory.

Cut the brain in half longitudinally along the sagittal fissure.

Sagittal View (figures 10.4–10.6)

As you examine the cat (or sheep) brain, compare the structures you find with those of the human illustrated in figure 10.5.

1. **CORPUS CALLOSUM** (K*O*R-pus ka-L*O*-sum). A band of nerve fibers connecting the cerebral hemispheres and forming the floor of the longitudinal fissure.
2. **PINEAL** (PIN-*e*-al) **BODY** (not illustrated). A short, stubby projection arising from the midbrain just caudal to the corpus callosum.
3. **Tectum mesencephali** (TEK-tum mez-en-SEF-a-l*i*). The tissue forming the roof of the third ventricle and expanded caudally into two pairs of lobes, the **CORPORA QUADRIGEMINA** (K*O*R-p*o*-ra qwod-ri-JEM-i-na), between the cerebrum and the cerebellum. The rostral pair of lobes are the **rostral colliculi** (ko-LIK-*u*-l*i*) and serve as optic reflex centers. The caudal pair are the **caudal colliculi** and serve as aural (auditory) reflex centers.
4. **FOURTH VENTRICLE.** The cavity of the medulla covered by a membrane, the **tela choroidea.**
5. **HYPOPHYSIS** (h*i*-POF-i-sis), or **PITUITARY** (pi-t*u*-i-t*a*r-*e*) **GLAND** (chapter 12). A complex endocrine gland situated within the sella turcica. The pituitary has two distinct parts: (1) the **adenohypophysis** (a-d*e*n-*o*-h*i*-POF-i-sis) (figure 10.6), derived from the embryonic pharynx and divisible into **pars intermedia, pars tuberalis,** and a **pars distalis,** and (2) the **neurohypophysis** (N*U*-r*o*-h*i*-POF-i-sis), an extension of the floor of the hypothalamus. The proximal portion of the neurohypophysis is the **infundibulum,** or stalk, and the distal portion is referred to as the **pars nervosa.** The **cavum** (K*A*-vum) **hypophysis** is a cleft between the pars intermedia and the pars distalis.

The Cranial Nerves (figure 10.3)

Number (Type)	Name	Origin on Brain (and Exit from Skull)	Distribution
I (sensory)	**OLFACTORY**	Olfactory bulb (olfactory foramina)	Nasal epithelium
II (sensory)	**OPTIC** (OP-tik)	Rostral colliculi and thalamus (optic foramen)	Retina of eye
III (motor)	**OCULOMOTOR** (ok-*u*-l*o*-M*O*-t*o*r)	Cerebral peduncle (orbital fissure)	Dorsal, ventral, and medial rectus and ventral oblique eye muscles
IV (motor) (not illustrated)	**TROCHLEAR** (TR*O*K-l*e*-ar)	Caudal to caudal colliculi (orbital fissure)	Dorsal oblique eye muscle
V (both)	**TRIGEMINAL** (tr*i*-JEM-i-nal) Divisions are ophthalmic (of-THAL-mik), maxillary, and mandibular	Pons (ophthalmic, orbital fissure; maxillary, foramen rotundum; mandibular, foramen ovale)	Ophthalmic to skin of face; maxillary to jaw muscles and upper teeth; mandibular to tongue and lower teeth
VI (motor)	**ABDUCENS** (ab-DOO-senz)	Rostral medulla oblongata (orbital fissure)	Lateral rectus and retractor bulbi eye muscles
VII (both)	**FACIAL**	Medulla oblongata near V (facial canal)	Muscles of the face: digastricus and stylohyoideus
VIII (sensory)	**AUDITORY** (AW-di-tory)	Medulla oblongata, caudal to VII (internal acoustic meatus)	Hair cells of inner ear
IX (both)	**GLOSSOPHARYNGEAL** (glos-*o*-f*a*h-RIN-j*e*-al)	Medulla oblongata near X (jugular foramen)	Pharynx and tongue
X (both)	**VAGUS** (V*A*-gus)	Medulla oblongata caudal to VIII (jugular foramen)	Larynx, heart, lungs, diaphragm, and stomach
XI (motor)	**SPINAL ACCESSORY**	Medulla oblongata and rostral end of spinal cord (jugular foramen)	Muscles of neck and pharyngeal viscera with vagus
XII (motor)	**HYPOGLOSSAL** (h*i*-p*o*-GLOS-al)	Medulla oblongata caudal to X (hypoglossal canal)	Tongue muscles

6. **Mammillary** (MAM-ih-ler-*e*) **bodies.** Located at the caudal end of the tuber cinereum and dorsal to the infundibulum. Nerve tracts connect these nuclear (gray matter) bodies with the thalamus or tegmental area.
7. **Fornix** (F*O*R-niks). A fiber tract connecting the hippocampus (most internal portion of cerebrum) with the mammillary body of the hypothalamus. The tract runs ventral to the third ventricle, then turn dorsad to reach the mammillary body.
8. **Septum pellucidum** (pel-LOO-sid-um). The neural area between the corpus callosum and fornix, separating the lateral ventricles.
9. **THIRD VENTRICLE.** A cavity just caudal to the fornix that extends ventrally into the infundibulum and caudally to the level of the corpora quadrigemina. The roof of the third ventricle is the **tectum mesencephalica.**

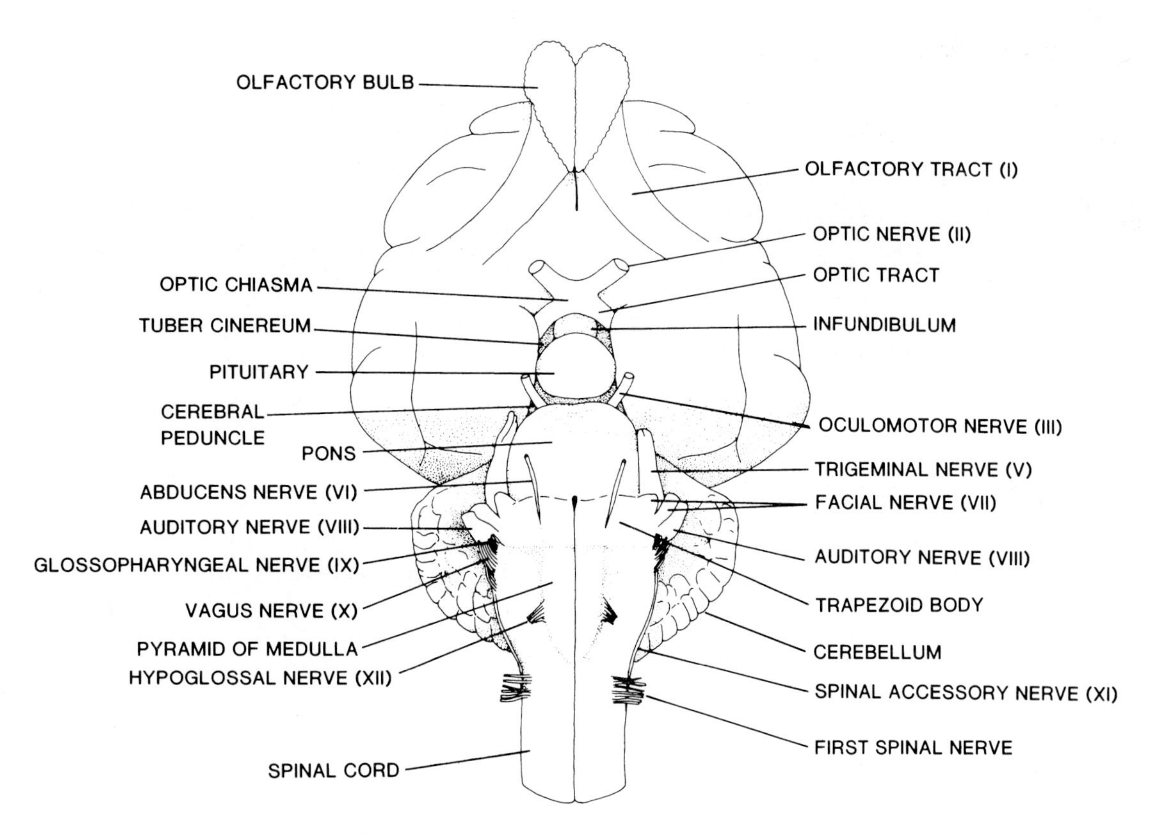

FIGURE 10.3. Ventral view of the cat brain.

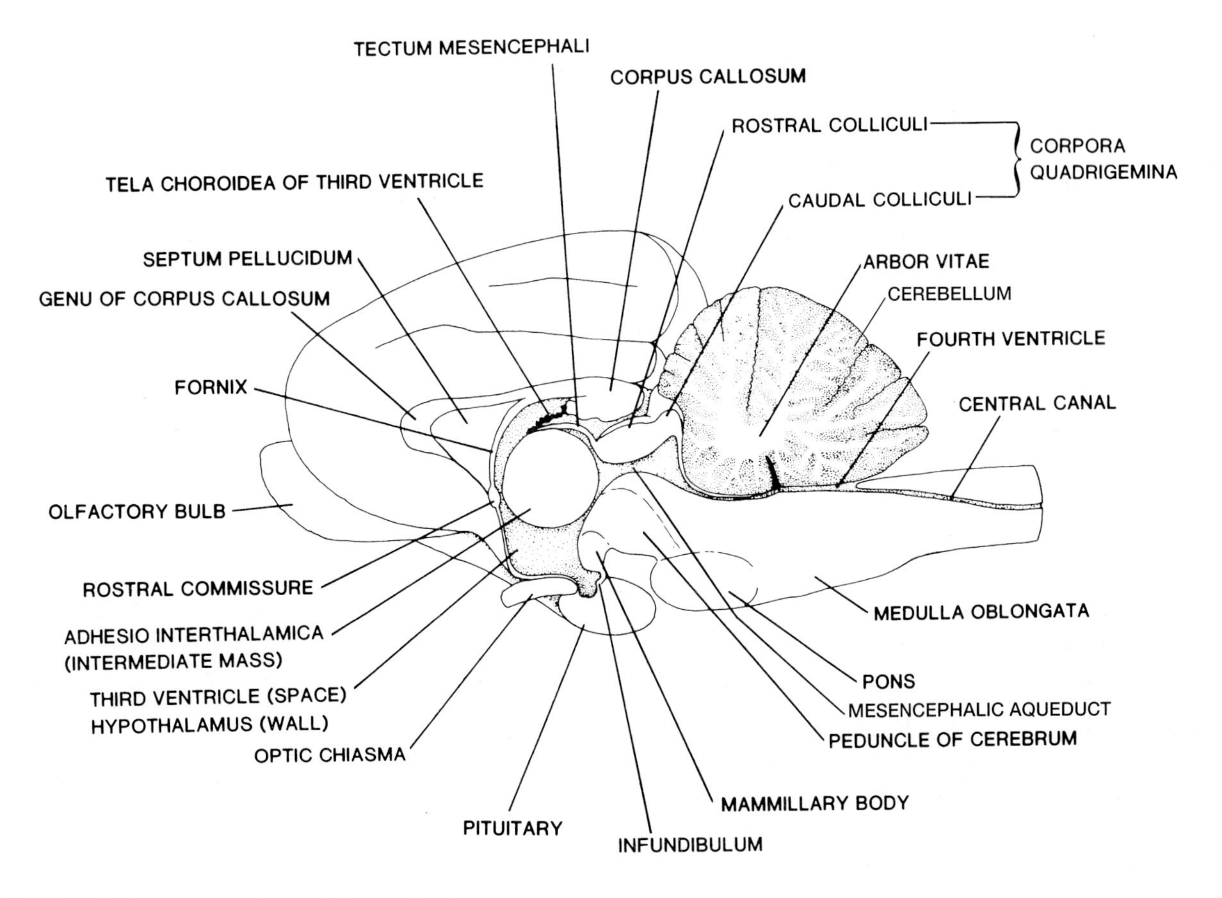

FIGURE 10.4. Sagittal section of the cat brain.

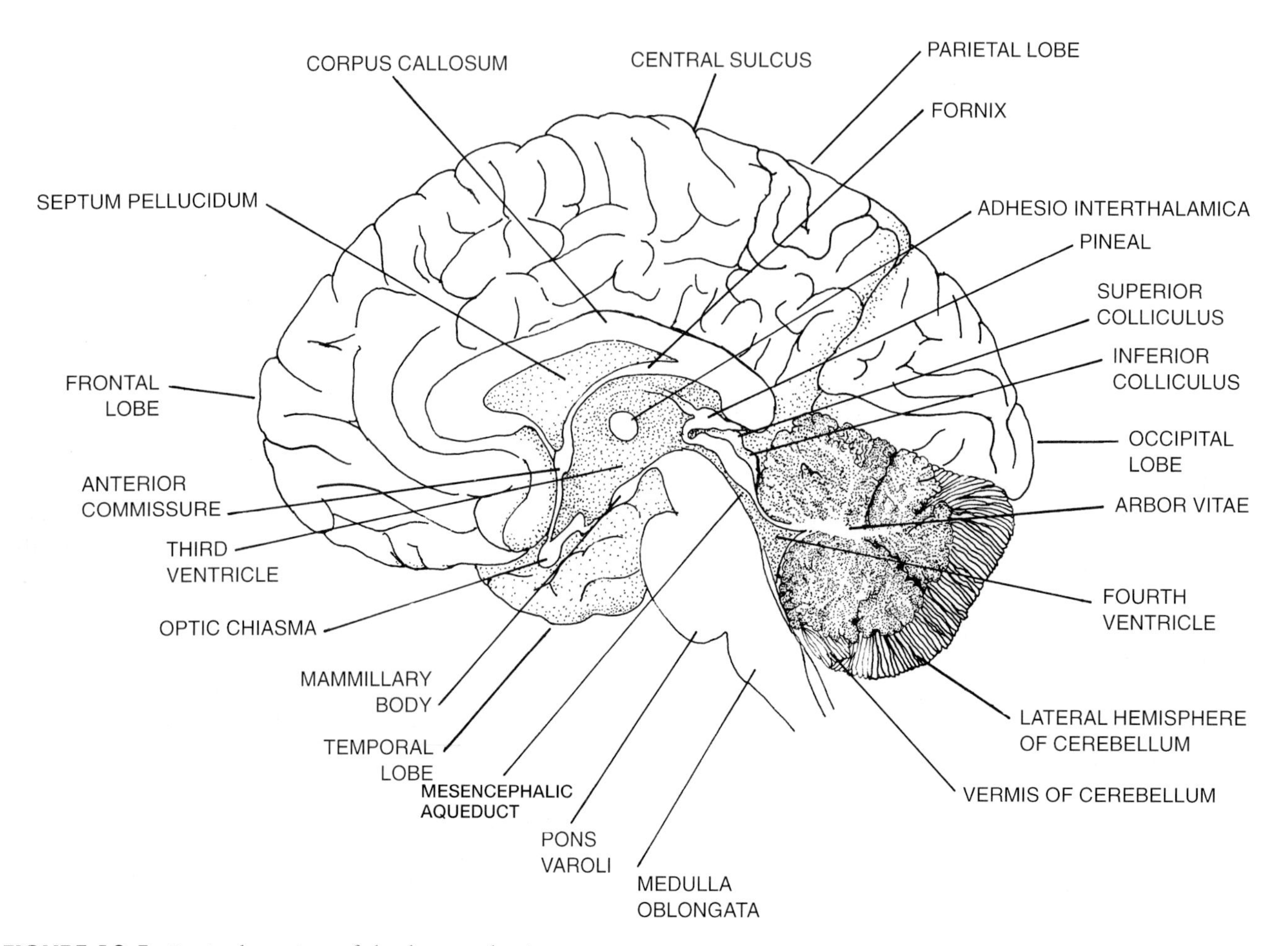

FIGURE 10.5. Sagittal section of the human brain.

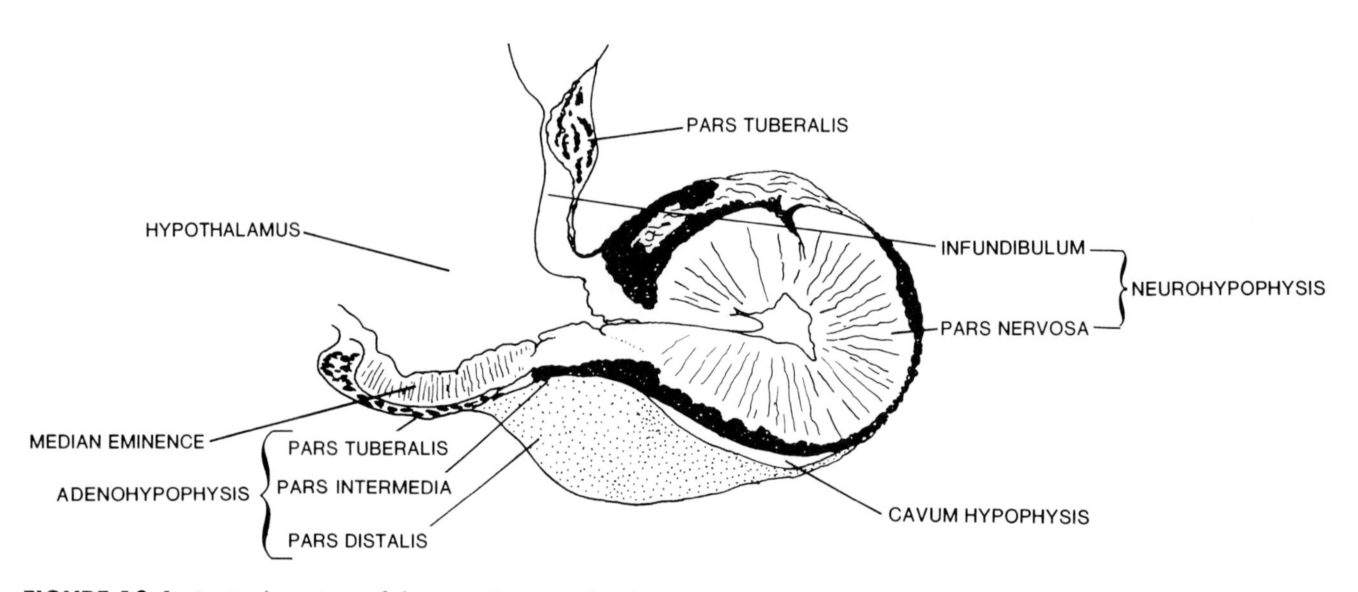

FIGURE 10.6. Sagittal section of the cat pituitary gland.

10. **ADHESIO INTERTHALAMICA (Intermediate mass).** Medial swellings of the walls of the third ventricle that meet in the center of the ventricle and give the *appearance* of a connection between the two halves of the thalamus.
11. **THALAMUS** (THAL-a-mus). Nuclei (see hypothalamus below) in the dorsal wall of third ventricle. The adhesio interthalamica contains the most ventral nuclei of the thalamus and roughly divides the thalamus and hypothalamus.
12. **Rostral commissure** (KOM-i-sh*ur*). Fiber tracts connecting right and left parts of the brain at the rostral end of the fornix. The anterior commissure of the human.
13. **HYPOTHALAMUS** (h*i*-p*o*-THAL-a-mus). Gray matter in the wall of the ventral half of the third ventricle. A concentration of neuron cell bodies in the hypothalamus is known as a **nucleus,** and each nucleus may be specifically named.
14. **MESENCEPHALIC AQUEDUCT,** or **AQUEDUCT OF SYLVIUS** (AK-w*e*-dukt, SIL-v*e*-us). The passageway connecting the third and fourth ventricles.
15. **Interventricular foramen,** (not illustrated) or **foramen of Munro** (MUN-r*o*). The connection between the first, second, and third ventricles just caudal to the fornix.
16. **CEREBELLUM.** The pattern of fiber tracts of the vermis is called the "tree of life" or **ARBOR VITAE** (AR-b*o*r V*I*-t*e*).

Spinal Cord (figures 10.7, 10.8)

The **SPINAL CORD,** like the brain, is surrounded by meninges. Paired **spinal nerves** emerge from the spinal cord through the intervertebral foramina (see figures 3.11, 3.12). Each nerve is formed of **dorsal** and **ventral roots.** The dorsal (sensory root) is enlarged with a **ganglion** (GANG-gl*e*-on) containing the nuclei and cell bodies of all the neurons in this root. The neurons of the dorsal root carry impulses to the spinal cord. The ventral root is motor, and its nuclei and cell bodies are located in the gray matter of the spinal cord. The ventral root neurons carry impulses away from the spinal cord.

The spinal cord is enlarged in the cervical and lumbar regions to accommodate the greater number of neurons and synapses associated with the limbs. Each of the enlargements is associated with nerve **plexi** (PLEKS-*i*) that serve the limbs. The cord terminates in a slender **filum** (F*I*-lum) **terminale** that together with the caudal spinal nerves forms the **cauda equina** (KAW-da e-KW*I*-na).

In humans, the spinal cord is shifted into a vertical position. This changes the relationships of the cerebellum and brain stem as well. Compare the midsagittal view of the human (figure 10.5) with that of the cat (figure 10.4) to get a better understanding of these differences.

The spinal nerves are designated according to the region in which they are found: **CERVICAL, THORACIC, LUMBAR,** and **SACRAL.** There are eight pairs of cervical nerves (C_1–C_8, figure 10.7). The first pair leaves the spinal canal via the foramina of the atlas. All other nerves make their exit through the intervertebral foramina. The first four cervical nerves serve the neck muscles, and the fourth also gives rise to the **phrenic** (FREN-ik) **nerve** that serves the diaphragm. The fourth (in part), fifth, sixth, seventh, and eighth cervical nerves and the first thoracic nerves contribute to the **BRACHIAL PLEXUS.** Major nerves from the brachial plexus to the breast and arm are the following:

1. **Suprascapular nerves.** Arise from the sixth and seventh cervical nerves (fifth and sixth in humans) and serves the infraspinatus and supraspinatus muscles.
2. **Subscapular nerves** (not illustrated). Arise from the sixth, seventh, and eighth cervical nerves (fifth and sixth in humans) and serve the subscapularis, latissimus dorsi, and teres major muscles.
3. **Cranial pectoral nerves** (not illustrated). From the seventh and eighth cervical nerves to the pectoral muscles (except pectoralis profundus). In humans, the pectoral branches include C_5–C_8, and T_1.
4. **Musculocutaneous nerves.** Formed by the sixth and seventh cervical nerves (and also C_5 in humans) and supplying the biceps, coracobrachialis, and brachialis muscles.
5. **Axillary nerves.** From the sixth and seventh cervical nerves (fifth and sixth in humans) on the caudal side of the humeri across the elbows to the lateral surface of the forearms. Cranial to the axillary are two branches to the medial surface of the shoulders.
6. **Long thoracic nerves.** From the seventh cervical nerves, (and, in addition, C_5 and C_6 in humans) to the serratus ventralis muscles.
7. **Lateral thoracic nerves.** From the seventh and eighth cervical and first thoracic nerves and serving the pectoral muscles.

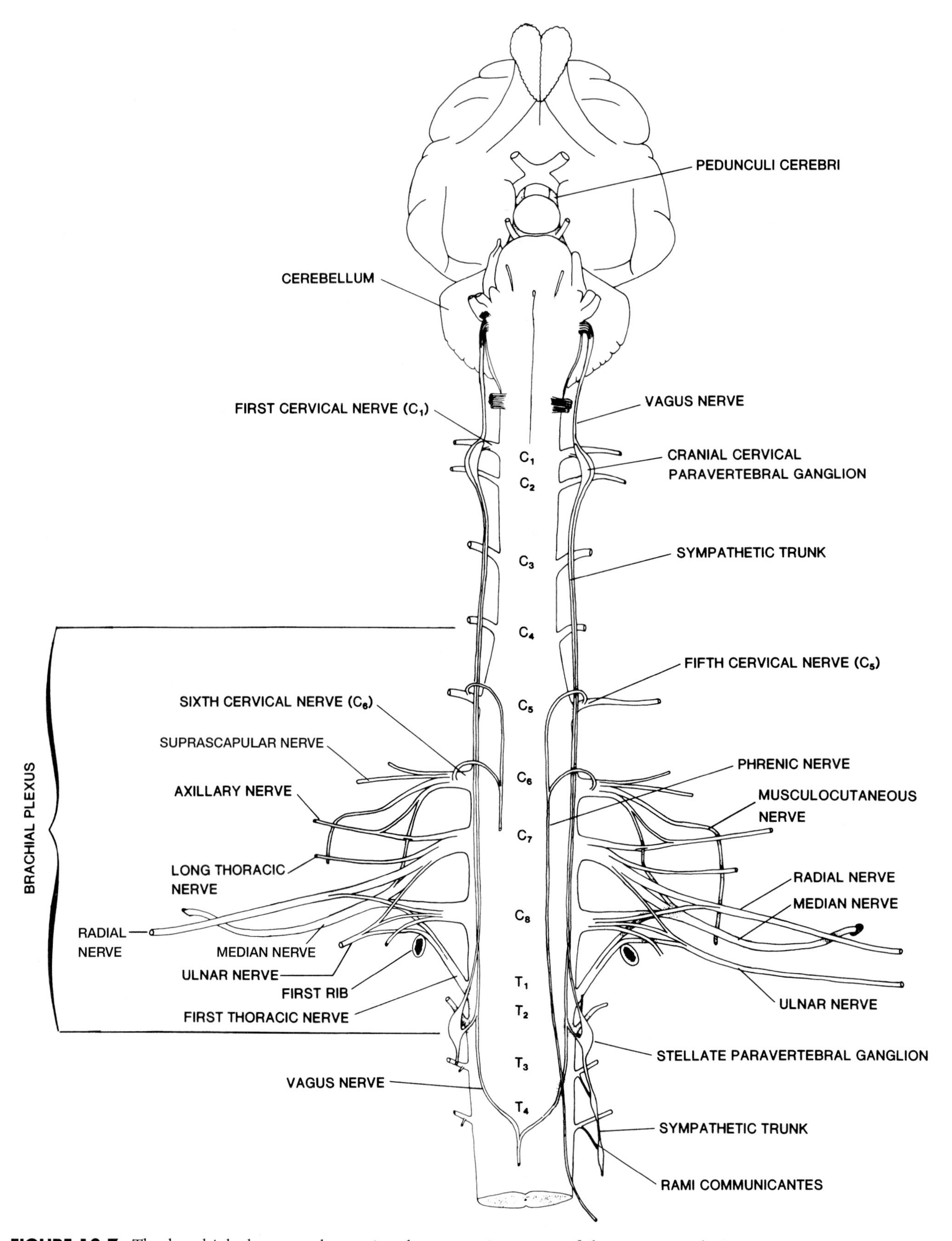

FIGURE 10.7. The brachial plexus and associated autonomic nerves of the cat, ventral view.

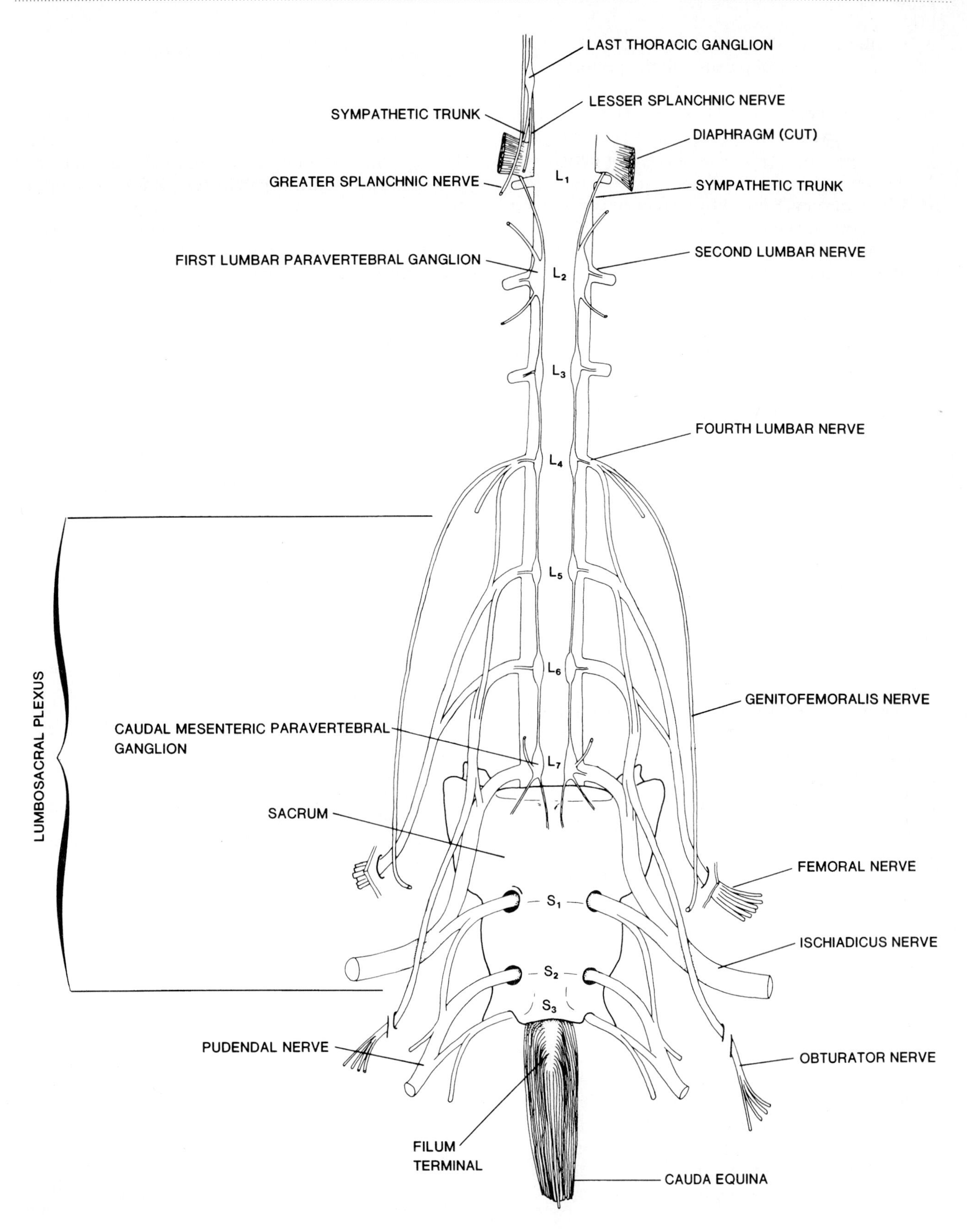

FIGURE 10.8. Lumbosacral plexus and abdominal autonomic nerves of the cat, ventral view.

8. **Caudal pectoral nerves** (not illustrated). From the caudal brachial plexuses to the pectoralis profundus muscle.
9. **Radial nerves.** Formed by the seventh and eighth cervical nerves (C_5–C_8, and T_1 in humans) to the triceps, supinator, and extensor muscles of the forearm.
10. **Ulnar nerves.** Formed by the eighth cervical and first thoracic nerves to flexor muscles of the forearms and digits.
11. **Median nerves.** Formed by the seventh and eighth cervical and first thoracic nerves (and also C_6 in humans) to the pronator and flexor muscles on the median surface of the forearms. These nerves cross the rostral surface of the elbows.

The second to the twelfth thoracic nerves and the first three pairs of lumbar nerves serve the dorsal trunk muscles. The fourth, fifth, sixth, and seventh lumbar (L_4–L_7) and the first two pairs of sacral nerves (S_1 and S_2) form the **LUMBOSACRAL** (lum-bo-*SA*-kral) **PLEXUS** (figure 10.8).

In humans, the lumbosacral plexus consists of the **lumbar plexus** (T_{12}, L_1–L_4), the **sacral plexus** (L_4 and L_5, S_1–S_3), and the **coccygeal nerves.**

The major divisions of the lumbosacral plexus are listed below.

1. **Genitofemoralis** (jen-i-t*o*-fem-*o*r-AL-is) **nerves.** Arise from the fourth and fifth lumbar nerves (first and second in humans) and serve the loin muscles (psoas and iliopsoas).
2. **Femoral nerves.** Arise from the fifth and sixth lumbar nerves (second, third, and fourth in humans) to the iliopsoas, sartorius, rectus femoris, and vastus medialis on the thigh and continue into the leg as the **saphenous** (sah-*FE*-nus) **nerves** (not illustrated), located superficially on the medial surface of the thighs.
3. **Obturator** (OB-tu-*ra*-tor) **nerves.** Arise from the sixth and seventh lumbar nerves (second, third, and fourth in humans) and pass through the obturator foramina to the hip muscles.
4. **Ischiadicus** (is-k*e*-AD-i-kus) (**sciatic** [s*i*-AT-ik]) **nerves.** From the sixth and seventh lumbar nerves and the first, second, and third sacral nerves to the muscles of the thighs (L_4 and L_5, S_1–S_3 in humans). They divide at the knee into the **peroneal** and **tibial nerves** to the muscles of the legs (not illustrated).

The Autonomic Nervous System

(figures 10.7, 10.8)

There are two longitudinal **SYMPATHETIC TRUNKS,** one on each side and ventral to the vertebral column. These trunks communicate with the spinal cord through **rami communicantes** (R*A*-m*i* ka-m*u*-ne-KAN-t*e*z) to the ventral root of the spinal nerves. The trunks are ganglionated, one **PARAVERTEBRAL** (p*a*r-a-VER-te-bral) **GANGLION** beneath each spinal nerve. **PREVERTEBRAL** (pr*e*-VER-te-bral) **GANGLIA** are associated with the visceral plexuses (figure 10.9).

The paravertebral ganglia lie in a column parallel to the vertebrae (the sympathetic trunk), as the name implies. Part of the sympathetic division nerves synapse in these ganglia.

The prevertebral ganglia are ventral to the vertebrae in the region of the abdominal aorta. The sympathetic nerves that do not synapse in the paravertebral ganglia synapse in prevertebral ganglia.

TERMINAL GANGLIA (not illustrated) are located at the visceral effector and are the site of the synapses in the parasympathetic division. Sympathetic nerves have their cell bodies located primarily in the cervical and thoracic regions of the spinal cord, but parasympathetic nerves have their cell bodies in the brain (cranial) and sacral region of the spinal cord.

Locate the following ganglia and plexi on the cat:

Cervical and Thoracic

1. **Cranial cervical ganglion.** A ganglion of the sympathetic trunk near the branching of the common carotid artery into internal and external carotid arteries.
2. **Middle cervical ganglion.** A ganglion of the sympathetic trunk between the first and second ribs. This ganglion is connected with the cervicothoracic ganglion by nerves that surround the subclavian artery.
3. **Cervicothoracic ganglion.** The combined last cervical and first thoracic ganglia.
4. **Vagus nerves.** These branch from their combination with the sympathetic trunk at different levels. The right vagus nerve separates from the sympathetic trunk just before the first rib and gives rise to a right recurrent laryngeal nerve, branching to the aorta and heart and then separating into dorsal and ventral branches. The left vagus branches from the sympathetic trunk at the level of the brachial plexus and gives off the left recurrent laryngeal nerve at the level of the aortic arch. The left vagus divides into dorsal and

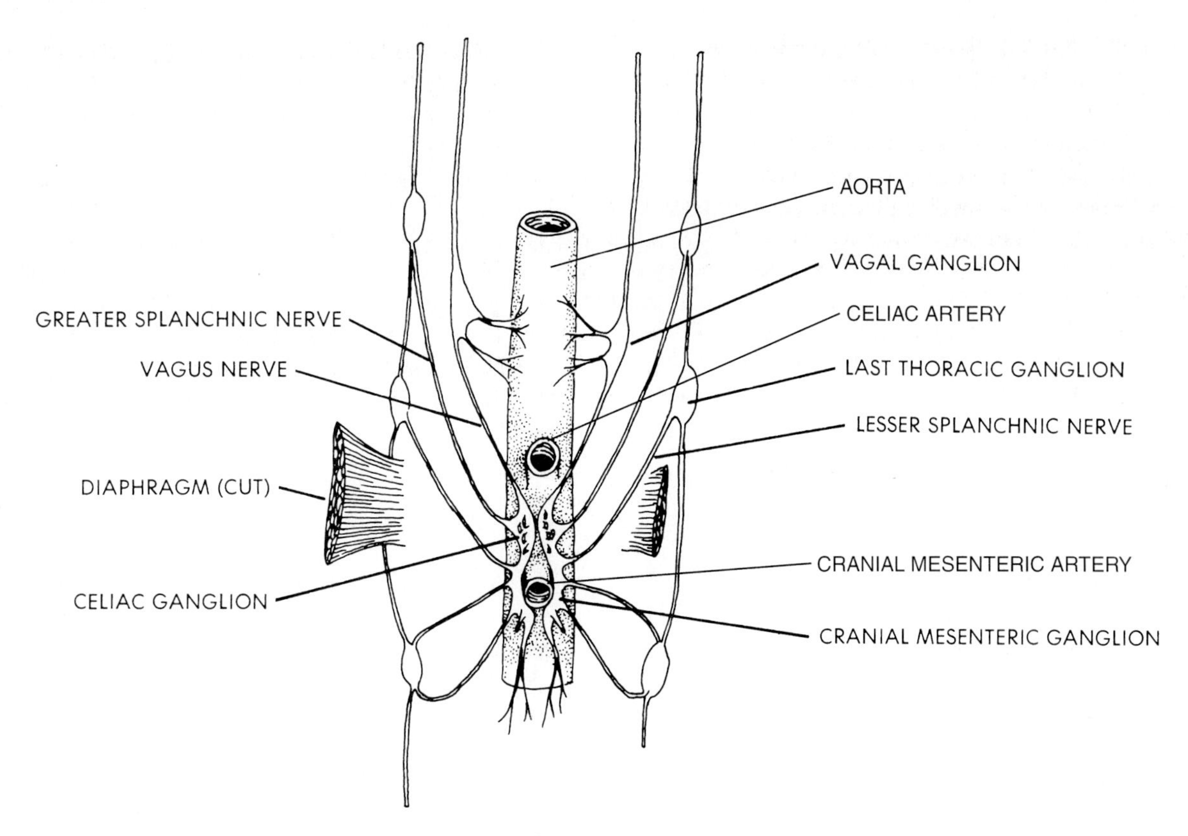

FIGURE 10.9. Celiac and cranial mesenteric plexi of the cat.

ventral branches caudal to the bronchi. The right and left dorsal and ventral branches join to form dorsal and ventral trunks to the abdominal cavity.

Abdominal

1. **Celiac plexuses** (figure 10.9). A pair of plexuses located on either side of the celiac and cranial mesenteric arteries and connected with the cranial mesenteric plexus that lies between them. The center of each celiac plexus is a **celiac ganglion** containing cell bodies and synaptic junctions.
2. **Cranial mesenteric plexus.** On the caudal surface of the cranial mesenteric artery with connections to the celiac plexuses. The **cranial mesenteric ganglion** is in the center of the cranial mesenteric plexus.
3. **Intermesenteric plexus** (not illustrated). On the ventral surface of the abdominal aorta and connecting the cranial mesenteric ganglion with the caudal mesenteric ganglion.
4. **Caudal mesenteric plexus** and **ganglion** (not illustrated). Found on the cranial surface of the caudal mesenteric artery, gives rise to a pair of **hypogastric nerves** and helps to form the **pelvic plexus.**
5. **Splanchnic nerves.** A major and minor branch from the sympathetic trunk and ganglia near the eleventh and twelfth ribs. Splanchnic nerves send branches to the coeliac ganglia.

Suggested Readings

Amat, P. 1978. Ultrastructural study of the median eminence of the cat. *Bull. Assoc. Anat.* 62(179):379–88.

Baker, M. A. 1979. A brain cooling system in mammals. *Sci. Am.* 240(5):130–9.

Greenfield, B. E., and B. D. Wyke. 1964. The innervation of the cat's temporomandibular joint. *J. Anat.* (London). 98:300.

Maillot, C., J. G. Koritke, and M. Laude. 1978. The vascular patterns in choroid tela of the 4th ventricle in the cat, *Felis domestica. Arch. Anat. Histol. Embryol.* 61:3–42.

Ranson, S. W., and S. L. Clark. 1959. *The Anatomy of the Nervous System: Its Development and Function*. Tenth Edition. Philadelphia: W. B. Saunders Company. (Although this book is old and now out of print, it contains an excellent description of the sheep brain and may be found in most large libraries.)

Snider, B. S., and W. T. Niemer. 1961. *A Stereotaxic Atlas of the Cat Brain*. Chicago: U. Chicago Press.

Voogd, J. 1964. *The Cerebellum of the Cat*. Philadelphia: F. A. Davis Co

CHAPTER 11

Receptors

The nerve cell that detects a stimulus does not conduct that impulse to the central nervous system (CNS) (with the possible exception of some pain endings). Instead, the impulse is transferred to one or more (usually more) intermediate cells before reaching the CNS. Within the CNS, the impulse activates neurons that (1) cause a response to the stimulus (contraction of muscles, secretion of glands, increase in heart or respiratory rate, etc.,) and (2) inform the higher centers of the brain (cerebral cortex) that a stimulus has been received. There are probably six to ten different nerve cells involved in the reception of and response to a single stimulus.

Receptors may be classified in several different ways. They are often grouped as **special sense receptors** (hearing, equilibrium, taste, smell or olfaction, and sight) and **somesthetic receptors** or **general sense endings.** There is considerable overlap in this classification.

Somesthetic receptors may be classified according to their location: **exteroceptors** (eks-ter-*o*-SEP-torz) are in or near the skin; **interoceptors** (in-ter-*o*-SEP-torz) (or **visceroceptors**) are on viscera and blood vessels; **proprioceptors** (pr*o*-pr*e*-*o*-SEP-t*o*rz) are in joints, tendons, and muscles. Also, they are classified by their structure (free nerve endings or encapsulated endings) or the kind of stimulus that activates them (mechanoreceptor, chemoreceptor, thermoreceptor, photoreceptor, or phonoreceptor).

Preparation of the Cat for Observation of the Eye

Before dissecting the eye, notice the upper and lower eyelids and a small fold of tissue, the **plica semilunaris** (PL*I*-ka sem-*e*-loo-N*A*-ris) (not illustrated), at the inner corner of the eye. A very thin membrane, the **CONJUNCTIVA** (kon-JUNK-ti-va) (see figure 11.4), covers the exterior of the eyeball and the inner margin of the eyelids.

Now remove the eyelids and the skin around the eye caudad as far as the ear.

The Eye

External Eye (figures 11.1, 11.2)

1. **LACRIMAL** (LAK-ri-mal) **GLANDS** (not illustrated). The intraorbital lacrimal gland is located at the caudal corner of the eye; the extraorbital lacrimal gland is found at the lower rostral base of the ear.
2. **EYE MUSCLES.** Find the ocular musculature, as listed in the following chart of cat ocular muscles (page 153). The cat and most other mammals have a **retractor bulbi** deep to the **rectus** muscles that retracts the eyeball. In addition, a **levator palpebrae** (PAL-pe-br*e*) **superioris** retracts the upper eyelid.

Remove the eyeball from the orbit by transecting the eye muscles and optic nerve. Leave as much muscle attached to the eyeball as possible.
Section the eyeball in a plane parallel to the cornea and optic nerve.

Internal Eye (figures 11.3, 11.4)

1. **SCLERA** (SKLE-ra). The tough, white, fibrous outer layer of the eyeball.
2. **CORNEA** (K*O*R-n*e*-a). The exposed transparent continuation of the sclera.
3. **ANTERIOR CHAMBER.** The chamber of the eye bordered externally by the cornea and internally by the iris and pupil.
4. **CHOROID** (K*O*-royd). The black tissue between the sclera and the retina.
5. **IRIS** (*I*-ris). Pigmented partition pierced by the pupil and separating the anterior and posterior chambers of the eye. The iris is a rostral continuation of the choroid.
6. **PUPIL** (PYOO-pil). Opening through the center of the iris. In the cat, the pupil is a vertical slit; in reduced light, it dilates to a large, round opening. An opening this large is not possible in a pupil that is round when contracted.

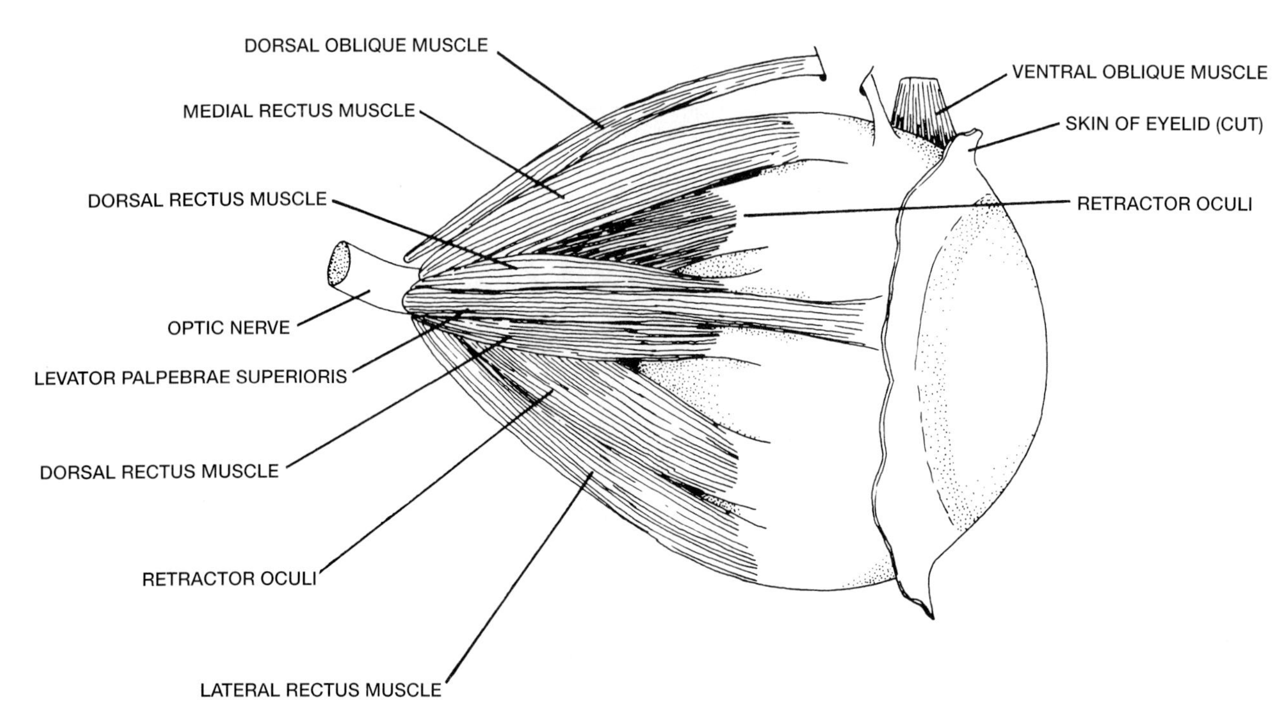

FIGURE 11.1. Muscles of the right cat eye, dorsal view.

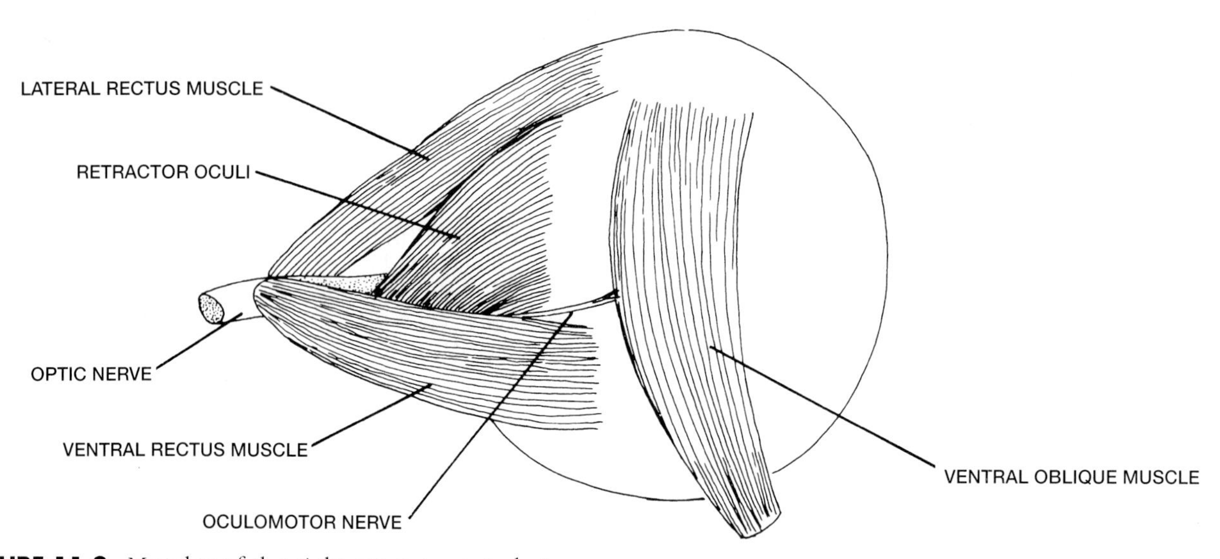

FIGURE 11.2. Muscles of the right cat eye, ventral view.

Cat Ocular Muscles (figure 11.2)

Name	Innervation	Attachment
LEVATOR PALPEBRAE SUPERIORIS	Oculomotor	Connective tissue of upper eyelid
DORSAL RECTUS	Oculomotor	Dorsal midline beneath levator palpebrae superioris
LATERAL RECTUS	Abducens	Lateral border of eyeball
VENTRAL RECTUS	Oculomotor	Ventral midline beneath ventral oblique
MEDIAL RECTUS	Oculomotor	Medial border of eyeball
DORSAL OBLIQUE	Trochlear	Rostral, dorsomedial area, beneath a large vein
VENTRAL OBLIQUE	Oculomotor	Rostral, ventrolateral border of eyeball
RETRACTOR BULBI (four parts)	Oculomotor	Deep to the rectus muscles

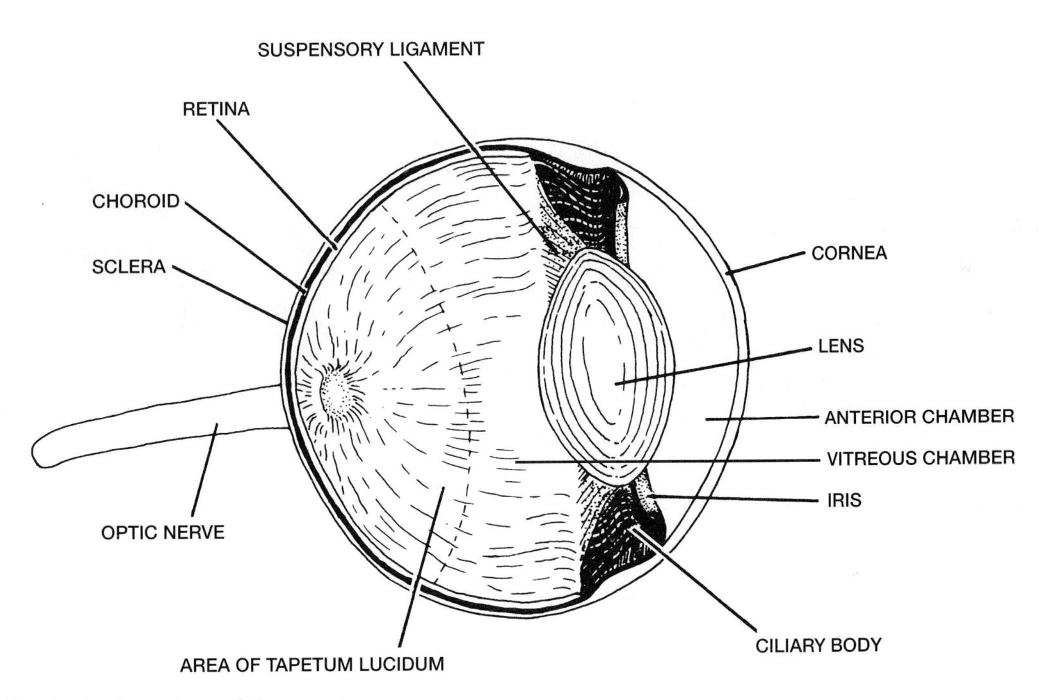

FIGURE 11.3. Sagittal section of the cat eye.

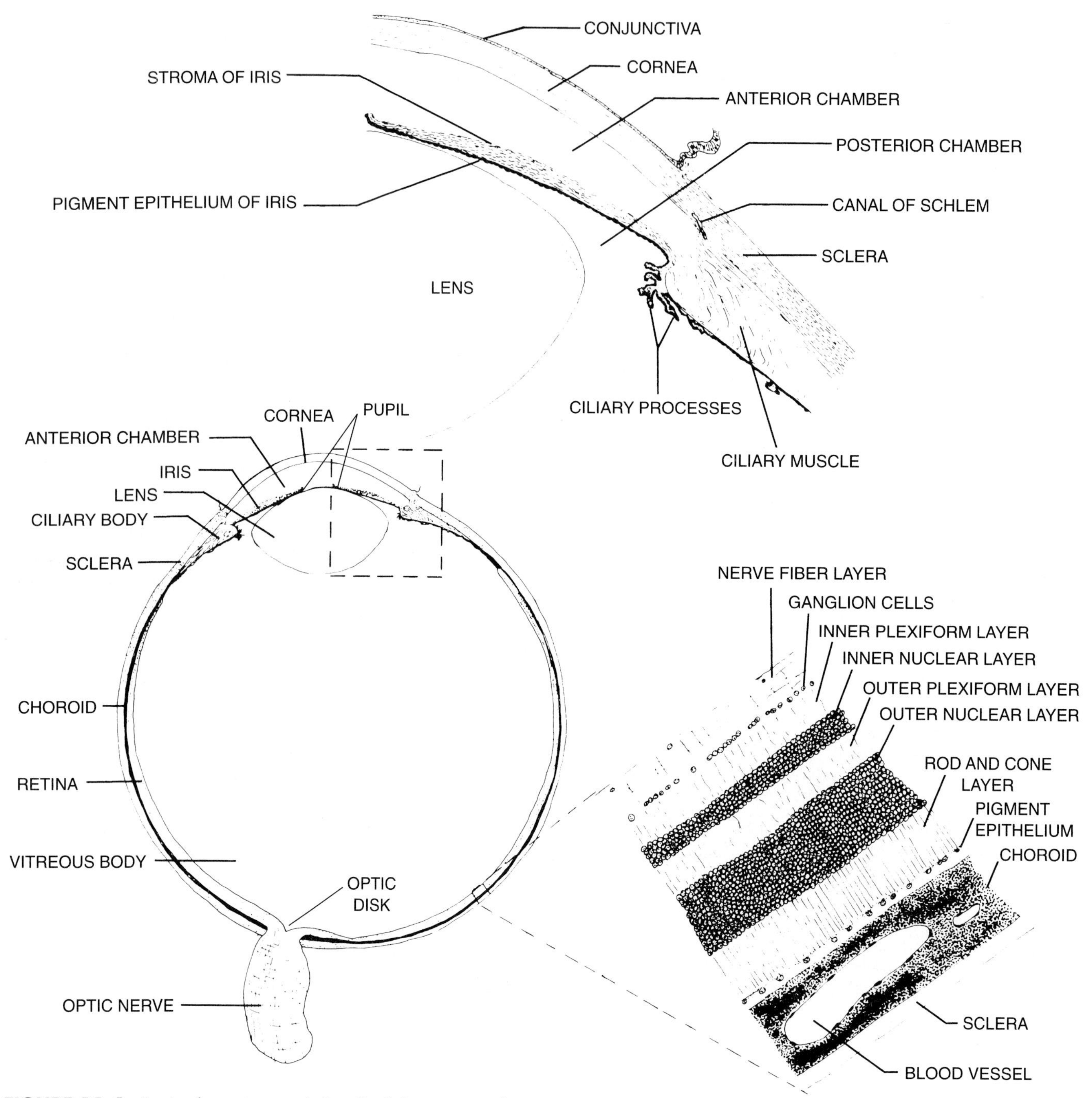

FIGURE 11.4. Sagittal section and detail of the mammalian eye.

7. **POSTERIOR CHAMBER.** Bordered rostraly by the iris and caudally by the lens, the anterior (and posterior) chamber is filled with **aqueous humor.**
8. **LENS** (lenz). The spherical body in the center of the eyeball. The lens is attached to the ciliary body by the **suspensory ligament.**
9. **Ciliary body.** An enlargement of the choroid at the base of the iris, containing muscles that control the lens.
10. **VITREOUS** (VIT-re-us) **CHAMBER.** The large cavity medial to the lens and filled with a gelatinous **VITREOUS BODY.**
11. **RETINA** (RET-i-na). The light-sensitive yellow tissue between the choroid and vitreous body.
12. **OPTIC** (OP-tik) **DISC.** The point of exit of the optic nerve.

The Ear (figure 11.5)

The following is a tedious dissection. Check with your laboratory instructor to determine if you are to complete this section.

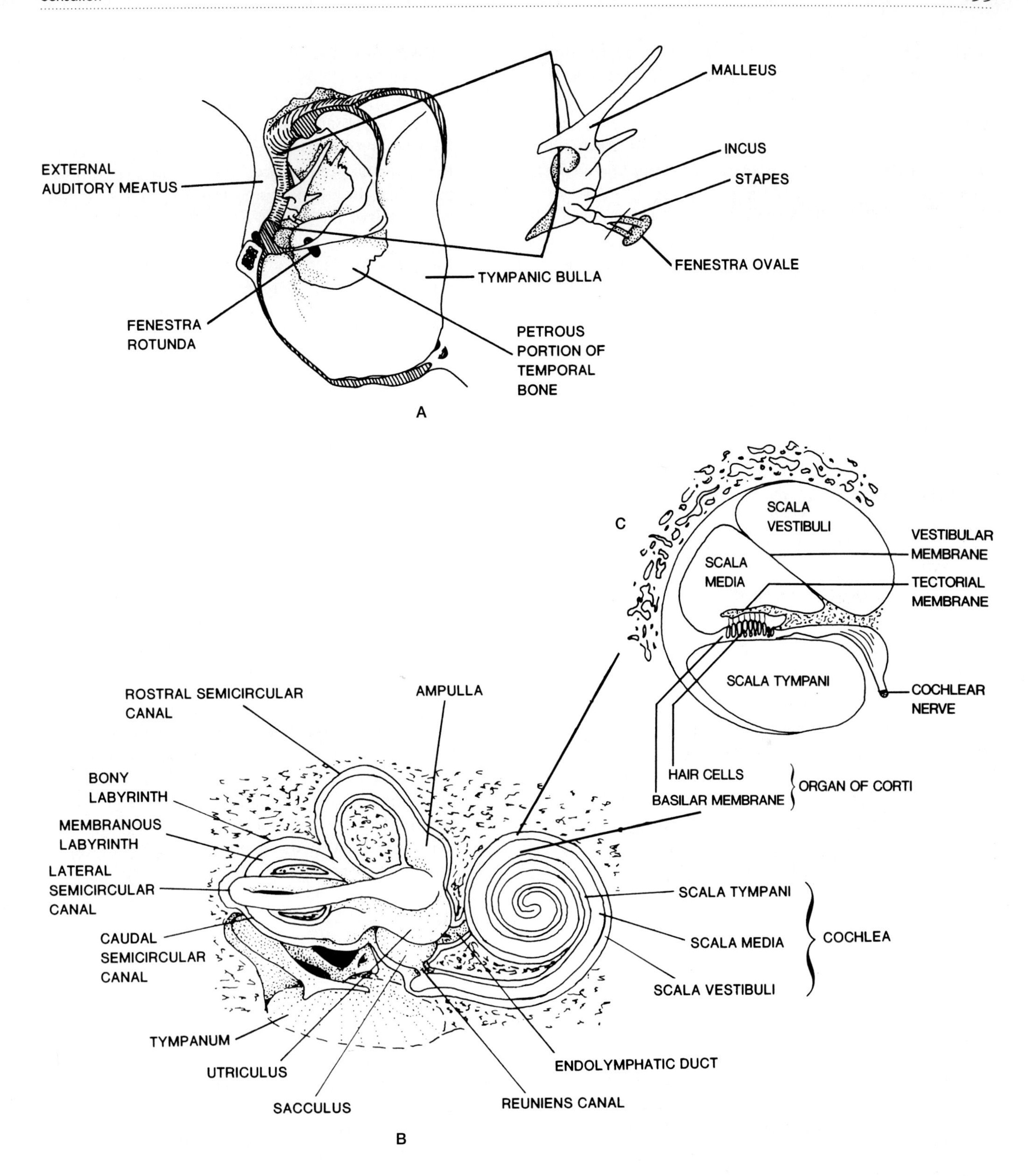

FIGURE 11.5. The middle ear of the cat. The ventral half of the bulla has been removed to expose the ossicles and part of the inner ear (A). The ossicles are enlarged to the right. Inner ear (B) with details of the cochlea shown enlarged (C).

The external ear consists of the **PINNA** (PIN-na) and a canal, the **EXTERNAL AUDITORY MEATUS** (me-*A*-tus), that leads to the **TYMPANUM** (tim-PA-num) or **EARDRUM** (figure 11.5B). *After removing the external ear, note the tympanum within the external auditory meatus. A dissecting microscope will be helpful for the following dissection.*

1. **TYMPANIC BULLA** (BUL-a). Forms the outer casing of the middle ear. The lateral margin of the bulla is perforated by a round opening that forms the innermost part of the external auditory meatus. The tympanum (tympanic membrane) separates the external auditory meatus from the middle ear.

   Before removing the bulla to observe the middle ear, carefully cut the tympanum loose from the bone. One of the ear bones is attached to the membrane, so you will need to cut around the attachment of the ossicle. Use a small blade and, with the aid of a dissecting microscope, cut the membrane. The ossicles are located in the upper rear corner of the membrane; this portion will stay with the cranium when you remove the bulla. Therefore, the ventral border of the membrane is all you need to cut. However, be certain that the membrane is not attached to the bulla when you remove this bone.
2. **AUDITORY OSSICLES** (AW-dih-to-re OS-si-kls). A chain of three bones extending from the tympanum to the **fenestra ovale** (*O*-val), or **oval window,** of the inner ear. The outermost bone is the **MALLEUS** ([MAL-*e*-us] hammer) with a manubrium that attaches the bone to the internal surface of the tympanum. The rounded head of the malleus articulates with the body of the **INCUS** (ING-kus). The incus (anvil) has a lenticular process attached to the head of the **STAPES** ([ST*A*-pez] stirrup). The base of the stapes is in the fenestra ovale of the inner ear. The **tensor tympani** (tim-PA-n*i*) **muscle** (not illustrated), innervated by the trigeminal nerve, attaches to the malleus and tenses the tympanum. The **stapedius** (sta-P*E*-de-us) **muscle** (not illustrated), innervated by the facial nerve, attaches to the stapes and helps maintain a tension on the fenestra ovale. The **fenestra rotunda** (r*o*-TUN-da), or **round window** is a thin membrane in the dorsal wall of the middle ear just caudal to the fenestra ovale.

   The cavity of the middle ear opens to the pharynx by means of the **auditory tube** (not illustrated), which passes through the ventrorostral corner of the tympanic bulla. The bulla has a short lip around the rostral bony canal that accommodates the auditory tube. The auditory tube is also called the eustachian (*u*-ST*A*-ke-an) tube.
3. **INNER EAR.** The inner ear is located within the petrosal (pe-TR*O*-sal) portion of the temporal bone and is not subject to gross dissection techniques. By removing the tympanic bulla, you were able to observe the lateral (outer) surface of the petrosal and the two fenestrae in its wall. If you open the cranium and remove the brain, you may observe the medial (inner) surface. Sagittally sectioned or disarticulated skulls will be good for this purpose. The medial surface of the petrosal region has a shallow recess with two foramina. The dorsal foramen is the **facial canal** (figure 3.7A) for the passage of the facial nerve. Ventral to this is the foramen for the trigeminal nerve. Caudal and ventral to the facial and trigeminal foramina is the **internal acoustic** (a-KOOS-tik) **meatus** (figure 3.7A,B) for passage of the auditory nerve.

   If they are available, prepared microscope slides of the inner ear should be used for study of the **COCHLEA** (K*O*K-le-ah), the sound receptor, and the **semicircular canals, sacculus,** and **utriculus,** the **VESTIBULAR** portion of the ear. Sound enters the cochlea through the fenestra ovale. Here the vibrations of the ossicles are converted to pressure waves in the perilymphatic fluid of the **scala** (SKA-la) **vestibuli.** As the wave passes along the perilymph, it is transferred through the **vestibular membrane** to the endolymphatic fluid in the **scala media.** This, in turn, perturbs the **basilar** (B*A*S-i-lar) **membrane.** Finally, the movements of the basilar membrane cause the **hair cells** to move against the **tectorial** (tek-T*O*-r*e*-al) **membrane** and produce impulses in the **cochlear nerve** (figure 11.5C).

Suggested Readings

Hogan, M. I., J. A. Alvarado, and J. E. Weddell. 1971. *Histology of the Human Eye.* Philadelphia: W. B. Saunders Company.

Prince, J. H., C. D. Diesem, I. Eglitis, and G. L. Russkell. 1960. *Anatomy and Histology of the Eye and Orbit in Domestic Animals.* Springfield, Ill.: Charles C. Thomas, Publishers.

Walls, G. L. 1942. *The Vertebrate Eye and its Adaptive Radiation.* Bloomfield Hills, Mich.: Cranbrook Institute of Science. (Reprint, New York: Hafner Publishing Company. 1963.)

Wersall, J., A. Flock, and Per-G. Lundquist. 1965. Structural basis for directional sensitivity in cochlear and vestibular sensory receptors. *Cold Spring Harb. Symp. Quant. Biol.* 30:115–32.

CHAPTER 12

Endocrine Glands

ENDOCRINE (EN-do-krin) **GLANDS** secrete substances called **HORMONES** (HOR-mons) directly into the blood stream. These hormones are transported by the blood to the **target organ** (the organ affected), where they usually produce their effects through some type of receptor.

The endocrine system is not a distinct system, but its components are often parts of another anatomical system. For example, the ovaries and testes are described with the reproductive system, although they contain hormone-producing cells. Similarly, the neurons of the nervous system all secrete substances, many of which are considered hormones, and the digestive system has hormone-secreting cells in its mucosa and in the islet cells of the pancreas.

Except for the pituitary, thyroid, and adrenal, it is difficult to observe these glands macroscopically. Figure 12.1 illustrates the locations of the endocrine glands in the cat.

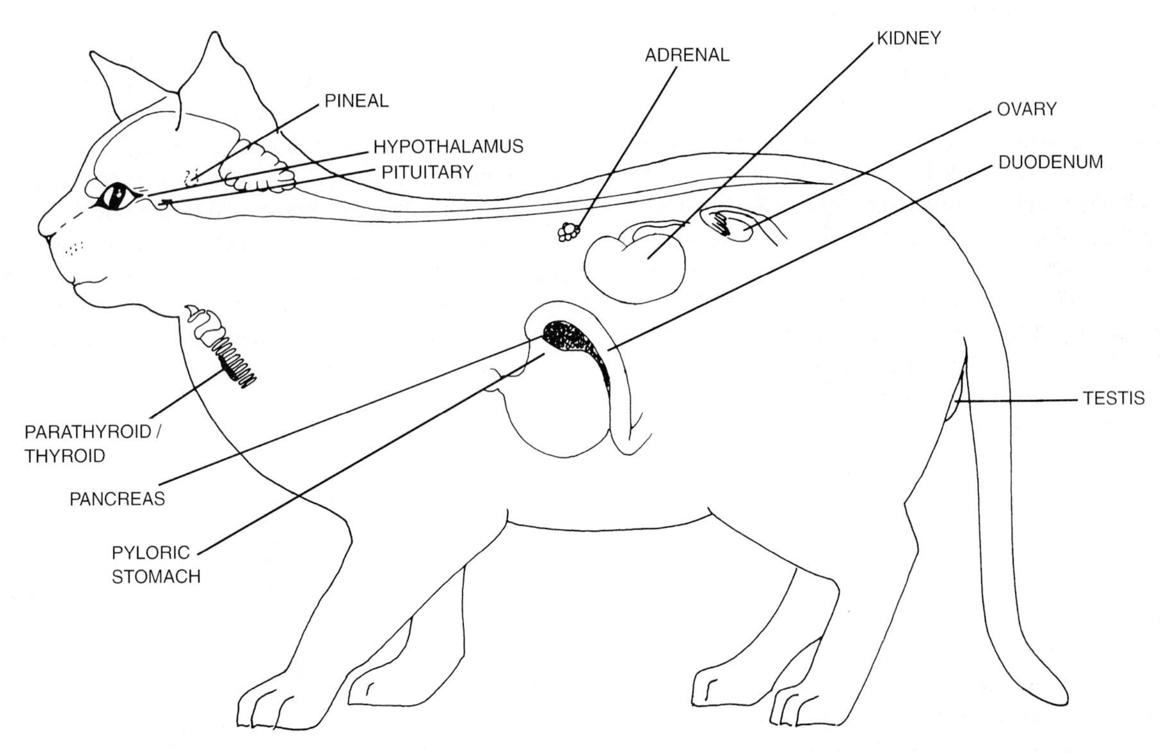

FIGURE 12.1. Endocrine glands of the cat.

Endocrine Glands and Their Secretions

Hypothalamus

The **hypothalamus** (h*i*-p*o*-THAL-a-mus) is the floor of the diencephalon. This region consists of the **hypothalamic nuclei** (not illustrated), which integrate the central nervous system with endocrine activity and, thus, with somatic functions. These nuclei produce **releasing factors (hormones)** that control the synthesis and release of hormones from the pituitary and make the hypothalamus the principal neurosecretory center of vertebrates. In mammals, the two principal nuclei are the **paraventricular** (p*a*r-a-ven-TRIK-*u*-lar) and **supraoptic** (soo-pra-OP-tik) **nuclei.**

Pituitary Gland

The **pituitary** (pi-T*U*-i-t*a*r-*e*) **gland,** or hypophysis (h*i*-POF-i-sis), is a ventral appendage of the hypothalamic diencephalon. Embryologically, this organ has both neural and epidermal origins. The neural components arise from the embryonic diencephalon, and the epidermal contribution is from the oral cavity of the embryo. In most tetrapods there is a stalk, or **infundibulum** (in-fun-DIB-*u*-lum), connecting the pituitary to the hypothalamus. Blood vessels and nerve axons emanating from the hypothalamus pass to the pituitary along this pathway. The vessels (called the **hypophyseal portal system**) carry the releasing hormones directly from the hypothalamic nuclei to the cells of the adenohypophysis.

The pituitary is subdivided into two anatomic divisions: neurohypophysis and adenohypophysis (see figure 10.6). The **neurohypophysis** (the neural component) releases neurosecretions passed via neuronal axons originating in the hypothalamus. The hormones are **oxytocin** ([ok-s*e*-T*O*-sin] all tetrapods) and **vasopressin** (vas-*o*-PRES-in). These hormones are polypeptides with nine amino acids (nonapeptides). Oxytocin causes uterine contractions and milk ejection in mammals. Vasopressin is an antidiuretic hormone (reduces water loss to the urine).

The **adenohypophysis** (a-d*e*n-*o*-h*i*-POF-i-sis), the epidermal division of the pituitary, consists of a **pars distalis** (parz DIS-ta-lis) and a **pars intermedia** (inter-M*E*-d*e*-a). The actual shape of the gland varies considerably among mammal species. However, molecular structures of the secretions are often identical in different species or differ by one or two amino acids in what are often quite large polypeptides. The table at bottom of page lists the adenohypophyseal hormones.

The only hormone of the pars intermedia is **melanotropin** (mel-a-n*o*-TR*O*-pin). It is involved in control of the cutaneous melanophores. This section is innervated by neurons from the hypothalamus, some of which are neurosecretory.

Organon Vasculosum Laminae Terminalis (OVLT)

The **OVLT** (not illustrated) is located in the vicinity of the third ventricle rostral to the infundibulum. Although an endocrine function is suspected for this organ, we know only that it acts as a shunt between the blood and the cerebrospinal fluid (CSF).

Pineal Gland

In the roof of the diencephalon lies the **pineal** (PIN-*e*-al) **gland** or **epiphysis** (*e*-PIF-e-sis). It is considered to be a gland rather than a photoreceptor because it is

Hormones of the Vertebrate Adenohypophysis

Hormone	Function
Growth hormone (GH)	Stimulation of somatic growth
Prolactin (pr*o*-LAK-tin) (PRL)	Milk secretion
Thyrotropic (th*i*-r*o*-TR*O*-pik) **hormone** (TSH)	Stimulation of the thyroid gland
Gonadotropins (g*o*n-a-d*o*-TR*O*-pins) (LH and FSH)	FSH stimulates gametogenesis in the testis and ovary; LH is involved in steroidogenesis
Adrenocorticotropic (ad-r*e*-n*o*-k*o*r-te-k*o*-TR*O*P-ik) **hormone**	Stimulation of the adrenal cortex in response to stress

extensively vascularized. The gland consists of the following parts (not illustrated). The **pineal stalk** extends dorsally just rostral to the optic lobe (tectum). The stalk is expanded into a **pineal vesicle,** with a central lumen just beneath the roof of the cranium. The ventral base of the stalk has a **caudal commissure.** A **subcommissural organ** on the caudal portion of the stalk is a membranous sack, the **dorsal sacculus** (SAK-*u*-lus). Between the base of the dorsal sacculus and the pineal stalk is a **habenular** (ha-BEN-*u*-lar) **commissure.** The tissues of the pineal vesicle consists of photoreceptive, supporting, and ganglion cells.

The pineal produces the hormone **melatonin** (mel-a-T*O*-nin) which is probably involved in the control of reproduction. Melatonin is known to exert suppressive effects on gonadal maturation and function.

Thyroid Gland

Embryologically, the **thyroid** (TH*I*-royd) **gland** is an outgrowth from the floor of the pharynx in all vertebrates. The form and final position of the gland in adults varies, but it is generally a diffuse organ arranged as follicles.

The thyroid hormones target the general body cells to increase oxygen consumption and carbohydrate metabolism. These hormones are also important in body growth and development of the central nervous system in mammals.

Ultimobranchial Gland

The **ultimobranchial** (ul-ti-mo-BRANG-k*e*-al) **gland** (not illustrated) becomes closely allied with the thyroid, where it occurs as scattered cells between the thyroid follicles. **Calcitonin** (kal-si-T*O*-nin) has been isolated from this gland. Calcitonin reduces levels of calcium in the blood by increasing bone deposition.

Parathyroid Glands

These glands are physically close to the thyroid gland in most mammals, and the name was derived because of this fact. The parathyroids secrete **parathyroid** (p*a*r-a-TH*I*-royd) **hormone,** which regulates the reabsorption of calcium from bone and thus increases blood calcium levels.

Endocrine Cells of the Pancreas

These cells, arranged as **islets of Langerhans** (LANG-er-hanz), are associated with the exocrine pancreas that, in turn, is associated with hepatic portal drainage. Islet cells produce **insulin** (IN-su-lin), **glucagon** (GLOO-ka-gon), **pancreatic polypeptide,** and **somatostatin** (s*o*-ma-t*o*-STAT-in). These hormones are important in the regulation of metabolism and the control of circulating plasma glucose levels.

Endocrine Cells of the Intestinal Mucosa

Mammals have a number of endocrine cells in the mucosal walls of the stomach and duodenum (see table at bottom of page).

Adrenal Gland

In mammals, the adrenal gland consists of **chromaffin** (kr*o*-MAF-in) **cells** (stain with chromates) in a central **medulla,** and **interrenal cells** surrounding the chromaffin cells as a **cortex.** The adrenal chromaffin tissue

Hormones of the Vertebrate Digestive Tract

Hormone	Source	Function
Gastrin (GAS-trin)	Pyloric stomach	Stimulates fundus to produce digestive enzymes
Secretin (s*e*-KR*E*-tin)	Duodenum	Stimulates release of pancreatic fluids
Pancreozymin (PAN-kr*e*-*o*-z*i*-min)	Duodenum	Stimulates release of pancreatic enzymes
Cholecystokinin (k*o*-l*e*-sis-t*o*-K*I*N-in)	Duodenum	Stimulates emptying of gallbladder
Enterogastrone (en-ter-*o*-GAS-tr*o*n)	Duodenum	Inhibits gastric activity

is under neural control and produces either **epinephrine** (ep-i-NEF-rin) or **norepinephrine** (nor-ep-i-NEF-rin), thus serving in conjunction with the sympathetic division of the autonomic nervous system. Steroids, including **mineralocorticoids** (min-er-al-o-KOR-ti-koyds), **glucocorticoids** (gloo-ko-KOR-ti-koyds), and **corticosteroids** (kor-ti-ko-STE-royds), are produced by the interrenal cortical cells. Mineralocorticoids regulate plasma sodium, glucocorticoids control plasma glucose, and corticosteroids are important in the response to stress. The chromaffin cells of mammals are arranged as cords of endocrine cells suspended in sinusoidal blood pools. The term **adrenal** implies a close association with the kidneys. In mammals, the adrenals lie just cranial to the kidneys.

Adrenal Hormones of Mammals

Interrenal cortical corticosteroids of the mammal	**Cortisol** (glucocorticoid) regulates blood sugar; controls metabolism of proteins and fats and inflammatory responses
	Aldosterone (mineralocorticoid) controls ion balance by regulation of kidney, intestine, and sweat and salivary glands
Chromaffin/medullary hormones of mammals	**Catecholamines** (kat-e-kol-AM-ens) (norepinephrine, epinephrine, and dopamine [DO-pa-men]) increase heart rate; control arterial constriction, dilatation, and venoconstriction; increase gastric motility; increase metabolic rate; other effects on lung, uterus, spleen, eye, skin, exocrine glands, skeletal muscle, pancreas, and behavior

Endocrine Cells of the Kidney

The hormone **renin** (RE-nin) is produced by **juxtaglomerular** (juks-ta-glo-MER-*u*-lar) **cells** of the mammalian kidney. This hormone promotes the conversion of angiotensinogen (from the liver) to angiotensin I, which is converted to angiotensin II. Angiotensin II has a steroidogenic action (aldosterone production) and may also act directly to increase blood pressure. The steroidogenic action of angiotensin II has been demonstrated in mammals. The kidneys are also the source of **erythropoietin** (e-rith-ro-POY-e-tin), the hormone that stimulates the proliferation and release of red blood cells from hemopoietic tissues into the blood stream. The endothelial cells of the peritubular capillaries are the source of erythropoietin in humans.

Endocrine Cells of the Gonads

In the female, the **gonads** (GO-nads) produce steroids **(estrogens** and **progestins)** that promote receptivity to the male and encourage the development of secondary sexual characteristics. Ovarian hormones are produced from the **interstitial** (in-ter-STISH-al) **cells, follicle,** and **corpus luteum.** In the male, the sex steroids **(androgens)** encourage male behavioral patterns, including territoriality, courtship, and sexual behavior. The androgens are produced from **Leydig's** (*LI*-digz) **cells,** found between the testis lobules that make up the bulk of the testes. There is some disagreement about the contribution of **Sertoli's** (ser-TO-lez) **cells** to androgen production.

Suggested Readings

Harris, G. W., and B. T. Donovan. 1966. *The Pituitary Gland.* Vols. 1–3. Berkeley, Calif.: U. California Press.

Pitt Rivers, R., and W. R. Trotter, eds. 1964. *The Thyroid Gland.* Vols. 1, 2. Washington, D.C.: Butterworths.

APPENDIX 1

Sheep Heart

The sheep heart will have been removed from the thoracic cavity, but the **mediastinal** (m*e*-d*e*-as-T*I*-nal) **membranes** may or may not have been removed. If the **pericardial** (per-i-KAR-d*e*-al) **membranes** are still intact, review the arrangement of the thoracic **coelomic** (s*e*-LOM-ik) **membranes** and then carefully separate the membranous walls of the **mediastinal septum** (SEP-tum). The outermost membrane lines the thoracic (**pleural** [PLOO-ral]) cavity but is usually regarded as the outer layer of the **pericardia** (per-i-KAR-d*e*-a). By tracing the evolution of these membranes, it is clear, however, that this outer membrane is actually the ventral **mesentery** (MES-en-ter-*e*) of the thoracic cavity. *Cut open the outer layer of the pericardium and note the fat, blood vessels, and any glandular tissue between the inner and outer layers of the pericardium. Now remove the inner layer of the pericardium. This is the true parietal pericardium. The visceral pericardium adheres to the heart surface. The blood vessels and fat that you see on the surface of the heart is actually beneath the thin visceral pericardium. The cavity between the visceral and parietal pericardial membranes is filled with pericardial (coelomic) fluid.* The pericardial fluid reduces friction and cushions the movements of the heart.

Ventral Superficial View of the Sheep Heart (figure H.1A)

ATRIA (*A*-tr*e*-a) (singular, **atrium** [a vestibule]) are the two small sacs on either side of the cranial portion of the heart. The free edge of each atrium (opposite the entrance of the large veins) is termed the auricle (little ear). The larger, caudal portion of the heart consists of two **VENTRICLES** (VEN-tri-kls) that superficially appear to be a single, large muscular structure. Ventrally, the main branch of the large **left coronary** (K*O*R-o-n*a*-re) **artery** and the **great coronary vein** (beneath the artery) lie in the **interventricular paraconal groove** (in-ter-ven-TRIK-*u*-lar p*a*ra-ah-K*O*N-al groov) and thus mark the division between the right and left ventricles. Branches of the vessels continue to the right side in the **coronary groove** between the atria and ventricles and medial to the **pulmonary** (PUL-m*o*-ner-*e*) **arch.** A small **marginal branch** occurs on the left margin of the left ventricle. *Use the left coronary artery as a guide to the interventricular septum (which lies beneath it) when dissecting the heart.*

Dorsal Superficial View of the Sheep Heart (figure H.2A)

The large vessels of the heart are best seen on the dorsal side. The cranial and caudal **VENA CAVA** (V*E*-na K*A*-va) enter the right atrium; the **coronary sinus** (S*I*-nus) enters the right atrium from the left, between the cranial and caudal vena cava. The **PULMONARY VEINS** enter the left atrium (two from the right lung and two from the left lung). The two pulmonary veins from the left lung usually join before entering the atrium and thus may appear to be a single vessel. The cut ends of the **pulmonary** and **aortic** (*a*-OR-tik) **arches** are cranial to the veins. The **PULMONARY ARCH** will be seen in the dissected ventral view of the heart.

The **right coronary artery** and the **middle cardiac vein** occur together in the **interventricular subsinosal** (sub-S*I*-n*o*-sl) **groove** and mark the division between the right and left ventricles on this aspect of the heart. Near the **apex** (*A*-peks) of the ventricles, an **anastomosing** (a-nas-t*o*-M*O*-sing) **branch** of the right coronary artery passes to the ventral surface of the heart to join the left coronary artery. The left and right

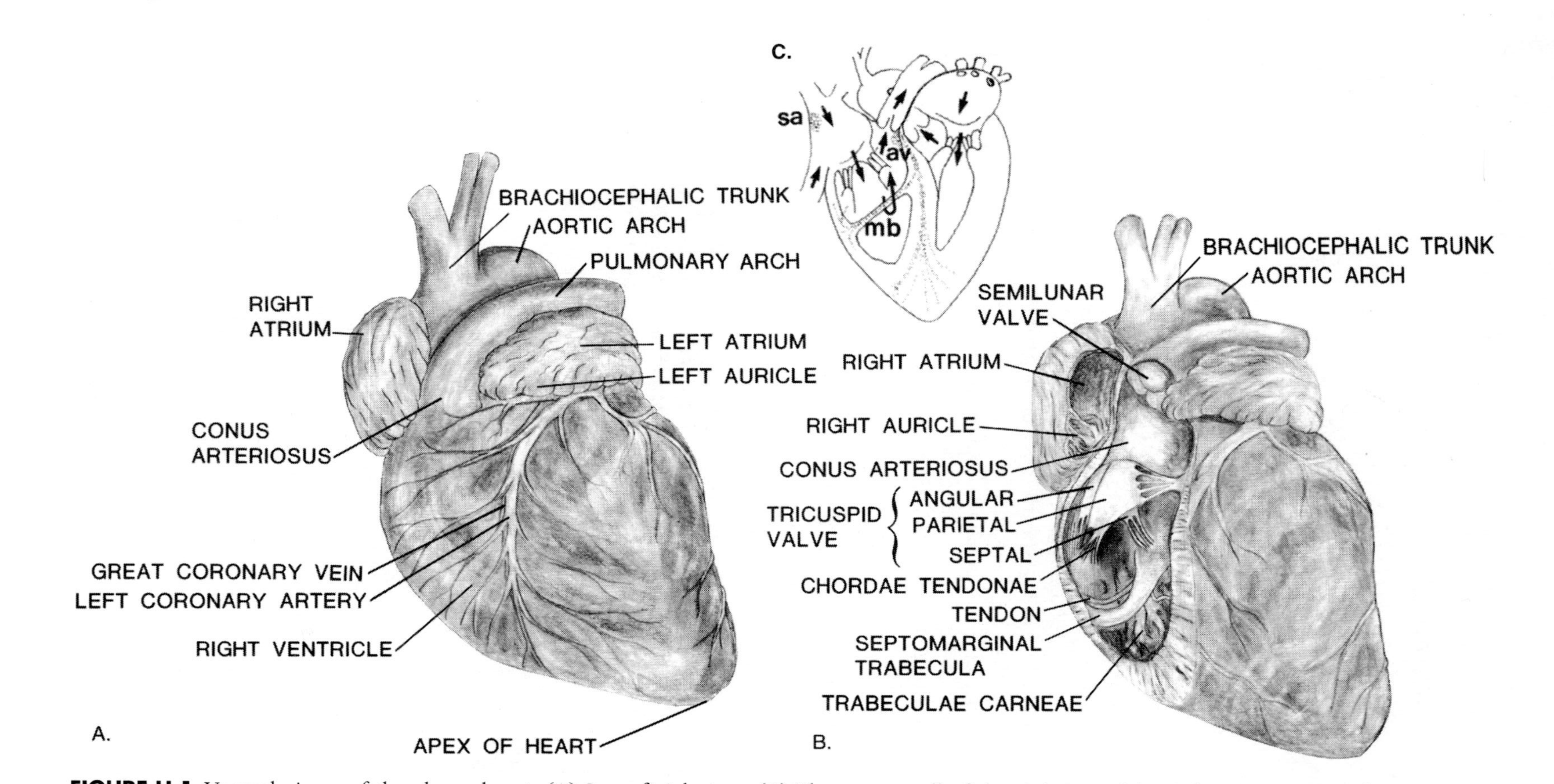

FIGURE H.1 Ventral views of the sheep heart. (A) Superficial view. (B) The outer wall of the right ventricle and atria is removed. (C) Diagram of blood flow in the sheep heart and conducting system (stippled). SA = sinoatrial node (pacemaker); MB = moderator band; AV = atrioventricular node.

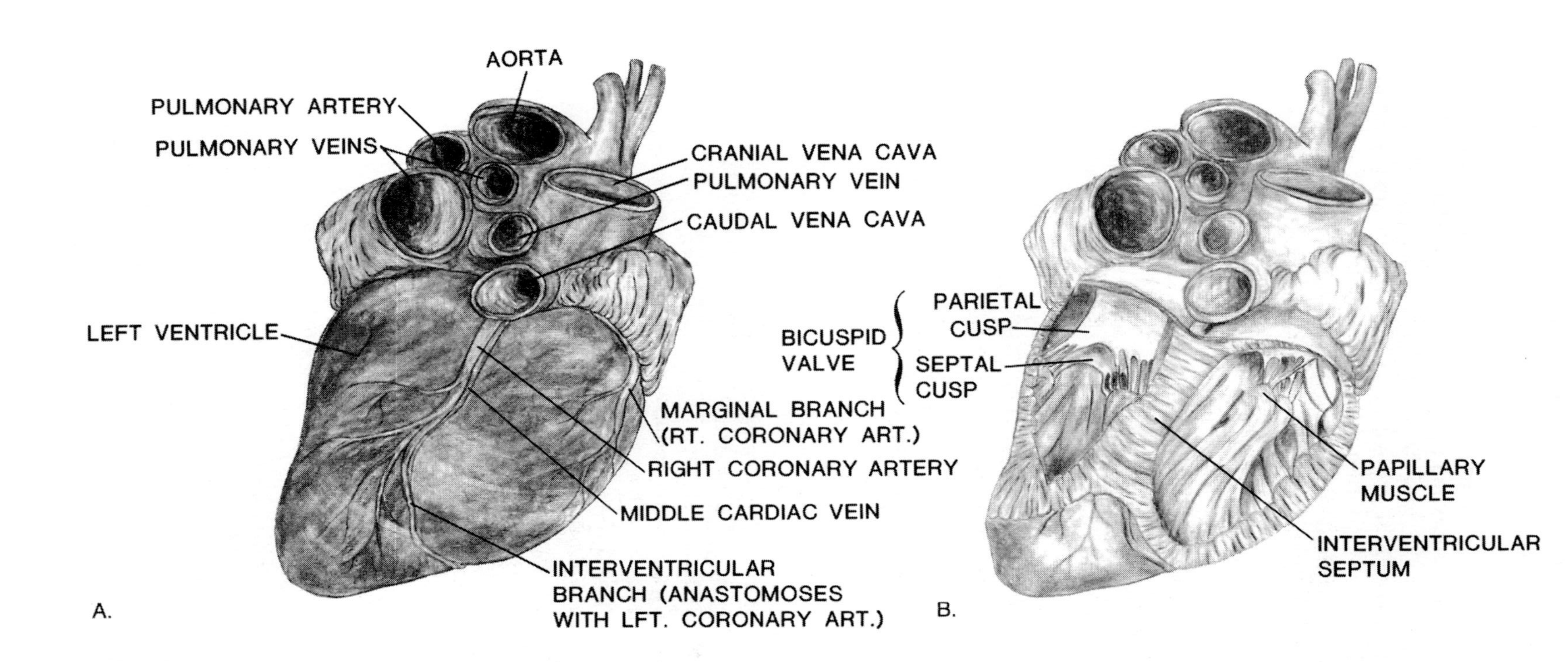

FIGURE H.2 Dorsal view of the sheep heart. (A) Superficial view. (B) The outer ventricular walls are removed.

coronary arteries originate in the right and left ventral **cusps** (kusps) of the **semilunar valve** (sem-*e*-LOO-nar valv) of the **aorta** (*a*-OR-ta).

The **coronary groove** is the division between the atria and the ventricles. The **great coronary vein** originates near the apex of the ventricles and passes through the **interventricular paraconal groove** to the **coronary groove,** where it receives the **middle cardiac** (KAR-d*e*-ak) **vein** and **marginal veins** from the ventricles and smaller veins from the atria. The great coronary vein enlarges to become the **coronary sinus** before opening to the right atrium.

Dissected Views of the Sheep Heart

Ventral (figure H.1B)

With a pair of scissors, cut off the medial half of the right auricle and atrium to expose the interior. Internally, the atrium is a large chamber, but the auricle has muscular trabeculae (pectinate muscles) extending between its inner and outer walls. Blood enters the right atrium from the cranial and caudal vena caval veins, which have drained blood from cranial and caudal parts of the body, and from the coronary sinus, which receives blood from the wall of the heart itself. The left atrium drains blood from the lungs via the pulmonary veins.

In cross section, the right ventricle is half-moon shaped and the left ventricle is round. This difference in appearance is due to the difference in thickness of the muscular walls. The right ventricle exerts enough force to drive the blood to the lungs (and back), but the left ventricle must pump the blood to all parts of the body. *With a sharp scalpel, cut through the wall of the right ventricle just to the right (your left) of the paraconal groove and the left coronary artery on the ventral surface of the heart. Start your cut at the apical end of the right ventricle, but note that the right ventricle does not reach the apex, so confine your cut to the right side of the left coronary artery. Before cutting all the way to the coronary groove, go back to the beginning of your incision and cut laterally along the right margin of the ventricle. This will provide a flap that you can lift up, thus allowing you to observe and avoid the internal structures as you continue to cut and remove the ventral wall of the right ventricle.*

Note the muscular bands on the inner wall of the ventricles; these are **TRABECULAE CARNEAE** (tra-BEK-*u*-le KAR-n*e*-*e*). The larger muscle papillae with cords of tendon attached to them are called **PAPILLARY** (PAP-i-l*ar*-*e*) **MUSCLES;** the tendons are **CHORDAE TENDINAE** (K*O*R-d*e* TEN-di-n*e*). The cranial ends of the chordae tendinae are attached to tendonous flaps of the **TRICUSPID** (tr*i*-KUS-pid) **VALVE** on the right, and to the **s** (b*i*-KUS-pid) or **MITRAL** (M*I*-tral), **VALVE** on the left. Blood passes to the pulmonary artery from the right ventricle and into the aorta from the left. One or more trabeculae extend from the interventricular septum to the base of the papillary muscle supporting the lateral flap of the tricuspid valve. These trabeculae are called moderator bands, or septomarginal (sep-t*o*-MAR-ji-nal) trabeculae, and contain branches of the interventricular conducting tract (Pur-kin-je fibers) in addition to cardiac muscle fibers.

Both the pulmonary artery and aorta are guarded by semilunar valves. Each valve consists of three cup-shaped flaps with the open end of the cup directed away from the heart. This arrangement allows the blood to flow away from the ventricles. If blood were to flow toward the heart, the cups would fill and close the opening between the ventricle and artery. Notice that the pulmonary semilunar valve is cranial to the level of the atrioventricular tricuspid valve and that the wall of the ventricle in this area is similar to the wall of the pulmonary arch. This portion of the ventricle leading to the pulmonary semilunar valve is called the **conus arteriosus** (K*O*-nus ar-te-r*e*-*O*-sus).

Dorsal (figure H.2B)

Use the right coronary artery and the interventricular subsinuosal groove to orient the interventricular septum and begin your cut (with a sharp scalpel) on the left side, just cranial to the apex of the heart. Again, make your first cut just to the left of the coronary artery and toward the coronary groove. Make a second cut from the beginning of the first cut along the left lateral margin. Be careful to avoid cutting the papillary muscles supporting the chordae tendinae of the bicuspid valve. Remove the wall of the left ventricle by cutting transversely just below the coronary groove. You may also wish to remove the dorsal wall of the right ventricle. If so, be sure to leave a marginal band intact to support the angular flap of the tricuspid valve.

The lateral flap of the bicuspid valve is now exposed in the left ventricle. Beneath the lateral flap is the **septal** flap. Both flaps are held in place by chordae tendinae attached to papillary muscles. The **AORTIC SEMILUNAR VALVE** is just cranial to and between the lateral flap of the bicuspid and the dorsal wall of the ventricle. You should be able to locate the valve with a blunt probe. *In order to see the valve you will need to* *dissect away most of the outer portion of the aortic arch.* Before doing this, identify the innominate (brachiocephalic), subclavian, and carotid arteries.

Heart Beat

The sequence of heart contraction begins with the right atrium and is immediately followed by contraction of the left atrium. Contraction of the atrial muscles and constriction of the atrial chambers is termed **atrial systole** (SIS-to-le). This constriction forces blood through the atrioventricular valves (bicuspid and tricuspid) into the two ventricles. Dilation of the ventricles [**diastole** (*di*-AS-t*o*-le)] accompanies atrial systole. During ventricular diastole, blood flows into the ventricles and through the semilunar valves, into the aorta and the pulmonary artery.

At the end of ventricular diastole, pressures in the atria, ventricles, and great arteries (aorta and pulmonary artery) are equal. As the ventricles contract, **ventricular systole** begins and the pressure in the ventricles and great arteries increases while that of the atria decreases.

Muscles at the apex of the ventricles contract first. The contraction continues toward the atria. The atrioventricular valves prevent the flow of ventricular blood into the atria so that the blood can only go into the great arteries through the semilunar valves.

At the end of ventricular systole, **atrial diastole** is also complete and atrial systole begins again. Semilunar valves prevent return blood flow from the great arteries into the ventricles so that the ventricles can only fill with blood from the atria.

APPENDIX 2

Sheep Brain

Preparation of the Sheep Brain for Dissection

Sheep brains completely removed from the cranium but with the dura mater in place are available from biological supply companies. Although other preparations are available, it is suggested that those with the dura mater intact be obtained. Rinse the brain thoroughly in tap water and KEEP IT MOIST DURING DISSECTION. Always wrap the brain in wet toweling when storing it in a plastic bag.

The brains of mammals are divisible into anatomical regions that correspond to the major embryonic portions as described for the cat.

Dorsal and Lateral Views (figures B.1, B.2)

1. **Meninges** (me-NIN-*jez*) (not illustrated). The membranes that cover the brain. Mammals have three layers of membranes; the **dura mater** (D*U*-ra M*A*-ter) is the outermost of the three. It is folded between the cerebral hemispheres as the

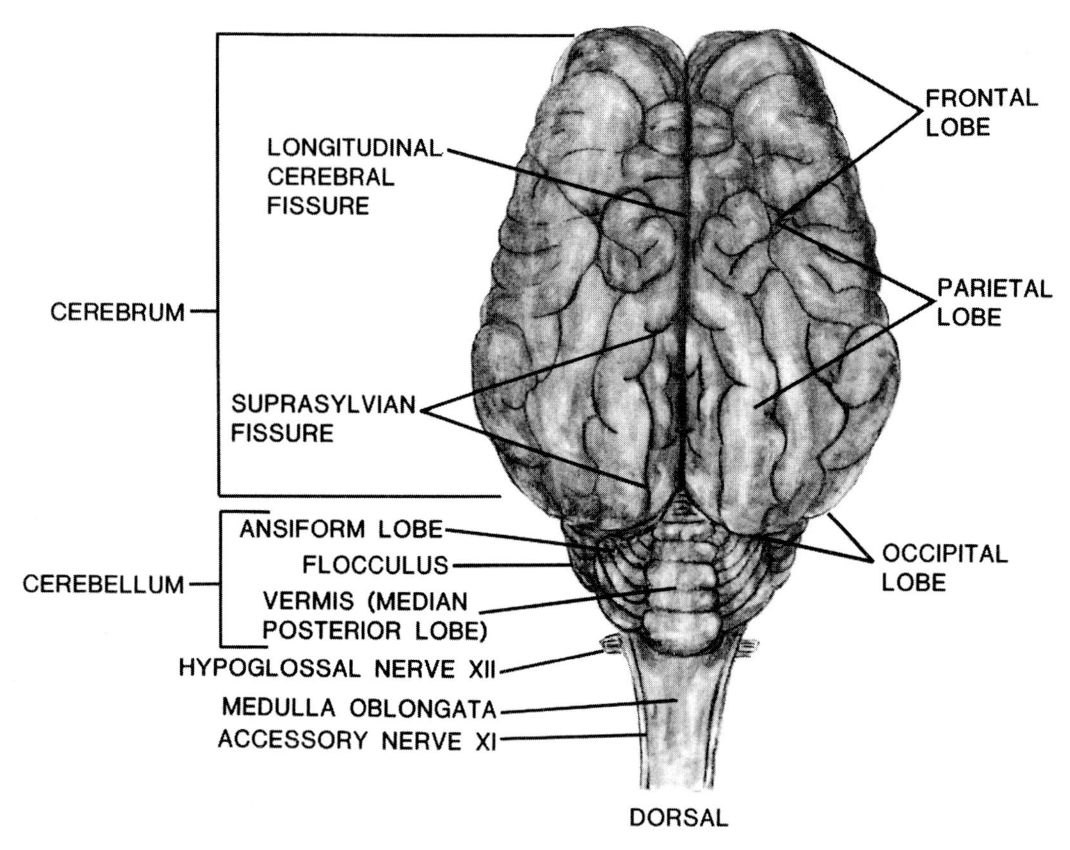

FIGURE B.1 Dorsal view of the sheep brain.

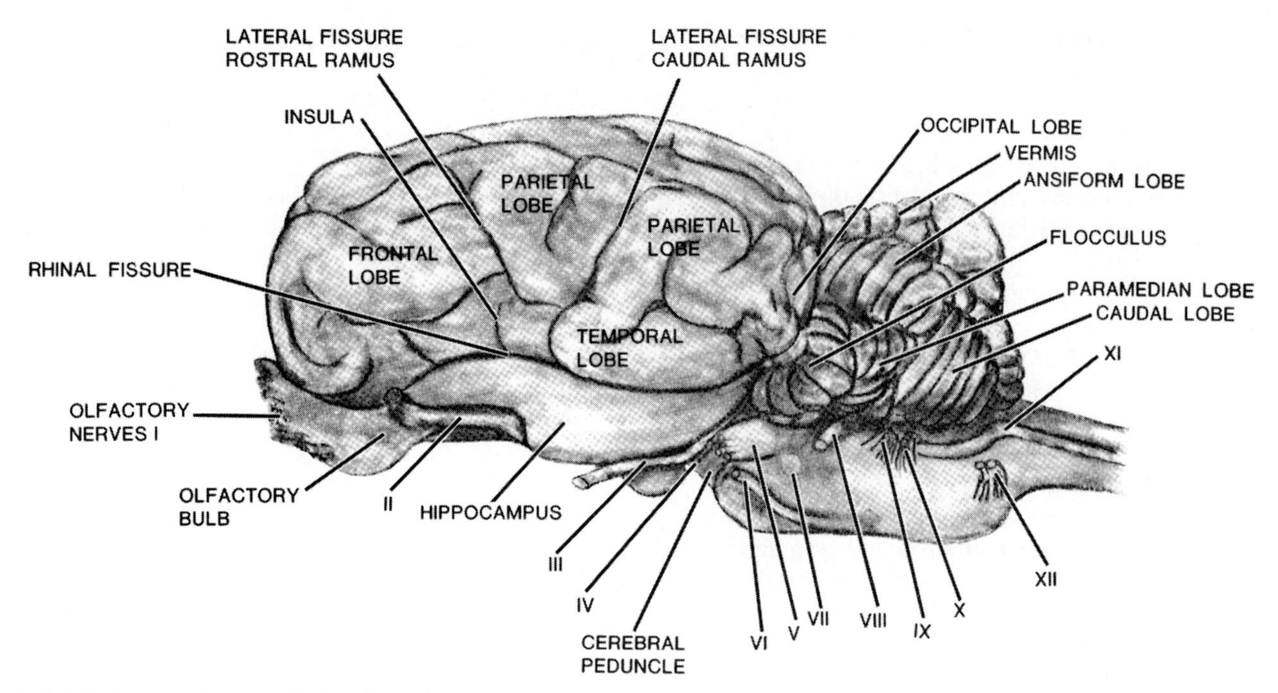

FIGURE B.2 Lateral view of the sheep brain.

falx cerebri (falks SER-e-br*i*) and between the cerebrum and cerebellum as the **tentorium cerebelli** (ten-T*O*-r*e*-um ser-e-BEL-l*i*). The innermost layer is the thin **pia** (P*I*-a) **mater,** which adheres to the surface of the brain. Between the dura mater and pia mater is a network of thin, delicate fibers called the **arachnoid** (a-RAK-noyd) **layer.** The spaces between the arachnoid fibers are the **subdural** and **subarachnoid** spaces. Most of the blood vessels of the brain occur in the subarachnoid space, but the large venous sinuses adhere to the inner surface of the dura mater, and arachnoid villi may project into these sinuses. *Carefully cut and remove the dura mater, if this has not been done previously,* and note the pattern of blood vessels on the surface of the cerebral hemispheres.

2. **CEREBRUM** (SER-e-brum). Consists of two halves or hemispheres divided by a deep longitudinal groove, the **LONGITUDINAL CEREBRAL FISSURE.** The major fissures and lobes of the sheep cerebrum are best seen in dorsal and lateral views. The deep fissures are called **SULCI** and the folds between sulci are called **GYRI.**

Each hemisphere is divided into four major lobes by the fissures and branches of the fissures. A more-or-less vertical **LATERAL FISSURE** divides the hemisphere into rostral and caudal halves. This is also referred to as the **fissure of Sylvius** (SIL-v*e*-us). In sheep there is a prominent **caudal ramus** (R*A*-mus) that partially subdivides the **PARIETAL LOBE** and sets off the **TEMPORAL LOBE** as a distinct portion. The **rostral ramus** of the lateral fissure separates the parietal and **FRONTAL LOBES.** The **INSULA** (IN-s*u*-la) is a discrete portion of the cerebrum between the frontal, parietal, and temporal lobes. The **OCCIPITAL LOBE** is not separated by distinct fissures.

The lobes of the cerebrum are usually associated with the broad functional categories that are listed here, but there is no attempt to list all the functions of the lobe or to describe the interrelationships of the lobes. Furthermore, the experiments that have identified the functions listed below were performed mainly on cats and monkeys, but not on sheep.

a. **FRONTAL LOBES.** The half of the lobes rostral to the lateral fissure relates to **primary motor** function in most mammals studied. Localized functional patterns correspond closely with localized sensory patterns of the sensory association cortex in the parietal lobes. Most motor functions are initiated here. The cortex rostral to the primary motor cortex has connections with the thalamus, and stimulation of this cortex in the monkey affects blood pressure, respiratory rate, intestinal tract motility, and (perhaps through the thalamus) emotional behavior.

b. **PARIETAL LOBES.** Narrow regions caudal to the lateral fissures are **sensory association** cortex, receiving input from all parts of the body. Caudal to the sensory association cortex is a **somatic** (touch, heat, cold, pressure) **area,** and caudal to this is a cortex concerned with **visual coordination** and **association.**
c. **OCCIPITAL LOBE.** In most animals, this area is associated with **visual projection** and **association.** Stimulation of this area in the monkey causes specific eye movements.
d. **TEMPORAL LOBES.** In humans, these lobes are concerned with **speech,** including production and association of speech with writing and vision. Although writing and speech are not performed by sheep, voice and visual association with voice may be centered in these lobes.
e. **INSULA.** Associated with **olfaction** through the piriform area and may also be involved in taste. This lobe is also thought to have a role in **visceral motor** function.

3. **CEREBELLUM** (ser-e-BEL-um). Consists of two lateral hemispheres, each with a laterally projecting **FLOCCULUS** (FLOK-*u*-lus), a median **VERMIS** (VER-mis), a dorsolateral **ANSIFORM** (AN-si-f*o*rm) **LOBE,** a ventrolateral **PARAMEDIAN LOBE,** and a **CAUDAL LOBE.**

   The cerebellum is principally a motor coordination center. The hemispheres are primarily responsible for coordinating movements of the limbs and digits; the vermis has areas for coordinating movements of the spinal column, shoulders, and hips.
4. **MEDULLA OBLONGATA** (ob-long-GA-ta). The most caudal part of the brain, which constricts imperceptibly into the spinal cord. This structure is better seen in lateral and ventral views. The regulatory centers in the medulla include respiration, heart rate, and blood pressure, but the medulla also plays an important role in regulating or mediating sensory impulses, endocrine secretions, and general awareness.

Ventral View (figure B.3)

1. **OPTIC CHIASMA** (k*i*-AS-ma). The crossing of the optic nerves between the caudal ends of the olfactory tracts.
2. **OPTIC TRACTS.** Continuations of the optic nerve fibers from the chiasma over the lateral surface of the diencephalon to the occipital lobe of the cerebrum, and to the rostral colliculi of the corpora quadrigemina.
3. **TUBER CINEREUM** (T*U*-ber si-NE-r*e*-um) (figure B.4). A prominence caudal to the optic chiasma and just rostral to the mammillary bodies. This is the area of attachment of the hypophyseal infundibulum to the brain. If the pituitary gland is not attached, the tuber cinereum will be seen as the area around a small hole, opening to the third ventricle.
4. **PONS VAROLII** (ponz va-R*O*-l*e*). Fiber tracts on the ventral rostral end of the medulla connecting the hemispheres of the cerebellum.
5. **CRANIAL NERVES.** See the cranial nerves (page 142).

Cut the brain in half longitudinally along the longitudinal cerebral fissure. You will need to spread the halves of the brain apart as you cut. You will most likely need to make more than one pass if your blade is not long enough to cut through the brain in one stroke. Keep your cuts as close to the midline as possible.

Sagittal View (figure B.4)

1. **OLFACTORY** (ol-FAK-t*o*-r*e*) **BULBS.** The extreme rostral end of the brain. The olfactory nerves terminate here after passing through the fenestrated ethmoid plate (olfactory foramina) from the nasal epithelium. Some of the olfactory nerves remain attached to the bulbs, giving them a brushlike appearance. Notice that the olfactory bulbs are two-lobed structures in the sheep, and that there is a dorsal and ventral grouping of the **OLFACTORY NERVES.** The sense of smell is a very primitive sense, and nearly all of the telencephalon of early vertebrates (i.e., dogfish shark) are devoted to this sense. The cerebral hemispheres make up a smaller part of their telencephalon.
2. **CORPUS CALLOSUM** (ka-L*O*-sum). A band of fibers connecting the gray matter areas of the cerebral hemispheres.
3. **CORPORA QUADRIGEMINA** (qwod-ri-JEM-i-na). Composed of two pairs of lobes between the cerebrum and the cerebellum. The rostral pair are

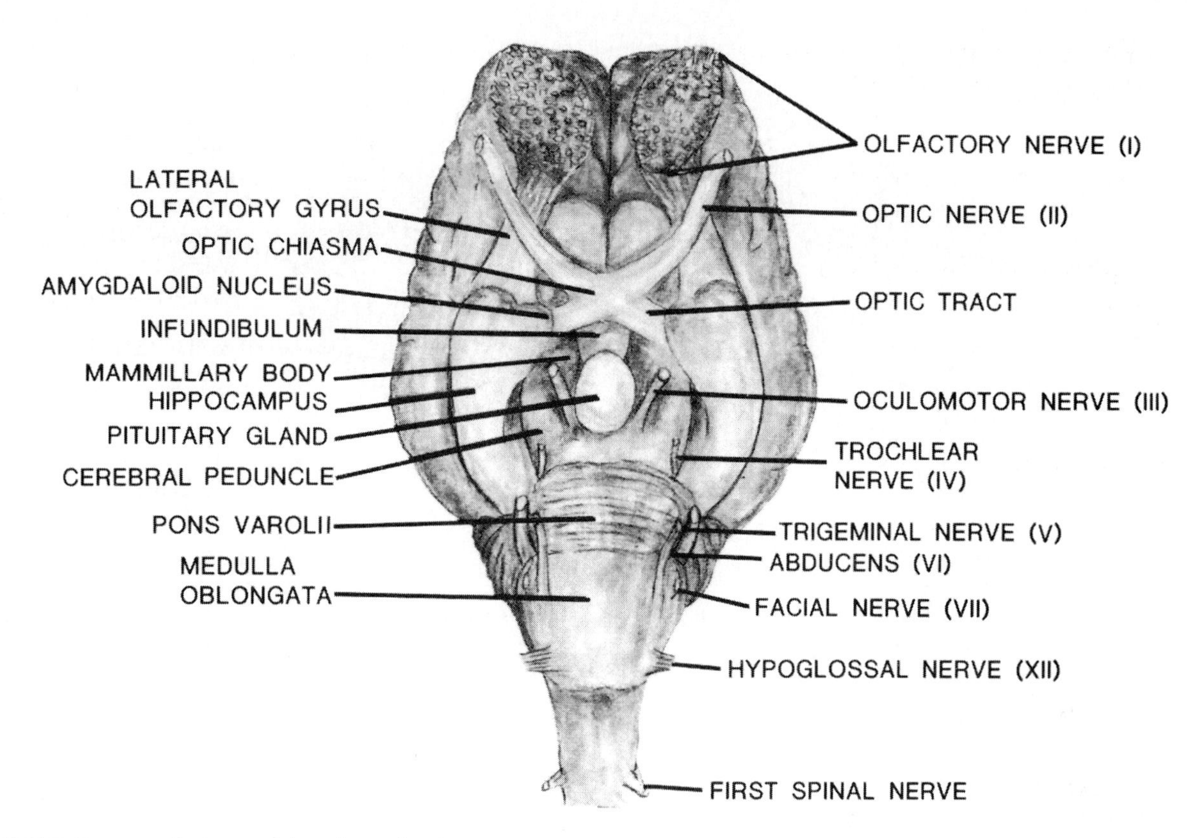

FIGURE B.3 Ventral view of the sheep brain.

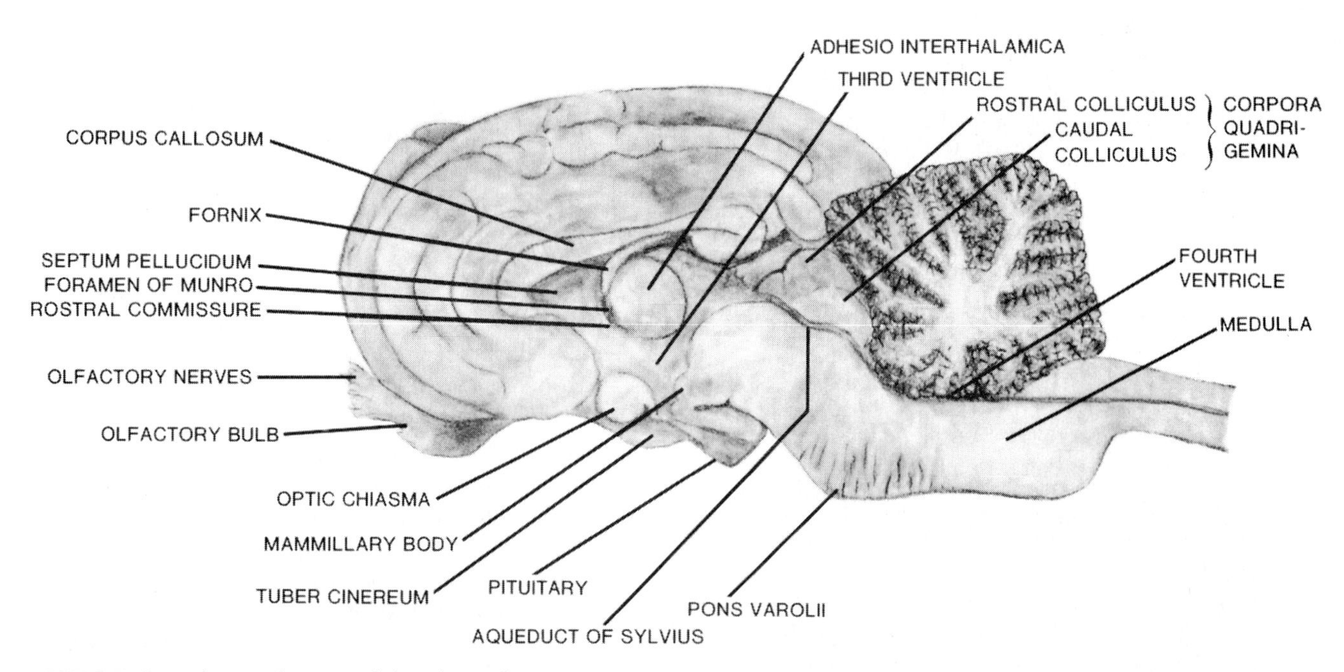

FIGURE B.4 Midsagittal view of the sheep brain.

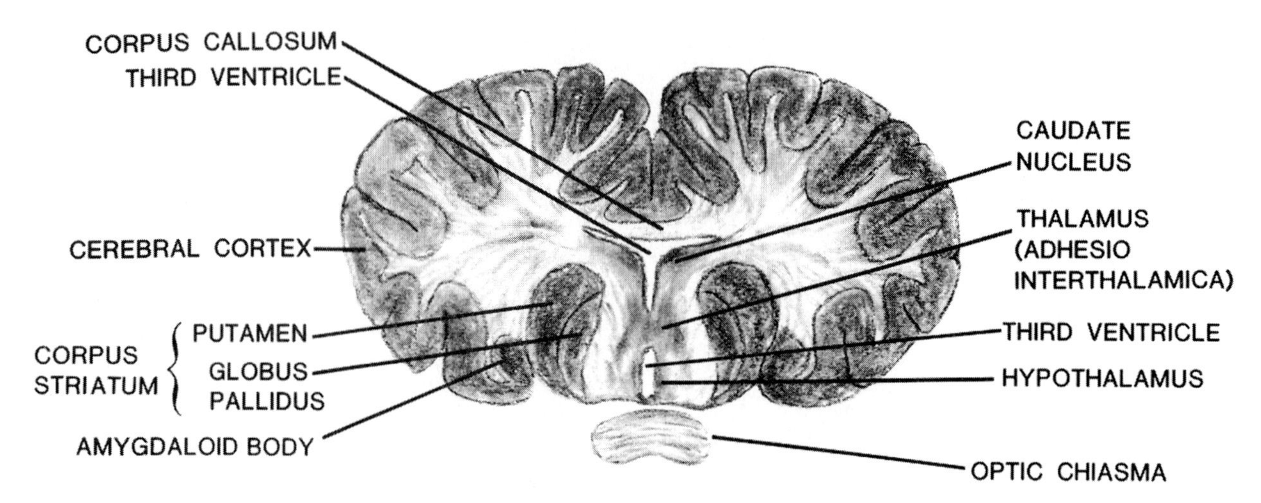

FIGURE B.5 Cross section of the sheep midbrain.

the **rostral colliculi** (ko-LIK-*u*-l*i*) and serve as optic reflex centers. The caudal pair are the **caudal colliculi** and serve as auditory reflex centers.

4. **FORNIX** (F*O*R-niks). A fiber tract connecting the hippocampus with the mammillary body (corpus mammillare). The tract runs ventral to the third ventricle, then turns dorsad to the mammillary body.
5. **SEPTUM PELLUCIDUM** (pel-LOO-sid-um). Consists of two narrow membranes of neural tissue extending between the corpus callosum and the fornix and separated by a narrow cavity. The septum pellucidum is located in the midline of the cerebrum and separates the lateral ventricles (I and II) in the cerebral hemispheres.
6. **THIRD VENTRICLE.** A cavity just caudal to the fornix that extends ventrally into the infundibulum and caudally to the level of the corpora colliculi.
7. **HYPOPHYSIS** (h*i*-POF-i-sis), or **PITUITARY** (pi-T*U*-i-t*a*r-*e*) **GLAND.** An important gland situated within the sella turcica and surrounded by a meningeal fold, which may prevent removal of the gland with the brain. See the discussion with the cat brain, sagittal section (page 141) and a discussion of hormones (page 158).
8. **PINEAL** (PIN-*e*-al) **BODY.** A gland arising from the midbrain just rostral to the corpora quadrigemina.
9. **ADHESIO INTERTHALAMICA** (ad-H*E*-z*e*-*o* in-ter-tha-LAM-ik-a) **(Intermediate mass).** Found in the center of the third ventricle and connecting the walls of the third ventricle. There are no fiber tracts in this body. The nuclei of the thalamus in this region bulges toward the midline on each side so the walls of the ventricle touch.
10. **Rostral commissure.** The rostral fiber tract connecting the basal nuclear areas of the right and left hemispheres near the rostral end of the fornix.
11. **AQUEDUCT OF SYLVIUS** (AK-w*e*-dukt, SIL-v*e*-us) or **Mesencephalic aqueduct.** The passageway connecting the third and fourth ventricles.
12. **Interventricular foramen** or **foramen of Munro** (MUN-r*o*). Connects the first, second, and third ventricles just caudal to the fornix.
13. **FOURTH VENTRICLE.** The cavity of the medulla oblongata just ventral to the cerebellum.
14. **Tela choroidea** (T*E*-la ko-ROY-d*e*-a) (not illustrated). Sheets of neural tissue and blood vessels covering the roof of each ventricle.

Cut a frontal section (at right angles to the sagittal section) through the optic chiasma.

Cross Section (figure B.5)

White matter consists of nerve processes (axons and dendrites) arranged in tracts; **gray matter** is composed of cell bodies with nuclei and nerve synapses. Gray matter occurs in layers (cortex) or in deep structures termed **nuclei.**

1. **CEREBRAL CORTEX.** The gray matter forming a thin layer on the outer surface of the cerebrum. Notice that the gray matter follows the folds (sulci and gyri) of the cerebral cortex. Together, the cortex and the underlying white matter fiber tracts form the **pallium** (PAL-*e*-um). The cells of the cortex are arranged in strata of three to six layers. All tetrapod vertebrates have a stratified cortex with at least three layers. This three-layered cortex is named the **archicortex**

(**archipallium** with its underlying white matter). The most specialized development is the six-layered **neocortex (neopallium)** that occurs in mammals. The archicortex is the **hippocampus** (hip-*o*-KAM-pus) (figure B.2). The archicortex and the paleocortex are part of the rhinencephalon, or "olfactory" brain. The neopallium and the **basal ganglia** (corpus striatum) form the nonolfactory portion of the telencephalon. The neocortex is roughly located on the dorsolateral surface of the cerebral hemispheres.

2. **CORPUS CALLOSUM.** You have already observed this in sagittal view; find it again in frontal view.
3. **CORPUS STRIATUM** (str*i*-a-tum). Consists of the basal nuclei **(globus pallidus, putamen** [p*u*-TA-men], and **caudate** [KAW-d*a*t] **nucleus)** and the fiber tracts. Phylogenetically, the putamen and caudate nucleus are the most recent portions of the corpus striatum and may be referred to as the **neostriatum.** The globus pallidus is an older feature; the **paleostriatum** and two additional nuclei, the **claustrum** (KLAWS-trum) and **amygdaloid** (a-MIG-da-loyd) **body,** form the archistriatum, or most primitive part of the subcortical gray matter. The archistriatum is separated from the rest of the corpus striatum by fiber tracts of the external capsule.
4. **THALAMUS** (THAL-a-mus). Consists of nuclei in the dorsal half of the diencephalon, or "between" brain. These nuclei are located medial and caudal to the basal nuclei.
5. **HYPOTHALAMUS.** Nuclei in the ventral half of the wall of the diencephalon.
6. **Internal capsules** (not illustrated). Fiber tracts connecting various areas of the cerebral cortex with the nuclei in the pons.
7. **Epithalamus** (not illustrated). A very small area dorsal to the thalamus and consisting of the **habenulae** (ha-BEN-u-l*e*) (nuclei), **stria medullaris** (fiber tracts), and **pineal body.**

APPENDIX 3

OX EYE

Most of the terms used in this description of the mammalian eye are in common use and recognized by *Nomina Anatomica Veterinaria* (NAV), but a few are not. In those instances in which the terms differ, the NAV terms are presented in bold type and in **(parentheses).**

External Features (figures E.1–E.3)

The ox or sheep eye has been removed with the attached extrinsic musculature, most of the lacrimal glands, much fibrous tissue, and some skin, including the eyelids. Before dissecting more of the eye, use figure E.1 and locate the following:

1. The **EYELIDS** (*I*-lidz), upper and lower.
2. **CILIA** (SIL-*e*-ah). Eyelashes. The cilia of the upper eyelid are much longer than those of the lower eyelid.
3. **PLICA SEMILUNARIS CONJUNCTIVA** (PL*I*-ka sem-*e*-loo-N*A*-ris kon-JUNK-tih-va). A fold of tissue at the inner or medial corner of the eye. This structure is also referred to as the third eyelid and is derived from the **CONJUNCTIVA,** which otherwise covers the exterior of the eyeball and the inner margin of the eyelids.
4. **CANTHUS** (KAN-thus). Fold of skin separating the upper and lower eyelids at both the medial **(medial canthus)** and lateral **(lateral canthus)** angles of the eye.

Knowing the position of the third eyelid (lower medial corner), you should orient the eye and determine whether the specimen is a right or left eye. Some other directions and orienting features of the eye are presented in figure E.2.

5. **Lacrimal** (LAK-ri-mal) **gland** (not illustrated). Located at the dorsal lateral corner of the eye. An additional lacrimal gland occurs at the rostral base of the ear of the ox.

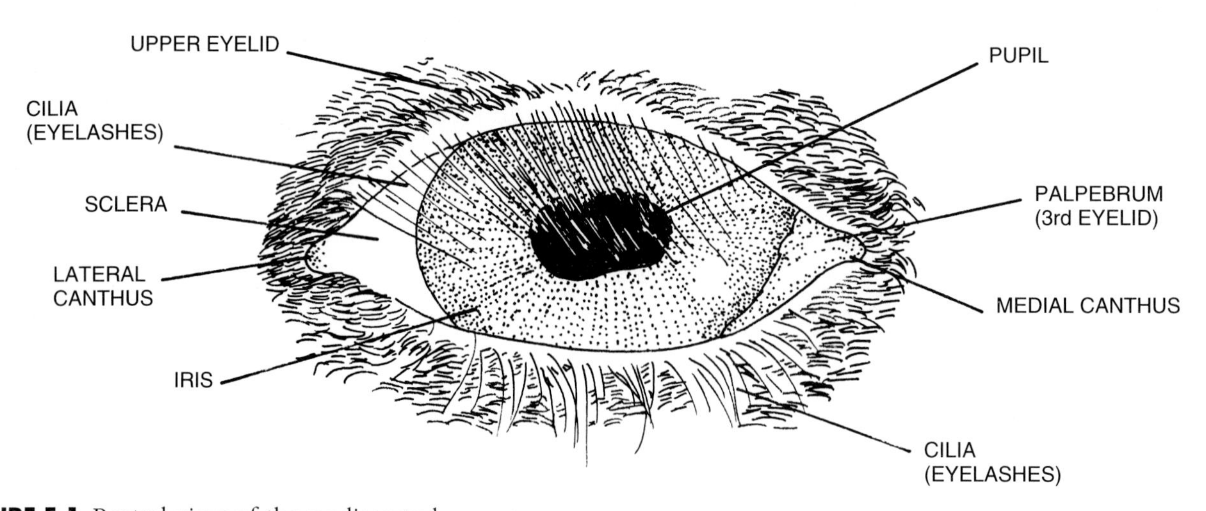

FIGURE E.1 Rostral view of the undissected ox eye.

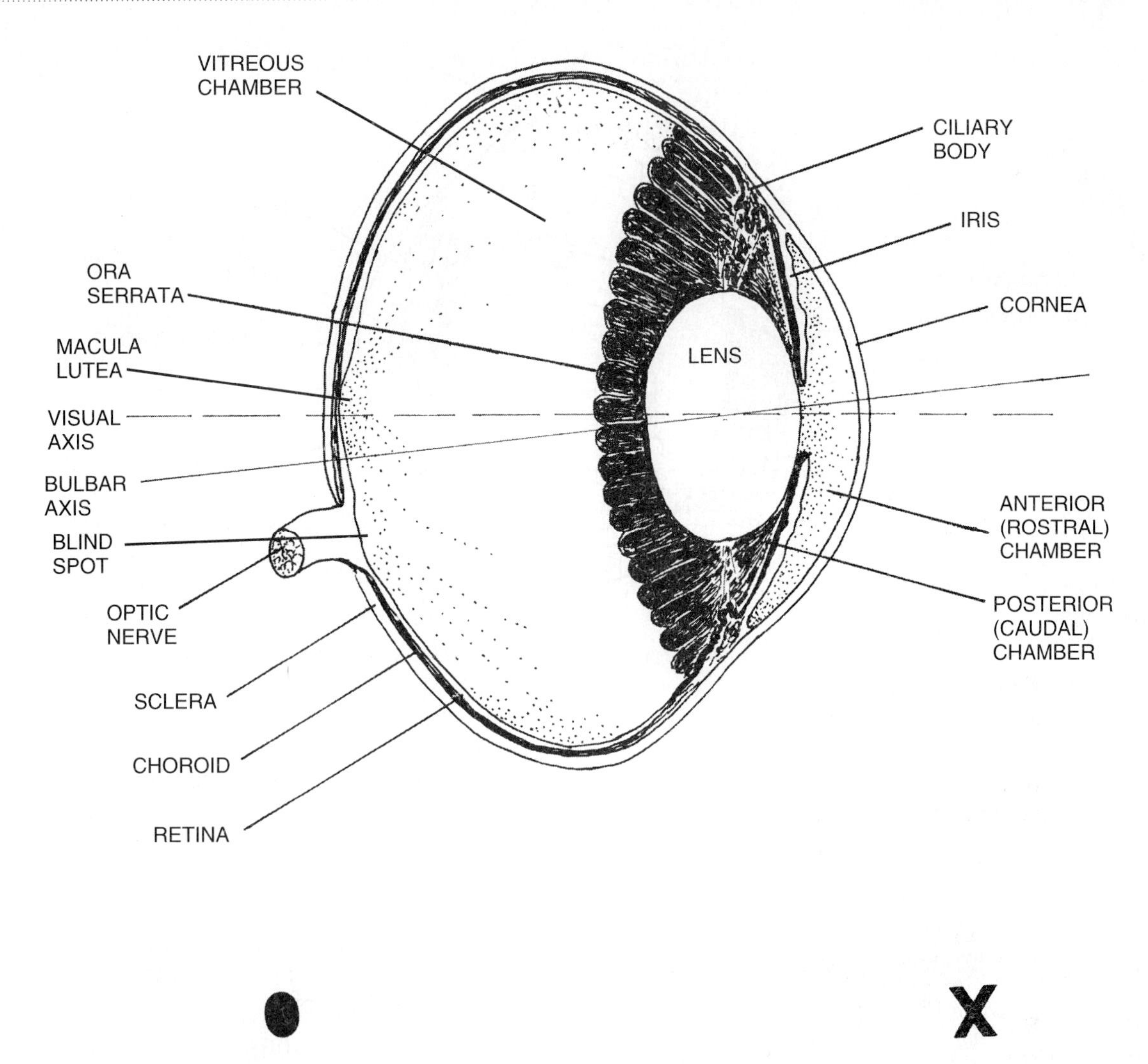

FIGURE E.2 Directions and planes of the mammalian eye. The **ANTERIOR POLE** is the center of the cornea. The **POSTERIOR POLE** is at the most posterior point of the eyeball. The **AXIS BULBI INTERNUS** (BUL-b*i* in-TER-nus), the bulbar axis, is a line drawn between the anterior and posterior poles. The **EQUATOR** is the greatest circumference of the eyeball. The **VISUAL AXIS** passes through the center of the lens to the **Fovea Centralis** (F*O*-v*e*-a sen-TRA-lis), which is the center of the **MACULA LUTEA** (MAK-*u*-lah LYOO-t*e*-ah). Note the relationship between the macula and the **BLIND SPOT.** Close one eye and focus the open eye on the dot on the bottom left of this figure. Next move the page forward or back until the X on the bottom right of the figure disappears. The dot is now focused on the macula of your eye and the X is focused on the blind spot.

6. **Harderian** (HARD-er-*e*-an) **gland** (not illustrated). C-shaped mucous gland in the skin surrounding the rostral part of the eyeball.
7. **EYE MUSCLES** (figure E.3). Find the extrinsic eye muscles listed in the following chart (top of page 175) and note their innervation. In addition to the six muscles that move the eyeball, a **levator palpebrae superioris** (l*e*-V*A*-tor PAL-peh-br*e* syoo-P*E*-r*e*-*o*-ris) draws the entire eyeball upward.

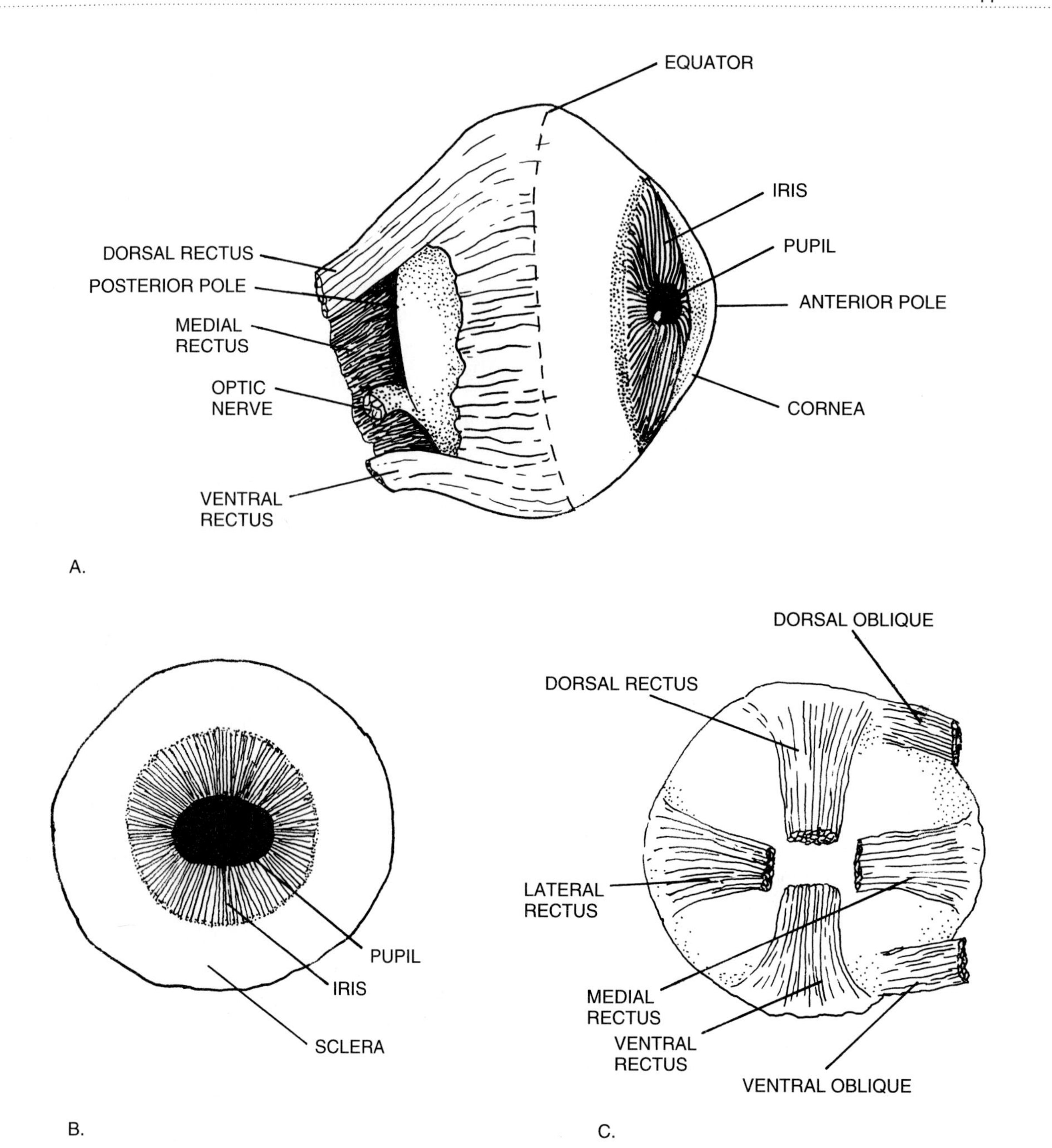

FIGURE E.3 Features of the ox eye. The skin has been removed and some fibrous tissue has been cleaned away. (A) Medial view of the left eye. (B) Rostral view. (C) Caudal view.

Chart of the Eye Muscles (figure E.3C)

Name	Innervation	Attachment	Action
DORSAL RECTUS (REK-tus)	Oculomotor	Dorsal midline beneath levator palpebrae superioris	Rotates eye (anterior pole) dorsad
LATERAL RECTUS	Abducens	Lateral border of eyeball	Rotates eye (anterior pole) laterad
VENTRAL RECTUS	Oculomotor	Ventral midline beneath ventral oblique	Rotates eye (anterior pole) ventrad
MEDIAL RECTUS	Oculomotor	Medial border of eyeball	Rotates eye (anterior pole) mediad
LEVATOR PALPEBRAE SUPERIORIS	Oculomotor	Connective tissue dorsal to eyeball	Elevates entire eyeball
DORSAL OBLIQUE (o-BLEK)	Trochlear	Anterior, dorsomedial area, beneath a large vein	Rotates eyebulb (bulbar) axis clockwise
VENTRAL OBLIQUE	Oculomotor	Anterior, ventrolateral border of eyeball	Rotates eyebulb (bulbar) axis counterclockwise

The Tunics and Internal Features

1. **SCLERA** (SKLE-ra). Tough, white, fibrous outer layer (tunic) of the eyeball. Also see figure E.1.
2. **CORNEA** (KOR-ne-a). Exposed transparent continuation of the sclera.

Section the eyeball (figure E.4) in a plane parallel to the cornea and optic nerve (as in figure E.2) and locate the following (figure E.5):

3. **ANTERIOR CHAMBER [camera anterior bulbi** (BUL-bi)]. The chamber of the eye, bordered anteriorly by the cornea and posteriorly by the iris and pupil.
4. **CHOROID** (KO-royd) **LAYER [choroidea** (KO-royd-e-a)]. Black tissue (tunic) between the sclera and retina.
5. **IRIS** (I-ris). Pigmented partition pierced by the pupil and separating the anterior and posterior chambers of the eye. The iris is an anterior continuation of the choroid.
6. **PUPIL** (PYOO-pil). Opening through the center of the iris. In the ox, sheep, and horse, the pupil is a horizontal opening that dilates to a large, round opening in the dark. This large an opening is not possible in a pupil that is round when contracted.
7. **POSTERIOR CHAMBER (camera posterior bulbi).** Bordered anteriorly by the iris and posteriorly by the lens. Both the anterior and posterior chambers are filled with **aqueous humor** (A-kwe-us HU-mor). The aqueous humor is a filtrate of blood that is released by the capillaries of the ciliary body and iris into the posterior chamber. The humor circulates through the pupil into the anterior chamber and reenters the circulatory system through the **canal of Schlemm** (schlem) (see histology of the eye and figure 11.4).
8. **LENS** (lenz). Spherical body in the center of the eyeball. The lens is attached to the ciliary body by the suspensory ligament.
9. **CILIARY** (SIL-e-er-e) **BODY.** Enlargement of the choroid between the iris and choroid proper and containing muscles that control the lens. The caudal edge of the ciliary body is serrated and is termed the **ORA SERRATA** (O-rah ser-RA-tah).
10. **VITREOUS** (VIT-re-us) **BODY [corpus vitreum** (KOR-pus VIT-re-um)]. Jellylike mass filling the cavity **(camera vitrea** (VIT-re-ah) **bulbi)** posterior to the lens.
11. **RETINA** (RET-i-na). Loose, folded yellow tissue between the choroid and vitreous body.
12. **OPTIC** (OP-tik) **DISK** (blind spot). Point of exit of the optic nerve from the retina.

Histology of the Eye (figure 11.4)

The following structures require a microscope for examination; the details cannot be seen in a gross dissection. Follow the instructions for the use of the microscope in Chapter 1 and learn the information regarding tissues before starting this exercise.

Slides of the eye are very difficult to prepare because the various layers have different consistencies, so the pressure required to cut through the lens will collapse or pull the retina out of position. For this reason, sections of the eyeball are rarely available commercially and those that are available usually have the lens removed. The following description assumes you have a perfect slide to study; you may not be able to find some of the structures, or the structures you observe may be in an unnatural position.

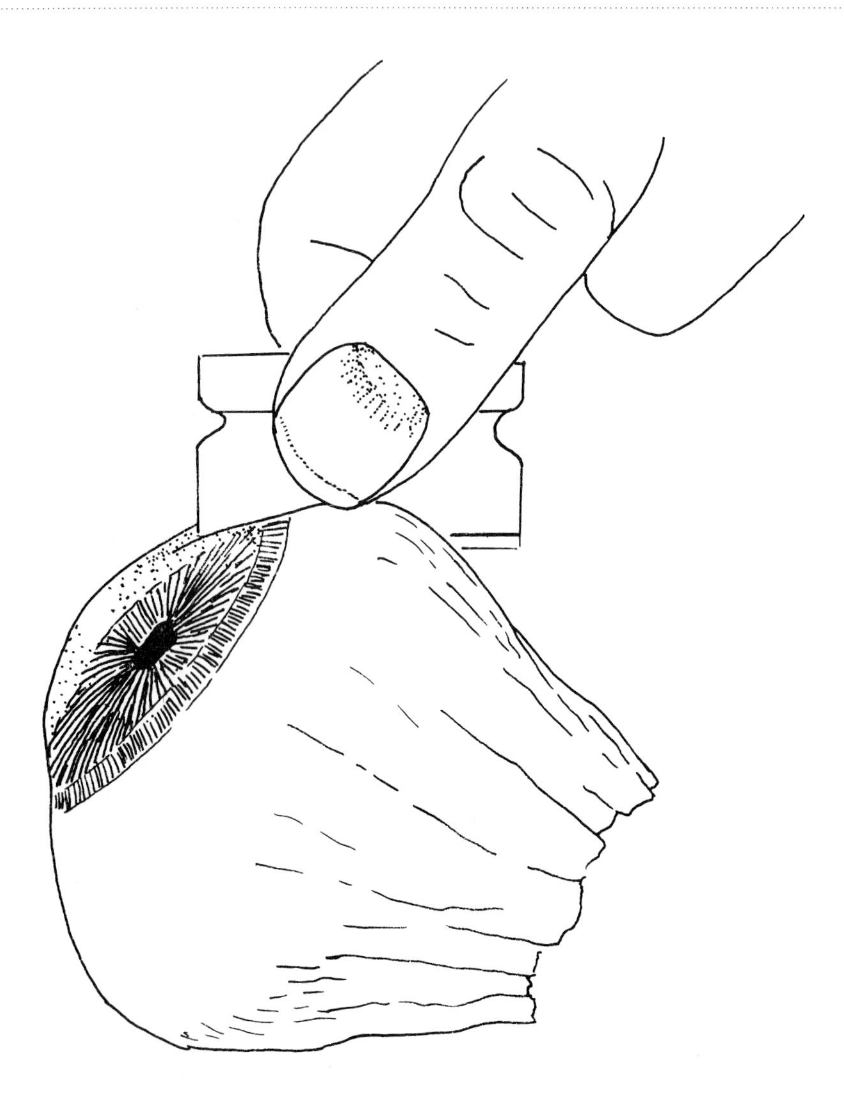

FIGURE E.4 Instructions for sectioning the ox eye. *Carefully clean the fat and some fibrous tissue from the eye (hemostats are a good tool to use for this purpose) in order to expose the glands and ocular muscles. Remove the skin (eyelids) and expose the skin musculature (orbicularis oculi) and lacrimal glands. Position the eye so the pupil is horizontal and start your cut through the cornea and sclera near the dorsal midline. Continue the cut through the tunics and completely around the eyeball with a sharp single-edge razor blade. After the tunics (including the iris) are cut, carefully remove the lens. Do not attempt to section the lens.*

Locate the following on your slide:

1. **SCLERA/CORNEA.** The outermost tunic. The sclera proper is the tough, white, fibrous outer layer of the eyeball and the cornea is the exposed transparent continuation of the sclera.
2. **CHOROID/CILIARY/IRIS** layer. The middle tunic. This layer of black pigmented tissue within the sclera is known as the **UVEA** (*U*-ve-ah).
3. **RETINA.** The third and innermost tunic. The three tunics enclose three fluid-filled chambers as follows:
   a. **ANTERIOR CHAMBER.** Bordered anteriorly by the cornea and posteriorly by the iris and pupil.
   b. **POSTERIOR CHAMBER.** Bordered anteriorly by the iris and posteriorly by the lens; the anterior and posterior chambers are filled with **aqueous humor.**
   c. **VITREOUS CHAMBER.** Between the lens and retina. Filled with a jellylike mass, the vitreous body.
4. **IRIS.** The pigmented partition pierced by the pupil and separating the anterior and posterior chambers of the eye. Verify that the iris is a rostral continuation of the choroid by tracing the tissue on the posterior side of the iris. The opening in the center of the iris is the pupil.

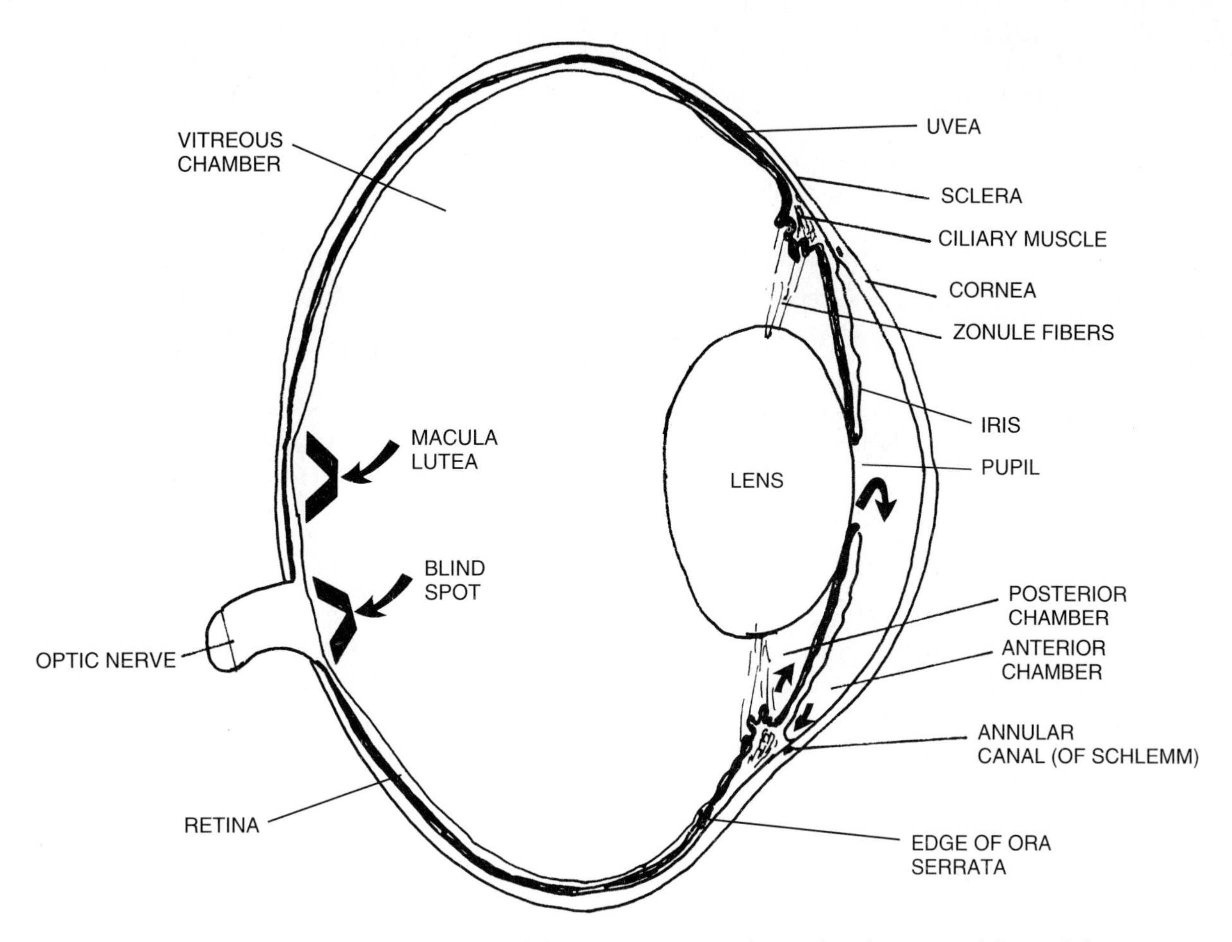

FIGURE E.5 Features seen in a gross dissection of the eye. Arrows indicate the direction of flow of the aqueous humor in the anterior and posterior chambers.

5. **CILIARY BODY.** An enlargement of the choroid layer which contains the **CILIARY MUSCLES** and suspends the lens by **suspensory** (sus-PEN-s*o*-r*e*) **ligaments** or **zonule** (ZON-*u*l) **fibers.** Contraction of the ciliary muscles change the shape of the lens in mammals (see accommodation of the eye).
6. **RETINA.** Contains the light-sensitive cells of the eye. There are four cell layers in the retina, but the parts of the cells, body, axons, and dendrites, are also arranged in layers. The layer of the retina adjacent to the choroid layer is a single layer of cuboidal cells called Bruch's membrane, or the **pigmented epithelium** (ep-i-TH*E*-l*e*-um). This layer is not photosensitive. The rod and cone cells form parts of three layers in the retina. The **RODS** and **CONES LAYER** of the retina are adjacent to the pigmented epithelium and contain the photosensitive chemicals that activate an impulse to the cell body [**OUTER NUCLEAR** (NYOO-kl*e*-ar) **LAYER**]. From the outer nuclear layer the axons extend into the **OUTER PLEXIFORM** (PLEK-si-f*o*rm) **LAYER,** where they synapse with dendrites of the **INNER NUCLEAR LAYER** cells. The inner nuclear layer cell bodies send axons to the **INNER PLEXIFORM LAYER.** These axons synapse with the **GANGLION** (GANG-gl*e*-on) cell dendrites. The ganglion cells send their axons to the brain via the **OPTIC NERVE.** In addition, the retina has several **glial** (GL*I*-al) **cells** that support the nerve cells. The point of exit of the optic nerve contains only the ganglion cell axons and therefore is not photosensitive.

Accommodation

Information regarding focus for near and far vision is presented in figures E.6 and E.7

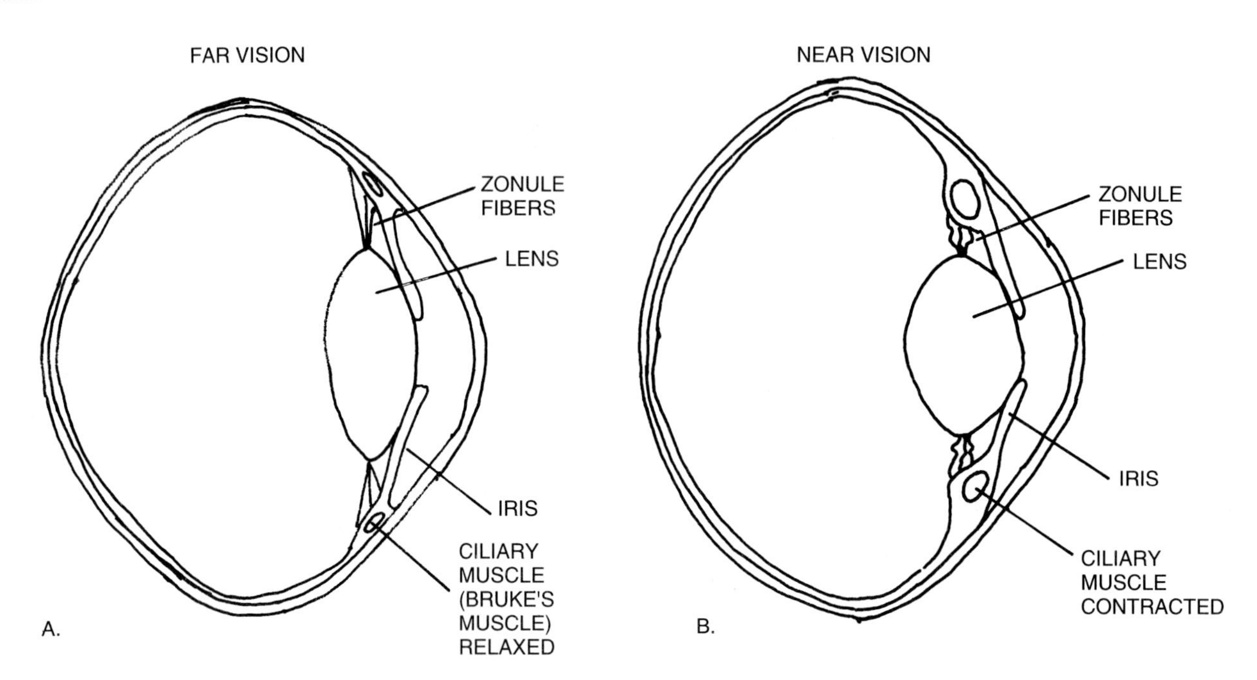

FIGURE E.6 Diagrams illustrating accommodation for far (A) and near (B) vision by the eye. Bruke's muscle of the ciliary body encircles the eyeball and acts as a sphincter muscle that constricts the eyeball in the plane of the lens' greatest diameter. This constriction allows the suspensory fibers between the lens and the ciliary body to slacken, thus releasing the tension that keeps the lens stretched when the muscle is not contracted (A). With the tension released, the natural elasticity of the lens causes the lens to "fatten" or become rounded (B).

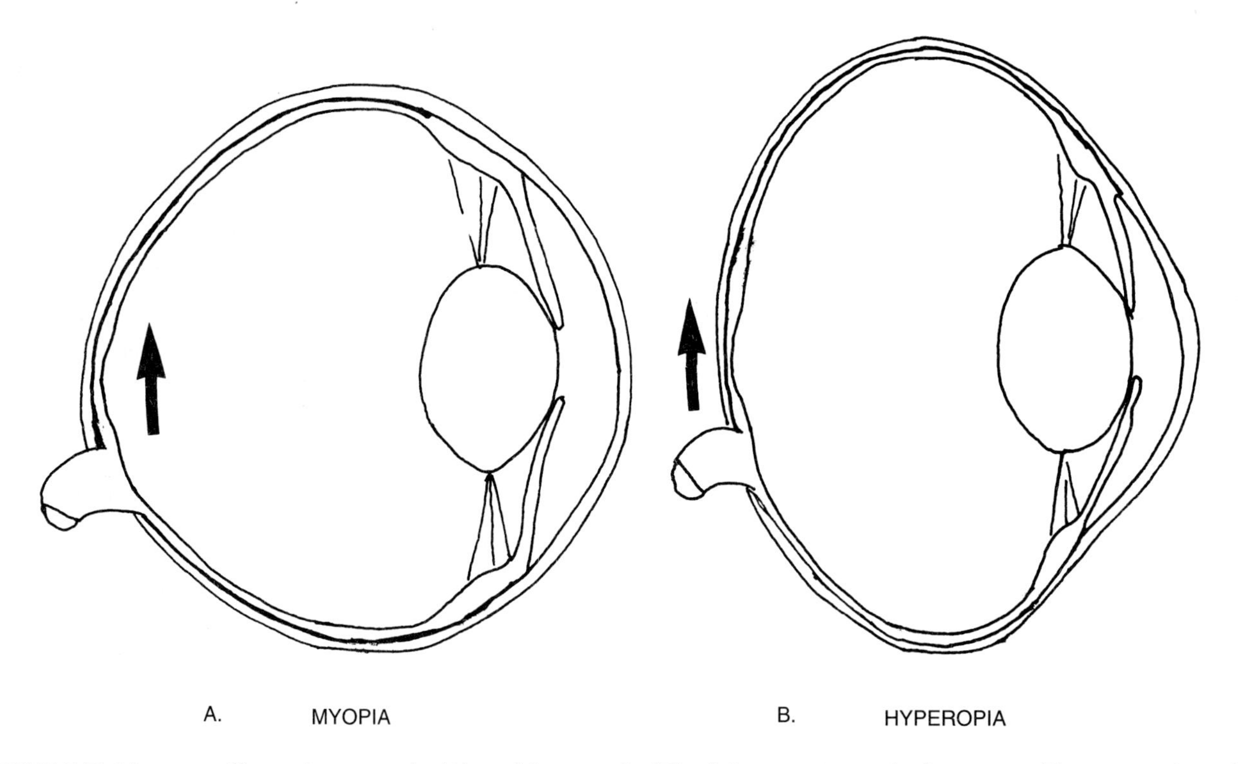

FIGURE E.7 Diagrams illustrating myopia (A) and hyperopia (B) of the eye. **Myopia** (m*i*-*O*-pe-ah) occurs when the image formed (in this case, the arrow) is in focus rostral to the retina. In effect the eyeball is too long or the lens is too round. This is corrected with a biconcave lens in front of the eye. **Hyperopia** (h*i*-per-*O*-pe-ah), occurs when the image (arrow) is focused behind the eye because the eyeball is too short or the lens is too flat. This condition is corrected with a biconvex lens.

GLOSSARY

The following is an abbreviated glossary of some of the terms used in this manual. Some word roots for structures are italicized when presented. Abbreviations: G. = Greek; L. = Latin.

A pronunciation guide is provided for the terms in this glossary. Each term is spelled phonetically in parentheses after its listing. The following rules were used in creating this guide:

1. The syllable with the strongest accent appears in capital letters; for example: a-NAT-*o*-m*e*.
2. Vowels to be pronounced in the long form are in italics, as in: bl*a*de and b*i*te.
3. Unmarked vowels are pronounced in the short form as in: mitt and drum.
4. Other indicators of sounds are: oo as in blue; oy as in foil.

Learning the correct spelling of new terms is easier if a few moments are spent studying the proper pronunciation.

Additional information regarding word origins, pronunciations, and definitions may be found in the following:

Blakiston's New Gould Medical Dictionary. (any recent edition). New York: McGraw-Hill Book Company.

Borror, D. J. 1960. *Dictionary of Word Roots and Combining Forms.* Palo Alto, Calif.: N-P Publications.

Dorland's Illustrated Medical Dictionary. 25th ed. 1980. Philadelphia and London: W. B. Saunders Company.

International Committee on Veterinary Gross Anatomical Nomenclature. 1983. *Nomina Anatomica Veterinaria.* 3d ed. International Committee on Veterinary Gross Anatomical Nomenclature, World Assoc. Veterinary Anat., Ithaca, N.Y.

A

Abducens (ab-DOO-senz) L. ab = away; duc = lead. A nerve.

Abduct (ab-DUKT) L. ab = away; duc = lead. An action.

Abductor (ab-DUK-tor) L. ab = away; duc = lead. A muscle that performs the action.

Acetabulum (as-e-TAB-u-lum) L. acetabul = vinegar cup. A joint socket.

Acinar (AS-i-nar) L. acin = a berry. Cellular arrangement of some glands.

Acoustic (a-KOOS-tik) G. acous = to hear.

Acromion (a-KRO-me-on) G. acro (akros) = top most; mion (omos) = point. A process on the scapula.

Adduct (ad-DUKT) L. adducere = to draw toward. An action.

Adenohypophysis (a-den-o-hi-POF-i-sis) G. adeno = gland; hypo = under; physis = to grow. An endocrine gland.

Adhesio Interthalamica (ad-HE-ze-o in-ter-tha-LAM-ik-a) L. adhaesio = to stick to; inter = between; L. & G. thalamica = chamber. Part of the brain.

Adrenal (ah-DRE-nal) L. ad = near; renes = kidneys. An endocrine gland.

Adrenocorticotrophic (ad-re-no-kor-te-ko-TROP-ik) L. adreno = near the kidneys; cortico = bark; G. trophikos = nourishing. A hormone secreted by the pituitary.

Afferent (AF-er-ent) L. afferens = to bear. Moving or conducting toward an area or point.

Agonist (AG-o-nist) G. ago = to lead, a chief. The "prime mover" or muscle of reference.

Alare (*A*-lar) L. ali = wing. Usually a bone process shaped like a wing.

Alisphenoid (al-e-SFE-noyd) L. ali = wing; G. spheno = wedge. A bone. (Note: Latin and Greek roots are sometimes combined in the same term.)

Alveoli (al-VE-o-li) L. alveol = pit or cavity [especially in the lungs].

Amnion (AM-ne-on) G. amnio = lamb. An embryonic membrane.

Ampulla (am-PUL-la) L. ampulla = a flask.

Amygdaloid (ah-MIG-dah-loyd) G. amygda = almond; loid = like. A part of the brain.

Anastomose (a-NAS-to-mos) G. ana = again; stomos = mouth. Coming together; usually of blood vessels.

Anconeus (an-KO-ne-us) G. ancon = elbow.

Angularis (ang-gu-LAR-is) L. angularis = sharply bent.

Antagonist (an-TAG-o-nist) G. antagonistes = an opponent. The muscle or muscles opposing the agonist.

Ante (AN-te) L. ante = before [a prefix].

Antibrachium (an-ti-BRA-ke-um) G. anti = against; brachium = the arm.

Antrum (AN-trum) G. antrum = a cave.

Aorta (a-OR-ta) G. aorta = great artery.

Aponeurosis (ap-o-nyoo-RO-sis) G. apo = from, away; neurosis = sinew, nerve, cord. Dense connective tissue forming a broad tendon.

Appendicular (ap-en-DIK-*u*-lar) L. append = hang to; cula = little. Arms and legs.

Arachnoid (a-RAK-noyd) G. arachn = spider, web; oid = like. Part of the membranes covering the brain.

Arbor (AR-bor) L. arbor = a tree.

Archinephric (ar-ki-NEF-rik) G. archi = beginning, original; nephros = kidney.

Arcuate (AR-ku-at) L. arcus = an arch, bow.

Arteriole (ar-TE-re-ol) G. arteria = to raise; iole = diminutive. A small artery that delivers blood to capillaries.

Artery (AR-ter-e) L. arteria; G. arteri (aer = air; terein = to keep) air pipe, artery.

Arthroidial (ar-THRO-de-al) G. arthrodia = gliding. A type of joint.

Articular (ar-TIK-u-lar) L. articularis = a joint.

Articulate (ar-TIK-u-l*a*t) L. articulatus = jointed.

Articulation (ar-tik-u-LA-shun) L. articlatio = to separate into joints. A joint or union of two bones.

Arytenoid (ar-e-TE-noyd) G. arytaina = ladle, jug; eidos = form. Cartilages in the larynx.

Atria (A-tre-a) L. atria = hall. Chambers of the heart.

Auricule (AW-re-kl) L. auricula = a little ear. Part of the atria of the heart.

Axilla (ak-SIL-ah) L. axillae = armpit; beneath the axis.

Axis (AKS-is) L. & G. axon = axle. The second cervical vertebra; any axial part.

Axon (AKS-on) G. axon = axis. Part of a nerve cell.

Azygous (AZ-i-gos) G. a = without; zygon = pair. A blood vessel.

B

Baculum (BAK-u-lum) L. baculum = rod. Penis bone.

Basihyo- (ba-se-HI-o) L. bas = foundation, step; G. hyo = Y-shaped. A hyoid bone.

Basilar (BAS-i-lar) L. bas = foundation. Forming a foundation.

Biceps (BI-ceps) L. bi = two; L. ceps = head.

Bicipital (bi-SIP-i-tal) L. bi = two; L. cipit = head.

Bicuspid (bi-KUS-pid) L. bi = two; cuspi = point.

Biventer (bi-VEN-ter) L. bi = two; L. vent = belly, underside.

Bone (bon) L. os, G. osteon = bone.

Brachial (BRA-ke-al) L. brachialis = arm.

Brachium (BRA-ke-um) L. The arm from shoulder to elbow.

Branchial (BRANK-e-al) L. branchiae = gills.

Bronchus (BRONG-kus) G. bronchos = trachea, windpipe.

Buccal (BUK-al) L. bucca = cheek.

Bulbo- (BUL-bo) L. A bulb.

Bulla (BUL-ah) L. A bubble.

Buttocks (BUT-oks) Prominence formed by the gluteal muscles.

C

Caecum (SE-kum) L. caecum = blind. Intestine; also see cecum.

Calcaneus (kal-KA-ne-us) L. calc = heel bone.

Canaliculus (kan-ah-LIK-u-lus) L. A small canal.

Canine (KA-nin) L. canis = a dog. A tooth.

Capitate (KAP-i-tat) L. caput = head; ate = like.

Capitulum (ka-PIT-u-lum) L. caput = head; ulum = little.

Carotid (ka-ROT-id) G. karos = stupor; L. carot = a carrot.

Carpal (KAR-pl) G. karpos = wrist.

Cartilage (KAR-ti-lij) L. cartilag = gristle.

Cauda (KAW-daw) L. Tail.

Cavum (KA-vum) L. cavus = hollow. Cavity.

Cecum (SE-kum) L. caecum = blind. Intestine; also see caecum.

Celiac (SE-le-ak) G. koilia = belly.

Centrum (SEN-trum) G. kentron = center.

Cephalic (se-FAL-ik) G. kephale = head.

Cerato- (ser-AH-to) G. keras = horn.

Cerebellum (ser-e-BEL-um) L. cere = brain; bellum = small, diminutive.

Cerebrum (se-RE-brum) L. Brain.

Cervicis (CER-vi-sis) L. cervix = neck; deer.

Choana (KO-a-nah) G. choane = funnel. Nasal passage.

Chorda (KOR-dah) G. A cord or string.

Chorion (KO-re-on) G. A membrane.

Choroid (KO-royd) G. Like a membrane.

Clavicle (KLAV-i-kl) L. Key, clavicle.

Cleido- (KLI-do) G. Key, clavicle.

Coccyx (KOK-siks) G. kokkyx = cuckoo. Anthropoid caudal vertebrae.

Cochlea (KOK-le-ah) G. kochlias = spiral shell; coiled chambers of the mammalian ear.

Coelom (SE-lom) G. koilos = hollow. Fluid-filled vertebrate body chambers.

Colliculus (ko-LIK-u-lus) L. collis = hill; ulus = diminutive. Eye and ear reflex centers in the brain.

Columnar (ko-LUM-nar) In the form of a column.

Conchae (KONG-ke) L. G. konche = shell, bivalve. Thin bones in the nasal chamber.

Condyle (KON-dil) G. kondylos = knuckle, knob.

Coracoid (KOR-a-koyd) G. korone = crow; oid = like.

Cornu (KOR-nu) G. Horn.

Coronary (KOR-o-na-re) Encircling as a crown. 1. Vessels encircling the heart. 2. A condition caused by a decreased blood flow to the heart muscle. 3. Specifically, a blocked coronary artery or a branch of a coronary artery.

Coronoid (KOR-o-noyd) L. corona = crown; oid = like.

Costal (KOS-tal) L. Pertaining to a rib.

Coxae (KOK-se) L. The hip.

Cremaster (kre-MAS-ter) G. Suspender.

Cribriform (KRIB-ri-form) L. crib = a sieve; form = shape.

Cricoid (KRI-koyd) G. A ring, circle.

Cruciate (KROO-she-at) L. cruci = a cross.

Crural (KROOR-al) L. cruri = leg.

Cuboidal (ku-BOYD-al) G. cubo = a cube.

Cuboid (KU-boyd) G. cubo = like a cube. Square.

Cutaneous (ku-TA-ne-us) L. cutane = skin.

Cystic (SIS-tik) G. kystis = a sack or pouch.

D

Deferens (DEF-er-enz) L. de = away from; ferens = carry.

Deltoid (DEL-toyd) G. delta [shaped like the Greek letter], triangular.

Diastema (di-as-STE-mah) G. A space, gap.

Digastricus (di-GAS-trik-us) G. di = two; gaster = stomach or belly.

Digitigrade (DIJ-i-ti-grad) L. digit = fingers or toes; grade = step, walk.

Duct (dukt) L. ductus = leading.

Duodenum (dyoo-o-DE-num) L. duodecem = twelve.

E

Ectoderm (EK-to-derm) G. Superficial primary germ layer of the embryo. This layer gives rise to the epidermis, nervous system, and mucous membranes.

Ectotympanic (ek-to-tim-PAN-ik) G. ecto = outer; tympani = a drum.

Effector (ef-FEK-tor) An organ that is innervated by a motor pathway.

Ellipsoidal (e-lip-SOYD-l) G. Elliptical shape.

Emulsify (e-MUL-si-fi) To produce small droplets of a liquid.

Endochondral (en-do-KON-dral) G. endo = within; chondros = cartilage.

Endocrine (EN-do-krin) A system of ductless glands secreting hormones into the blood stream and targeting another organ to produce a response.

Endoderm (EN-do-derm) G. Deep primary germ layer of the embryo. It gives rise to the linings of the pharynx, respiratory tree, digestive tract, urinary bladder, and the urethra.

Endometrium (en-do-ME-tre-um) The lining of the uterus; during the reproductive years, this layer is sloughed each month during menstruation.

Endothelium (en-do-THE-le-um) A layer of epithelial cells forming a tissue that lines the circulatory system and body cavities; derived from the mesoderm of the embryo.

Epi (EP-i) G. A prefix meaning on, upon, or over.

Epicondyle (ep-i-KON-dil) G. epi = above; condyl a knuckle.

Epidermal (ep-i-DER-mal) G. epi = above; derma = skin.

Epiphyseal (e-PIF-e-sel) **disk.** A cartilaginous growth line at the ends of long bones.

Ethmoid (ETH-moyd) G. Like a sieve.

Exteroception (eks-ter-o-SEP-shun) Perception of information provided by the exteroceptors.

Exteroceptor (eks-ter-o-SEP-tor) Sensory nerve terminals stimulated by the external environment.

Extrinsic (eks-TRIN-sik) On the outside; related to some region other than the organ with which it is associated.

F

Facet (FAS-et) L. faci = face; et = small. A small, facelike articular surface.

Facial (FASH-e-al) L. faci = the face.

Falciform (FAL-si-form) L. falci = a sickle; form = shape.

Femur (FE-mur) L. fem = thigh.

Fenestra (fe-NES-trah) L. fenestr = window.

Fetus (FE-tus) L. Postembryonic unborn. From the seventh or eighth week of gestation.

Fibula (FIB-u-lah) L. fibul = buckle. A leg bone.

Flexion (FLEK-shun) L. flect = bend.

Foramen (fo-RA-men) L. foram = an opening.

Fossa (FOS-a) L. foss = depression or trench.

Fovea (FO-ve-a) L. fovea = a pit.

Frontal (FRUN-tal) L. frons = brow, foliage.

G

Genu (JE-nyoo) L. Knee.

Ganglion (GANG-gle-on) G. gangli = swelling. A group of nerve cell bodies.

Genome (JE-nom) The haploid complement of hereditary factors.

Gluteal (GLOO-te-al) G. The rump.

Gonad (GO-nad) An ovary or testis.

H

Hallux (HAL-uks) L. Digit number one of the foot.

Hamulus (HAM-u-lus) L. hamul = little hook.

Haploid (HAP-loyd) A cell such as a gamete with half of the somatic chromosome number (23 in humans).

Hematopoiesis (hem-a-to-poy-E-sis) G. hema = blood; poie = produce. Blood forming.

Hepatic (he-PAT-ik) G. hepar = the liver.

Histology (his-TOL-o-je) G. hist = web; ology = study [of tissues].

Homeostasis (ho-me-o-STA-sis) G. The maintenance of body metabolism within set boundaries.

Hormone (HOR-mon) G. A chemical messenger secreted by a gland or tissue.

Humerus (HU-mer-us) L. humer = shoulder.

Hyoid (HI-oyd) G. Y-shaped.

I

Iliac (IL-e-ak) L. ilia = flank, loin.

Implantation (im-plan-TA-shun) L. im = into; plantare = to set. Attachment of the blastocyst to the endometrium.

Interosseus (in-ter-OS-e-us) L. inter = between; os = bones.

Interstitial (in-ter-STISH-al) L. inter = between; sistere = to set. Between cells of a tissue, as in the interstitial cells of the testis that produce testosterone.

Intrinsic (in-TRIN-sik) L. intrinsecus = found on the inside.

Incisive (in-SI-siv) L. Cut into.

Infra- (IN-fra) L. Below.

Inter- (IN-ter) L. Between.

Ischia (IS-ke-a) G. Hip.

J

Jejunum (je-JOO-num) L. jejun = hunger, dry.

Jugal (JOO-gal) L. jugum = yoke.

Jugular (JUG-u-lar) L. jugulum = the neck.

K

Keratin (KER-ah-tin) G. kera = horn.

Kinetic (ki-NET-ik) G. kinesis = movement.

L

Labialis (la-be-AL-is) L. labi = a lip.

Lacerum (las-ER-um) L. lacer = torn.

Lacrimal (LAK-ri-mal) L. lachr = tears.

Lacteals (LAK-te-als) L. lact = milk.

Lambdoidal (lam-DOYD-al) G. lambda = like the Greek letter lambda.

Latissimus (la-TIS-i-mus) L. later = side.

Lien (LI-en) L. The spleen.

Ligament (LIG-ah-ment) L. liga = bound or tied.

Lingual (LING-gwal) L. lingua = the tongue. Pertaining to the tongue.

Lumbosacral (lum-bo-SA-kral) L. lumb = the loin; sacrum = sacrum.

Lympho- (LIM-fo) L. lymph = water.

M

Malar (MA-lar) L. mala = the jaw, cheek.

Malleus (MAL-e-us) L. A hammer.

Mammalian (mah-MA-le-an) L. mamm = a teat. Pertaining to animals feeding their young with milk from a breast.

Mandible (MAN-di-bl) L. A jaw.

Manubrium (ma-NYOO-bre-um) L. A handle.

Masseter (mas-SE-ter) G. A chewer.

Mastication (mas-ti-KA-shun) G. Chewing.

Mastoid (MAS-toyd) G. mastos = breast; eidos = shape. A bony process of the skull.

Maternal (mah-TER-nal) L. Pertaining to the mother.

Maxilla (mak-SIL-ah) L. Jawbone.

Meatus (me-A-tus) L. A passageway.

Mediastinum (me-de-as-TI-num) L. media = middle; stin = membrane. The membranes enclosing the tissues and organs between the two lungs.
Medulla (me-DUL-a) L. Marrow.
Meiotic (mi-OT-ik) G. meio = less. Cell division that reduces the number of chromosomes from diploid to haploid.
Meissner's (MIS-ners) **corpuscles** [Meissner, German physiologist, 1829–1901] L. corpusc = a small body. Tactile sensory receptors in the superficial dermis.
Meninges (me-NIN-jez) G. Membranes covering the brain.
Menisci (men-IS-si) G. meniskos = crescent. A crescent-shaped fibrocartilage.
Mesentery (MES-en-ter-e) G. A membrane [usually double] attaching the viscera to the internal body wall.
Mesoderm (MES-o-derm) G. The middle primary germ layer of the embryo. This layer gives rise to bone, cartilage, blood, and other connective tissues as well as muscle, blood vessels, lymphatics, and the pleural, pericardial, and peritoneal membranes, kidneys, and gonads.
Molar (MO-lar) L. molaris = a grinder.
Muscle (MUS-el) L. musculus = a contractile organ.

N

Nephric (NEF-rik) G. nephros = kidney.
Natal (NA-tal) Pertaining to birth.
Navicular (na-VIK-u-lar) L. navicul = a little ship. A carpal bone shaped like a boat.
Neurotransmitter (nyoo-ro-TRANS-mit-er) G. neuro = nerve; trans = across. Hormones that bridge the synapse or myoneural junction.
Nuchal (NYOO-kal) **ligament** L. The back of the head or neck; ligare = to bind.

O

Obturator (OB-tyoo-ra-tor) L. A disk covering an opening.
Olecranon (o-LEK-ra-non) G. olekranon = elbow.
Omentum (o-MEN-tum) L. omentum = fat skin. A folded peritoneal membrane from the stomach to the dorsal body wall.
Oogenesis (o-o-JEN-eh-sis) The ovarian process that produces ova.
Optic (OP-tik) G. The eye.
Orbito- (OR-bi-to) L. Circle, track.
Organelle (or-gan-EL) Subcellular structures, including the mitochondria, lysosomes, and endoplasmic reticulum.
Ossicle (OS-si-kl) L. A little bone.

P

Pacinian (pa-SIN-e-an) **corpuscles** [Filippo Pacini, Italian anatomist, 1812–1883] L. corpusc = a small body. Pressure receptors found deep in the dermis and in visceral organs, especially the liver.
Palatal (PAL-a-tal) L. palatum = the palate [the roof of the mouth].
Palpebrae (PAL-peh-bre) L. Eyelid.
Pampiniform (pam-PIN-i-form) L. pampin = a tendril. A threadlike plexus.
Pancreas (PAN-kre-as) G. pan = all; kreas = flesh. A large gland near the stomach and duodenum.
Papilla (pa-PIL-a) L. Pimple.
Para- (PAR-ah) G. Beside or near.
Parietal (pa-RI-e-tal) L. pariet = wall. Bone of the cranium or coelomic membrane, as parietal peritoneum.
Parotid (pa-ROT-id) G. para = near; -otis = ear. The parotid gland at the base of the ear.
Parturition (par-tyoo-RISH-un) L. parturitio = birth.
Patella (pa-TEL-a) L. patera = plate. Knee bone.
Peduncle (pe-DUNG-kl) G. ped = foot; unculus = diminutive, thus a small support.
Peristalsis (per-ih-STAL-sis) G. peri = around; stalsis = contraction. Movement provided by the smooth muscles to the contents of the digestive tract, ureters, urethra, uterine duct, and vas deferens.
Pes (pes) L. A foot.
Petrous (pe-TROS) G. petro = a rock.
Pineal (PIN-e-al) L. pinea = a pine cone.
Pituitary (pi-TU-i-tar-e) L. pituitar = mucus secretion. An endocrine gland at the base of the brain.
Placenta (pla-SEN-ta) L. A flat cake. A vascular structure through which the developing embryo and fetus derive nourishment and dispose of waste in conjunction with the maternal endometrium.
Plantar (PLAN-tar) L. The sole of the foot.
Plantigrade (PLANT-i-grad) L. plantae = the sole of the foot; grade = walk.
Plasma (PLAZ-ma) G. plasma = form. An undifferentiated tissue from which another tissue or tissues may form.
Plexus (PLEK-sus) L. A network. G. Percussion.
Pollex (POL-eks) L. The thumb.
Popliteal (pop-LIT-e-al) L. poples = the back of the knee.
Portal (POR-tal) L. porta = gate. A pathway or entrance.
Process (PROS-es) L. To project from.
Prostate (PROS-tat) G. "One who stands before." The prostate gland is "before" the bladder.
Protract (pro-TRAKT) A muscle action that draws an element anteriorly.
Proximal (PROX-si-mal) L. Nearest.
Psoas (SO-as) G. The loin.
Pterygoid (TER-i-goyd) G. pterygion = wing; oid = like.
Pubis (PYOO-bis) L. The hair that appears at puberty.
Pudendo- (pyoo-DEN-do) L. pudenda = to be ashamed. External female genitalia.
Pulmonary (PUL-mo-ner-e) L. pulmo = lung. Pertaining to the lungs.

Q

Quadriceps (KWOD-ri-ceps) L. quadr = four; caput = head.
Quadrupeds (KWOD-roo-peds) L. quadr = four; ped = foot.

R

Radial (RA-de-al) L. A spoke or ray.
Renal (RE-nal) L. ren = kidney.
Repolarization (re-po-lar-ih-ZA-shun) Return of a neuron to a resting potential.
Retroperitoneal (re-tro-per-i-to-NE-al) External to the peritoneum.

S

Sacrum (SA-krum) L. Sacred.
Saphenous (sah-FE-nus) G. saph = clear, apparent. A large superficial vein of the leg.
Scaphoid (SKAF-oyd) G. scaph = a boat or a bowl.
Scapula (SKAP-u-la) L. The shoulder blade.
Sciatic (si-AT-ik) L. The hip.
Serous (SE-rus) L. Pertaining to serum.

Sesamoid (SES-ah-moid) L. Resembling a sesame grain. A small bone formed in tendons.

Simple (SIM-pl) Without complexity. In the case of a tissue, without layers.

Soleus (SO-le-us) L. solea = sandle. The sole of the foot.

Somatic (so-MAT-ik) G. Either pertaining to body cells (as opposed to gametes) or to the body wall (as opposed to the viscera).

Spermatogenesis (sper-ma-to-JEN-e-sis) G. sperm = seed, the male gamete; genesis = origin. The testicular process that produces sperm.

Sphenoid (SFE-noyd) G. sphen = wedge. The sphenoid bone.

Sphincter (SFINK-ter) G. To squeeze. A ring or rings of smooth muscle that constrict a passage.

Spine (spin) L. A thorn. A sharp projection from the body or main part of a bone.

Squamous (SKWA-mus) L. A scale. Flat or scalelike.

Stapes (STA-pez) L. Stirrup.

Stratified (STRAT-ih-fid) L. stratum = a bed cover, blanket. In layers.

Styloid (STI-loyd) G. stylos = pillar or stalk.

Sulcus (SUL-kus) L. A furrow or groove.

Suture (SOO-chur) L. sutur = a seam. An immovable joint.

Synergist (SIN-er-jist) G. syzyg = yoked together. Acting together.

Systemic (sis-TEM-ik) G. Pertaining to, or acting on, the body as a whole.

T

Tarsal (TAR-sl) G. tarsus = the ankle.

Tentorium (ten-TO-re-um) L. tendere = to stretch or extend. An extention of the parietal bone between the cerebral and cerebellar hemispheres; actually an ossification of the meninges.

Thyroid (THI-royd) G. thyreos = a shield. An endocrine gland near the shield-shaped cartilage of the larynx.

Tibia (TIB-e-a) L. The shin.

Trachea (TRA-ke-a) L. The windpipe.

Trapezium (tra-PE-ze-um) G. trapez = a table.

Trigeminal (tri-JEM-i-nal) L. tri = three; geminus = twin.

Triquetrum (tri-KWE-trum) L. triquetr = triangular. A carpal bone.

Trochanter (tro-KAN-ter) G. trochanter = runner. Two processes on the proximal femur.

Trochlea (TROK-le-a) L. trochlea = a pulley.

Tuberculum (too-BUR-qu-lum) L. A little knob. Process on some bones.

Tuberosity (too-be-ROS-i-te) L. tuber = swollen. A normal raised area on a bone, such as the deltoid tuberosity, which is the site of insertion of the deltoid muscle.

Turbinate (TUR-bi-nat) L. turba = a crowd. Folded parts of the ethmoid bone.

U

Ulna (UL-na) L. Elbow.

Ultimobranchial (ul-ti-ma-BRANG-ke-al) L. ultim = the last; branch = a gill. The last embryonic gill slit [often a pouch].

Urethra (u-RE-thra) G. ureo = urine. A duct for urine.

Uterus (U-ter-us) L. The womb.

Utriculus (u-TRIK-u-lus) L. A bottle.

Vas (vas) L. Duct or vessel.

Ventral (VEN-tral) L. venter = the belly or towards the belly. In human anatomy, anterior.

Ventriculi (ven-TRIK-u-li) L. ventricul = belly.

Venule (VEN-yool) L. ven = vein; ule = small. Small vessels that join to form veins.

Vermis (VER-mis) L. verm = worm.

Vertebrata (ver-te-BRA-ta) L. Having vertebrae.

Villi (VIL-i) L. villus = hair.

Visceral (VIS-er-al) L. viscus = organs of the thoracic, abdominal, and pelvic cavities.

Vomer (VO-mer) L. Plough.

W X

Xiphoid (ZIF-oyd) G. xiphos = sword.

Y Z

Zygoma (zi-GO-ma) G. zygon = yoke.

Index

Abdomen, xi, 11, 78, 121
Abdominal, 78, 80, 86, 90, 91, 96, 113, 116, 117, 119, 120, 121, 125, 127, 128, 133, 147–49
Abdominis, 60, 70, 76, 78–80, 87
Abducens, 27, 28, 142, 143, 153, 169, 175
Abduct, 45, 57, 59, 66–68, 81–83, 86
Abductor cruris, 83
Abductor digiti, 68, 69
Accessory, 23, 27, 34, 37, 55, 67, 68, 92, 108, 110, 140, 143, 166, 169
Accommodation, 177, 178
Acetabular, 39, 45
Acetabulum, 39, 41, 45, 86
Acinar, 92, 102
Acoustic, 15, 19–22, 27, 28, 142, 156
Acromialis, 56, 57, 63
Acromion, 34, 35, 37, 38, 57
Actin, 4
Adduct, 45, 57–59, 67, 68, 81, 86
Adductor, 48, 58–60, 65, 67, 68, 70, 80, 84–87, 119, 121
Adductor brevis, 70, 87
Adductor femoris, 48, 60, 80, 85, 119
Adductor longus, 70, 86, 87
Adductor magnus, 70, 86, 87
Adenohypophysis, 141, 144, 157
Adhesio interthalamica, 143–45, 169, 170
Adipose, 12, 111
Adrenal, 117, 120, 125, 133, 136, 156, 158, 160
Adrenocorticotropic, 158
Adrenolumbar, 117, 125, 126
Adventitia, 110
Afferent, 134, 139
Ala, 23
Alba, 60, 78
Aldosterone, 160, 161
Alisphenoid, 27
Allantoin, 132
Alveolar, 12, 27, 108, 109,
Alveoli, 109, 110
Alveolus, 109
Amnion, 137
Ampulla, 96, 101
Amygdaloid, 139, 168, 170, 171
Amygdaloideum, 171
Anal, 11
Anastomose, 116, 120, 128
Anastomosing, 53, 115, 120, 129, 161
Anconeus, 56, 64, 65, 80
Angular, 15, 19, 162, 164
Ansiform, 166–68
Antebrachialis, 61, 64, 69
Antebrachii, 64
Antebrachium, 10, 57, 64, 66
Anterior chamber, 151–54, 175, 176
Antrum, 137
Anus, 90, 101, 135
Aorta, 91, 113, 115–17, 119–21, 131, 136, 148, 149, 163, 164, 165
Aortic, 90, 95, 112–14, 116, 148, 162, 164, 165
Apex, 113, 133, 161, 162, 164, 165
Apical, 164
Aponeurosis, 67, 78, 86
Appendage, 10, 34, 39, 58, 81, 127, 157
Appendicular, 14, 34
Appendix, ix, 139, 161, 166, 172
Aqueduct of Sylvius, 145, 169, 170
Aqueous, 154, 175, 176
Arachnoid, 139, 167
Arbor vitae, 143–45
Arch, 16, 23, 24, 28, 51, 53, 112–14, 116, 148, 149, 162, 164, 165
Archicortex, 170, 171
Archinephric, 132, 136, 137
Archinephric duct, 136, 137
Archistriatum, 171
Arcuate artery, 133
Arrector pili, 11, 12
Arteries, 27, 110, 113–17, 119–21, 124, 127, 131, 134, 148, 149, 164, 165
Arteriosus, 162, 164
Artery
- afferent, 134, 139
- aorta, 91, 113, 115–17, 119–21, 131, 136, 148, 149, 163, 164, 165
- aortic, 90, 95, 112–14, 116, 148, 162, 164, 165
- arcuate, 133
- arteriosus, 162, 164
- axillary, 114, 116, 124
- basilar, 116
- brachial, 116, 124, 125
- carotid, 20, 112, 114, 115, 123, 148
- caudal mesenteric, 117, 120, 149
- celiac, 116–18, 120
- circumflex, 119, 121, 124, 127
- conus, 162, 164
- conus arteriosus, 162, 164
- coronary, 112, 113, 162, 161–64
- cranial mesenteric, 117, 118, 120, 129, 149
- dorsal aorta, 113, 116, 121, 131
- epigastric, 119, 121, 127
- external carotid, 115
- femoral, 86, 87, 117, 119, 121
- gastric artery, 116, 118, 120
- gastrosplenic, 96, 128, 129
- genus descendens, 121, 127
- gonadal, 119
- hepatic, 101, 118, 120
- iliac, 117, 119–21, 136
- iliolumbar, 117, 119, 121, 136
- interlobar, 133
- internal carotid, 20
- pancreaticoduodenal, 120, 128, 129
- pancreaticoduodenalis, 118, 128
- phrenic, 117
- pudendal, 119, 121
- pulmonary, 112, 113, 163–65
- renal, 117, 133, 134
- saphenous, 60, 119
- subclavian, 90, 107, 113–17, 148
- thyrocervical trunk, 115, 116
- thyroid, 114, 115
- umbilical, 119

Arthroidal, 40
Articular, 20, 28, 33, 34, 39, 45, 76
Articularis, 34
Articulate, 20, 23, 24, 34, 39
Articulation, 20, 24, 28, 34, 39, 90
Arytenoid, 105
Aspera, 44, 86, 88
Atlas, 20, 28, 31, 40, 53, 58, 71, 116, 131, 145, 150
Atria, 112, 113, 161, 162, 164, 165
Atrial, 165
Atrioventricular, 112, 113, 164, 165
Atrium, 112–14, 125, 161, 162, 164, 165
Auditory, 20, 27, 28, 95, 115, 141–43, 155, 156, 170
Aural, 141

Auricle, 112, 161, 162, 164
Auricular, 115, 123
Autonomic, 138, 146–48, 160
Axial, 14, 58
Axilla, 11, 116, 124, 125
Axillary artery, 114, 116, 124
Axillary nerve, 146
Axillary vein, 122, 125
Axis, 28, 31, 40, 45, 49, 71, 77, 118, 173, 175
Axon, 8
Axons, 157, 177
Azygous vein, 127

Baculum, 135
Basal, 3, 4, 12, 170, 171
Basale, 11
Basihyal, 24, 25, 51, 52, 96
Basilar, 116, 155, 156
Basilar artery, 116
Basilar membrane, 155, 156
Basilic vein, 125
Basioccipital, 19, 20, 28, 58, 71
Basisphenoid, 19, 20, 22, 23, 27, 28, 51
Biceps, 51, 58, 61, 64, 69, 70, 79–84, 116, 125, 145
Biceps brachii, 61, 64
Biceps femoris, 51, 79–84
Bicipital, 58, 59
Bicuspid, 112, 113, 163–65
Bile, 96, 101, 102, 118
Bile duct, 99, 101, 102, 118
Bilobed, 101
Biventer, 73, 74
Bladder, 4, 87, 91, 96, 117, 121, 129, 132, 133
Blood vessel
- anastomose, 116, 120, 128
- anastomosing, 53, 115, 120, 129, 161
- antrum, 137
- corpus cavernosum, 135
- glomerulus, 133, 134
- hepatic portal, 101, 118, 120
- microvascular, 110
- pampiniform, 135
- portal, 128
- rete, 115, 124, 135
- sinusoid, 101
- sinusoids, 128
- vasa, 133, 135, 136
- vasa recta, 133, 135, 136
- vascularized, 95, 158
- vasculature, 6, 40

Bone, 18, 23
- alisphenoid, 27
- angular, 15, 19, 162, 164
- atlas, 20, 28, 31, 40, 53, 58, 71, 116, 131, 145, 150
- axis, 28, 31, 40, 45, 49, 71, 77, 118, 173, 175
- baculum, 135
- basihyal, 24, 25, 51, 52, 96
- basioccipital, 19, 20, 28, 58, 71
- basisphenoid, 19, 20, 22, 23, 27, 28, 51
- calcaneus, 39–43, 88
- canine, xiv, 16, 19, 23, 26, 27, 51, 93
- capitate, 34, 67
- carnassial, 47
- carpale, 34, 39
- carpals, 10, 14, 29, 30, 34–40, 43, 44, 67, 68
- centrale, 39
- ceratohyal, 24, 25, 52
- cervical, 14, 16, 28–31, 43, 44, 57–59, 71, 73, 76, 114, 115, 116, 123, 145, 146, 148
- clavicle, 14, 30, 34–38, 44, 52, 53, 57
- clavicula, 34
- clavicular, 57
- coracoid, 34, 35, 37, 38, 43, 58
- cuboideum, 40
- cuneiform, 40, 41, 89
- dermal, 10, 12, 13, 14, 23, 24
- ectotympanic, 20
- endotympanic, 20
- epihyal, 25
- ethmoid, 18, 19, 21, 24, 27, 139, 168
- exoccipital, 19, 20, 71
- femur, 10, 14, 29, 30, 39, 41, 42, 45, 46, 81–83, 86, 88, 89
- fibula, 10, 29, 30, 39, 41, 42, 46, 88, 89
- frontal, xii, 15–18, 21, 22, 24, 30–32, 35–37, 41, 43, 49, 51, 62, 144, 167
- hamate, 34, 68
- humerus, 10, 14, 29, 30, 34, 40, 43, 44, 57–59, 64, 66, 67
- hyoid, 13, 14, 20, 24, 25, 29, 49, 51, 52, 106
- ilium, 30, 39, 41–44, 48, 58, 76, 78, 81–83, 86
- incisive, 16–19, 21–24, 27, 95
- incus, 20, 24, 155, 156
- interparietal, 15–17, 20
- ischium, 30, 39, 41, 42, 81, 83, 86
- jugal, 24
- lacrimals, 23
- lunate, 34
- malleus, 20, 24, 155, 156
- mandible, 15, 16, 18, 20, 21, 24, 27, 30, 49, 51–53, 92
- manubrium, 14, 29, 33, 35, 38, 43, 52, 53, 59, 156
- maxilla, 15, 16, 18, 19, 21–24, 27, 30, 92
- maxillae, 23
- mediale, 69, 88
- metacarpal, 10, 14, 29, 30, 35, 37–39, 43, 44, 67, 68
- metatarsal, 41, 88, 89
- metatarsals, 10, 29, 30, 40–42
- molar, 24, 26, 51, 92
- navicular, 39–41
- occipital, xii, 15–17, 19, 20, 22, 27, 28, 30, 44, 53, 57, 73, 76, 115, 124, 139, 167, 168
- orbitosphenoids, 23
- os coxae, 14, 29, 39
- ossicle, 155, 156
- osteocyte, 6
- palatine, 19, 21, 23, 24, 27, 51, 92, 95
- parietals, 16
- patella, 29, 30, 39, 41–43, 47, 81–83, 88
- petromastoid, 20, 27
- petrosa, 20–22, 28
- petrosal, 20, 23, 28, 156
- petrous, 20, 23
- phalange, 10, 30, 33, 35, 37–44, 67
- phalanx, 66–68, 88, 89
- pisiform, 34, 39, 67, 68
- premolars, 16, 23, 24, 26, 51, 92, 93
- presphenoid, 19, 20, 22–24, 27
- pterygoid, 16, 19, 22, 23, 51, 53, 95
- pubic, 39, 121, 127
- radiale, 34, 39
- radius, 10, 14, 29, 30, 34–40, 43, 44, 64, 66, 67
- sacrum, 14, 29, 30, 33, 39, 43, 44
- scaphoid, 34
- scapula, 29, 30, 34–38, 44, 55, 57–59, 64
- scapulae, 34, 55, 56, 58, 59, 125
- sesamoid, 34, 39
- squamosal, 15–18, 22, 24, 51
- stapes, 20, 24, 155, 156
- sternal, 77, 90
- sternebrae, 14, 29, 33, 59
- sternebrium, 33
- sternum, 14, 28, 29, 33, 34, 36, 43, 47, 53, 59, 76–78, 90, 91, 121
- suprascapular, 145
- talus, 39–41
- tarsal, 10, 29, 30, 39–42, 88
- temporal, 15, 16, 18, 20, 22–24, 27, 30, 51, 76, 115, 144, 156, 167, 168
- thyrohyal, 24, 25
- tibia, 10, 29, 30, 39–42, 46, 81–83, 86, 88, 121
- trapezium, 34, 67
- trapezoid, 34, 140, 143, 169
- triquetrum, 34
- turbinate, 21, 22, 24
- tympanohyal, 20, 24, 25, 52
- ulna, 10, 14, 29, 30, 34–40, 43–45, 57, 64, 66, 67
- ulnare, 34, 37, 39, 45
- zygoma, 15, 24, 30
- zygomatic, 16–19, 22–24, 51, 53

Brachial, 64, 116, 122, 124, 125, 145, 146, 148
Brachial artery, 116, 124, 125
Brachialis, 48, 51, 56, 58, 63, 64, 145
Brachii, 58, 61, 63, 64, 80
Brachiocephalic, 107, 112–16, 121–23, 126, 162, 165
Brachiocephalicus, 55, 57, 58, 61, 69
Brachioradialis, 63, 66, 69, 70, 80
Brachium, 10, 55–59, 63–65, 116, 124

Brain
- adhesio interthalamica, 143–45, 169, 170
- amygdaloid, 139, 168, 170, 171
- amygdaloideum, 171
- aqueduct of Sylvius, 145, 169, 170
- arbor vitae, 143–45
- archicortex, 170, 171
- archistriatum, 171
- callosum, 141–44, 168–71
- cerebelli, 139, 167
- cerebellum, 16, 139–41, 143–46, 150, 166–70
- cerebral, 16, 20, 115, 116, 139–42, 144, 151, 166–71
- cerebral hemisphere, 140
- cerebri, 139, 146, 167
- cerebrospinal fluid, 139, 158
- cerebrum, xiii, 139, 141–43, 166–68, 170
- cinereum, 141, 142, 168, 169
- commissure, 143–45, 158, 169, 170
- corpora quadrigemina, 141, 142, 168–70
- corpus callosum, 141, 143, 144, 169, 170
- crus, 139
- diencephalon, 139, 157, 158, 168, 171
- dura mater, 16, 139, 166, 167
- eminence, 39, 143, 144, 149
- falx cerebri, 139, 167
- flocculus, 167
- fornix, 142–45, 169, 170
- geniculate, 168, 171
- globus pallidus, 170
- gyri, 141, 167, 168, 170
- gyrus, 140, 167–69
- hemisphere, 140, 167
- hippocampus, 139, 142, 168, 170, 171
- hypothalamus, 139, 141–45, 157, 158, 170
- infundibulum, 141–44, 157, 158, 168–70
- insula, 167, 168
- interthalamica, 143–45, 169, 170
- mammillary body, 142–44, 169, 170
- median eminence, 143, 144, 149
- medulla oblongata, 139, 140, 142–44, 166, 168–70
- meningeal, 139, 170
- meninges, 139, 145, 166
- mesencephali, 141
- mesencephalic, 145, 170
- mesencephalon, 138, 139
- metencephalon, 139
- munro, 145, 170
- myelencephalon, 139
- neopallium, 139, 171
- neurohypophysis, 141, 144, 157
- optic lobe, 158
- pallium, 170
- paraflocculus, 28
- paramedian, 167, 168
- paraventricular, 157
- parietal lobe, 167, 168
- peduncle, 141–43, 167, 169
- pellucidum, 142–44, 169, 170
- pons, 139–41, 143, 144, 168, 169, 171
- prosencephalon, 138, 139
- putamen, 170, 171
- rhinencephalon, 139, 171
- rhombencephalon, 138, 139
- supraoptic, 157
- tectum, 141–43, 158
- tegmen, 170
- tela choroidea, 141, 143, 170
- telencephalon, 139, 168, 171
- tentorium, 16, 21, 22, 139, 167
- thalamic, 171
- tuber cinereum, 141, 142, 168, 169
- varoli, 144, 168, 169

Brevis, 66–70, 84, 86, 87, 89
Bronchi, 108, 110, 131, 149
Bronchiole, 109
Bronchioles, 110
Bronchopulmonary, 110
Bronchus, 108, 109
Bruke's muscle, 178
Buccal, 115
Bulbi, 142, 152, 153, 173, 175
Bulbocavernosum, 135
Bulbourethral, 135, 136
Bulbus, 168
Bulla, 16, 19, 20, 24, 27, 28, 155, 156
Bullae, 28, 115

Caecum, 22, 96, 101, 102
Calcaneus, 39–43, 88
Calcitonin, 159
Callosum, 141–44, 168–71
Canaliculi, 101
Canine, 16, 19, 26, 27, 51
Canines, xiv, 23, 93
Canthus, 172
Capitate, 34, 67
Capitis, 57, 58, 71, 73–76, 80
Capitulum, 34, 43
Capsule, 12, 45, 64, 83, 131–34, 171
Caput, 28, 34, 136
Cardiac, 4, 7, 48, 95, 161, 164
Carnae, 112, 162
Carnassial, 47
Carneae, 113, 164
Carnivora, xiii, xiv
Carnivores, xiv, 131
Carotid artery, 20, 112, 114, 115, 123, 148
Carpal, 10, 14, 29, 30, 34–40, 43, 44, 67, 68
Carpale, 34, 39
Carpi, 45, 51, 61, 63, 66, 67, 69, 79
Carpi radialis, 51, 61, 63, 66, 67, 69, 79
Cartilage, 6, 13, 14, 20, 23–25, 33, 35, 37, 38, 40, 45, 52, 77, 78, 90, 106, 109, 110
- ala, 23
- arytenoid, 105
- costal, 28, 52, 77, 78
- cricoid, 25, 106
- epiglottis, 25
- glottis, 52, 95, 96, 106
- hyoid, 13, 14, 20, 24, 25, 29, 49, 51, 52, 106
- menisci, 46
- meniscus, 46, 47, 88
- thyroid, 24, 25, 52, 106
- tracheal, 25

Cartilago, 33
Cauda equina, 145
Caudal mesenteric artery, 117, 120, 149
Caudal vein, 126, 127
Caudate, 101, 170, 171
Cavernosum, 135
Cavum, 141, 144
Cecum, 129
Celiac, 116–18, 120, 149
Celiac artery, 116–18, 120
Cell, 3, 4, 8, 96, 139, 145, 148, 149, 151, 170, 177
Cenozoic, xiii
Central canal, 139, 143
Centrale, 39
Centralis, 173
Centripetal, 135
Centroacinar, 91, 92
Centrum, 28, 33, 71
Ceratohyal, 24, 25, 52
Ceratohyoideus, 52
Cerebelli, 139, 167
Cerebellum, 16, 139–41, 143–46, 150, 166–70
Cerebral, 16, 20, 115, 116, 139–42, 144, 151, 166–71
Cerebral hemisphere, 140
Cerebri, 139, 146, 167
Cerebrospinal fluid, 139, 158
Cerebrum, xiii, 139, 141–43, 166–68, 170
Cervical, 14, 16, 28–31, 43, 44, 57–59, 71, 73, 76, 114–16, 123, 145, 146, 148
Cervicalis, 55–57, 79
Cervicis, 59, 73–75
Cervicothoracic, 148
Cervix, 137
Chamber, 23, 24, 45, 91, 95, 106, 115, 151–54, 164, 173, 175, 176, 177
Chiasma, 141, 143, 144, 168, 169
Choanae, 19, 24, 95, 106
Cholecystokinin, 160
Choledochus, 101, 102
Chordae tendineae, 113
Chordata, 139
Chorion, 137
Choroid, 149, 151–54, 175–77
Choroidea, 141, 143, 170
Choroid layer, 177
Chromaffin, 160, 161
Cilia, 3, 4, 172
Ciliary, 152–54, 173, 175–78
Ciliary body, 152–54, 173, 175, 177, 178
Cinereum, 141, 142, 168, 169
Circumduct, 45
Circumferentia, 34
Circumflex, 119, 121, 124, 127

Circumvallate, 95
Cisterna chyli, 131
Clavicle, 14, 30, 35–38, 44, 52, 53, 57
Clavicula, 34
Clavicular, 57
Cleidobrachialis, 34, 54–57, 63
Cleidocephalicus, 51, 54, 55, 57, 72, 79, 80
Cleidomastoideus, 34, 53, 54, 72, 80
Cleidotrapezius, 34
Clinoid process, 22
Clitoris, 121, 127, 137
Cloaca, 132
Coccygeal, 148
Coccyx, 30, 33, 34, 43
Cochlea, 155, 156
Coelomic, 91, 105, 137, 161
Colic, 118, 120, 129
Collecting tubule, 133, 134
Colli, 71, 72
Colliculi, 139, 141–43, 168, 170
Colliculus, 144, 169
Colon, 96, 97, 101, 102, 120, 129
Columnar, 4, 5, 96
Commissure, 143–45, 158, 169, 170
Communicantes, 148
Complexus, 73, 74
Condyle, 15, 16, 19, 20, 27, 28, 39, 44, 46, 53, 71, 88
Condyloid, 15, 19, 21, 22, 24, 27, 39
Conjunctiva, 151, 154, 172
Conus, 162, 164
Conus arteriosus, 162, 164
Convoluted, 132–34, 141, 168
Coracobrachialis, 58, 61, 69, 145
Coracoid, 34, 35, 37, 38, 43, 58
Cornea, 151–54, 173–77
Corneum, 11, 12
Cornu, 137
Coronary artery, 112, 113, 162, 161–64
Coronoid, 15, 24, 51
Corpora, 135, 141, 142, 168–70
Corpora quadrigemina, 141, 142, 168–70
Corpus, 135, 136, 137, 139, 141–44, 161, 168–71
Corpus callosum, 141, 143, 144, 169, 170
Corpus cavernosum, 135
Corpus luteum, 161
Cortex, 11, 12, 120, 132–34, 141, 151, 158, 160, 168, 170, 171
Cortical, 134, 141, 160, 171
Corticosteroids, 160
Cortisol, 160
Costal, 28, 52, 77, 78
Costarum, 58, 60, 75
Costocervical, 114, 116, 122, 123
Cowper's, 135
Coxae, 14, 29, 39
Cranialis, xi, xii, 58, 59, 79, 80, 84–86
Cranial mesenteric artery, 117, 118, 120, 129, 149
Cranium, xii, 23, 125, 139, 156, 158, 166
Cremaster, 135
Cribriform, 21, 22, 24, 27, 28
Cricoid, 25, 106
Cristae, 3
Cruciate, 46, 47
Crural, 78
Cruris, 83, 84
Crus, 139
Cubital, 124, 125
Cuboidal, 4, 5, 177
Cuboideum, 40
Cuneiform, 40, 41, 89
Curvature, 91, 96, 116, 120, 128, 129, 131
Cusp, 47, 113, 163
Cuspid, 26
Cusps, 113, 164
Cutaneous, 49, 55, 158
Cuticle, 11, 12
Cystic, 101, 129
Cytoplasm, 3, 11

Deferens, 127, 135, 136
Deltoid, 34, 37, 57, 125
Deltoideus, 51, 55–57, 60, 63, 65, 70, 79, 80
Dendrites, 177
Dental, 26
Dentition, xiii, 47
Dermal, 10, 12, 13, 14, 23, 24
Dermis, 4, 11, 12
Diaphragm, 1–3, 77, 78, 90, 91, 107, 116, 125, 131, 142, 145, 147, 149
Diastema, 51
Diastole, 165
Diencephalon, 139, 157, 158, 168, 171
Diffusion, 105
Digastricus, 51, 53, 54, 72, 142
Digit, 10, 39, 40, 66–68, 88, 89
Digital, 84, 85, 127, 168
Digiti, 65–69, 88
Digitigrade, 10
Digiti minimi, 68
Digitorum communis, 51, 61, 63, 66, 68, 69, 79
Digitorum longus, 47, 79, 80, 84, 89
Digitorum profundus, 65, 67–69, 79
Dogfish, 168
Dopamine, 161
Dorsal aorta, 113, 116, 121, 131
Dorsal mesentery, 90, 91, 129
Dorsal root, 145
Duct
  bile, 96, 101, 102, 118
  collecting tubule, 133, 134
  convoluted, 132–34, 141, 168
  cystic, 101, 129
  deferens, 127, 135, 136
  ductules, 101
  endolymphatic, 155, 156
  epididymis, 135–37
  excretory, 9, 132, 136, 137
  fallopian, 91, 137
  fimbriae, 137
  haversian, 6
  henle, 132–34
  hypoglossal canal, 21, 22, 142
  incisive, 95
  incisive canal, 19
  oviduct, 4, 137
  schlemm, 175
  ureter, 132–34, 136
  urethra, 4, 11, 132, 135–37
  vasa efferentia, 135, 136
  vas deferens, 127, 135, 136
Ductules, 101
Duodenal, 129
Duodenum, 91, 95, 96, 101, 102, 118, 120, 128, 129, 159, 160
Dura mater, 16, 139, 166, 167

Ectoderm, 11
Ectomesenchymal, 110
Ectotympanic, 20
Effector, 138, 139, 148
Efferent, 138, 139, 148
Ellipsoidal, 40, 45
Eminence, 39, 143, 144, 149
Endochondral, 14, 24
Endocrine, 9, 95, 102, 139, 141, 156–61, 168
Endolymphatic, 155, 156
Endoplasmic reticulum, 3
Endothelium, 4, 110
Endotympanic, 20
Enterogastrone, 159
Epicondyle, 34, 35, 37, 38, 64, 66, 67, 83, 88, 89
Epidermal, 10–12, 157
Epidermis, 4, 11, 12
Epididymis, 135–37
Epigastric, 119, 121, 127
Epiglottis, 25
Epihyal, 25
Epinephrine, 160, 161
Epiphysis, 158
Epithelial, 4, 96
Epithelium, 24, 95, 101, 109, 115, 139, 142, 168, 177
Equina, 145
Erythropoietin, 161
Esophageal, 95
Esophagus, 4, 56, 71, 72, 90–92, 95, 96, 106, 107, 115, 116, 129
Estrogen, 137
Ethmoid, 18, 19, 21, 24, 27, 139, 168
Eustachian, 20, 27, 115, 156
Eutheria, xiii, xiv
Excretory, 9, 132, 136, 137
Exoccipital, 19, 20, 71
Extensor carpi, 51, 61, 63, 66, 67, 69, 79
Extensor digitorum, 47, 51, 61, 63, 66–69, 79, 84, 89
External carotid artery, 115
External oblique, 51, 60, 70, 74–76, 79, 87
Exteroceptors, 139, 151

Eye, xii, 1, 10, 13, 24, 49, 115, 142, 151–54, 156, 161, 168, 172–78
Eyelid, 10, 151–53, 172, 176

Facet, 28
Facial, 20–23, 27, 28, 49, 92, 115, 123, 142, 143, 156, 169
Facial canal, 21, 22, 27, 28, 142, 156
Facialis, 122
Facial nerve, 20, 27, 28, 92, 156
Facial vein, 123
Falciform, 91, 101
Fallopian, 91, 137
Falx cerebri, 139, 167
Fascia, 49, 51, 53, 57–59, 61, 67, 69, 75–79, 81–83, 85, 88, 127, 135
Faucium, 96
Femoral artery, 86, 87, 117, 119, 121
Femoris, 39, 48, 51, 60, 70, 79–87, 119, 121, 148
Femur, 10, 14, 29, 30, 39, 41, 42, 45, 46, 81–83, 86, 88, 89
Fenestra, 155, 156
Fetus, 121
Fibrous, 6, 12, 13, 40, 109, 151, 172, 174–76
Fibula, 10, 29, 30, 39, 41, 42, 46, 88, 89
Filaments, 4
Filiform, 95, 96
Filum, 145
Fimbriae, 137
Fissure, 16, 18, 19, 22, 23, 27, 28, 91, 110, 141, 142, 166, 167, 168
Flexion, 40, 45, 49, 88
Flexor carpi, 61, 67, 69
Flexor digitorum, 48, 61, 65, 67–69, 79, 80, 88
Flexor muscles, 148
Flexors, 64
Flocculus, 167
Foliate, 95, 96
Follicle, 11, 12, 137, 161
Follicular, 168
Footpad, 12
Foramen, 15–24, 27, 28, 39, 41, 42, 44, 95, 115, 116, 142, 145, 156, 170
  choanae, 19, 24, 95, 106
  facial canal, 21, 22, 27, 28, 142, 156
  infraorbital, 15, 17–19, 22–24
  internal acoustic meatus, 20–22, 28, 142
  intervertebral, 28, 30, 116, 123, 145
  jugular, 19–22, 28, 142
  lacerum, 19, 23, 115
  magnum, 19, 20, 22, 27, 116
  mandibular, 21, 24
  mastoid, 19
  mental, 18
  munro, 145, 170
  obturator, 39, 41, 42, 44
  optic, 16, 18, 21–23, 142
  orbital fissure, 16, 18, 19, 22, 23, 28, 142
  rostral palatine, 19, 95
  rotundum, 19, 22, 142
  spinosum, 11, 19
  stylomastoid, 19, 20, 27
Forefoot, 10
Fornix, 142–45, 169, 170
Fossa, 19, 20, 23, 34, 37–39, 43–45, 51, 57, 58, 83, 86, 115
Fovea, 45, 173
Frenulum, 92
Frontal, xii, 15–18, 21, 22, 24, 30–32, 35–37, 41, 43, 49, 51, 62, 144, 167
FSH, 158
Fundic, 91, 95, 96, 102
Fundic stomach, 96
Fundus, 160
Fungiform papillae, 95, 96

Gallbladder, 4, 129
Ganglia, 148, 149, 155, 171
Ganglion, 145–49, 154, 158, 177
Ganglionated, 148
Gastric, 96, 116, 118, 120, 128, 129, 160, 161
Gastric artery, 116, 118, 120
Gastric vein, 128, 129
Gastrin, 160
Gastrocnemius, 51, 70, 79, 84, 85, 88
Gastroduodenal vein, 118, 120, 128
Gastroepiploic, 118, 120, 128, 129
Gastrosplenic, 96, 128, 129
Gemellus, 85, 86
Geniculate, 168, 171
Genioglossus, 25, 52, 96
Geniohyoideus, 25, 52, 96
Genital, 11, 136
Genitofemoralis, 147
Genu, 143
Genus descendens, 121, 127
Germinal, 131
Gill slits, xiii
Ginglymus, 40
Gland
  acinar, 92, 102
  adenohypophysis, 141, 144, 157
  adrenal, 117, 120, 125, 133, 136, 156, 158, 160
  bulbourethral, 135, 136
  centroacinar, 91, 92
  chromaffin, 160, 161
  cowper's, 135
  digital, 84, 85, 127, 168
  endocrine, 9, 95, 102, 139, 141, 156–61, 168
  goblet, 96, 101
  gonad, 125
  gonads, 161
  harderian, 173
  hypophyseal, 157, 168
  hypophysis, 141, 144, 157, 170
  hypothalamus, 139, 141–45, 157, 158, 170
  infraorbital, 92
  interrenal, 160
  isles of langerhans, 102
  juxtaglomerular, 161
  lacrimal, 151, 172
  leydig's, 161
  liver, 91, 96, 101, 120, 125, 128, 161
  mucous, 91, 92, 173
  neurohypophysis, 141, 144, 157
  pancreas, 91, 96, 101, 102, 116, 118, 120, 128, 129, 156, 159, 161
  parathyroid, 159
  paraventricular, 157
  parietal cell, 96
  parotid, 4, 79, 91, 92
  pars distalis, 141, 143, 144, 157
  pars nervosa, 141, 143, 144
  pineal, 141, 144, 158, 170, 171
  pituitary, 23, 28, 136, 139, 141, 143, 144, 156, 157, 168, 169, 170
  preputial, 135
  prostate, 121, 127, 135, 136
  salivary, 49, 51, 91, 92, 102, 116, 160
  sebaceous, 11, 12
  sertoli's, 161
  spigelian, 101
  sublingual, 91, 92
  submaxillary, 91, 92
  sudoriferous, 11, 12
  testes, 11, 120, 135–37, 156, 161
  testicles, 91
  testis, 136, 137, 158
  thymus, 90, 95, 107, 129, 131
  thyroid, 115, 116, 123, 158, 159
  ultimobranchial, 95, 159
Glans, 135, 136
Glenoid, 34, 37, 43, 57, 64
Glial, 4, 8, 138, 177
Globus pallidus, 170
Glomerular, 134
Glomeruli, 120
Glomerulus, 133, 134
Glossopharyngeal, 23, 27, 143, 169
Glottis, 52, 95, 96, 106
Glucagon, 102, 159
Glucocorticoids, 160
Gluteal, 121, 127
Gluteus, 48, 70, 80–86
Goblet, 96, 101
Golgi, 3, 4
Gonad, 125, 161
Gonadal, 119, 120, 125, 127, 158
Gonadal artery, 119
Gonadal vein, 119
Graafian follicle, 137
Gracilis, 60, 70, 86, 87, 121
Granulosa, 137
Granulosum, 11
Groove, 10, 28, 38, 58, 59, 113, 141, 161, 162, 164, 167, 168
Gyri, 141, 167, 168, 170
Gyrus, 140, 167–69

Hallux, 40
Hamate, 34, 68
Hamulus, 51

Harderian, 173
Haversian, 6
Helicotrema, 155
Hemifacet, 28
Hemisphere, 140, 167
Hemopoietic, 129, 161
Henle, 132–34
Hepatic, xiii, 91, 96, 101, 110, 116, 118, 120, 125, 126, 128, 129, 159
Hepatic artery, 101, 118, 120
Hepatic portal vein, 101, 118, 120
Hepatoduodenal, 91
Hilus, 120, 133
Hippocampus, 139, 142, 168, 170, 171
Histology, 1, 12, 110, 132, 134, 156, 175
Hormone
- adrenocorticotropic, 158
- aldosterone, 160, 161
- calcitonin, 159
- cholecystokinin, 160
- corticosteroids, 160
- cortisol, 160
- enterogastrone, 159
- epinephrine, 160, 161
- erythropoietin, 161
- estrogen, 137
- FSH, 158
- gastrin, 160
- glucagon, 102, 159
- glucocorticoids, 160
- insulin, 102, 159
- luteinizing, 136
- melanotropin, 158
- melatonin, 158
- norepinephrine, 160, 161
- oxytocin, 157
- pancreozymin, 160
- PRL, 157
- prolactin, 157
- renin, 161
- secretin, 160
- testosterone, 136
- thyrotropic, 158
- TSH, 158
- vasopressin, 157

Humerus, 10, 14, 29, 30, 34, 40, 43, 44, 57–59, 64, 66, 67
Humor, 154, 175, 176
Hyaenidae, xiv
Hyoglossus, 52, 96
Hyoid, 13, 14, 20, 24, 25, 29, 49, 51, 52, 106
Hyperopia, 178
Hypoglossal, 20–22, 27, 140, 142, 143, 166, 169
Hypoglossal canal, 21, 22, 142
Hypophyseal, 157, 168
Hypophysis, 141, 144, 157, 170
Hypothalamus, 139, 141–45, 157, 158, 170

Ileal, 129
Ileocolic valve, 97, 120
Ileocolic vein, 129
Ileum, 96, 97, 101, 129
Iliac artery, 117, 119–21, 136
Iliac crest, 76, 78, 81, 86, 120
Iliac spine, 39, 42, 81
Iliacus, 70, 86, 87, 120
Iliac vein, 119, 126, 127
Iliocostalis, 48, 74, 76, 80
Iliofemoral, 45
Iliolumbar artery, 117, 119, 121, 136
Iliolumbar vein, 119, 126
Iliopectineal, 39, 86
Iliopsoas, 86, 148
Iliosacral, 39
Ilium, 30, 39, 41–44, 48, 58, 76, 78, 81–83, 86
Illuminator, 1, 2
Incisive, 16–19, 21–24, 27, 95
Incisive canal, 19
Incisive duct, 95
Incisor, 16, 19, 23, 26, 93
Incus, 20, 24, 155, 156
Inferior oblique, 152
Inferior rectus, 152
Infraorbital foramen, 15, 17–19, 22–24
Infraorbital gland, 95
Infraspinatus, 34, 56, 57, 65, 79
Infundibulum, 141–44, 157, 158, 168–70
Inguinal, 119, 130
Insula, 167, 168
Insulin, 102, 159
Integument, 4
Integumentary, 12
Intercondyloid, 39
Intercostal, 77, 116, 117, 121, 123
Intercostales, 76
Intercostalis, 70, 75
Interlobar artery, 133
Interlobar vein, 133
Intermedia, 141, 143, 144, 157, 158, 170
Intermesenteric, 149
Intermesenteric plexus, 149
Internal acoustic meatus, 20–22, 28, 142
Internal carotid artery, 20
Internal oblique, 60, 70, 76
Interossei, 67
Interosseus, 67, 89, 125
Interosseus membrane, 67, 89
Interparietal, 15–17, 20
Interrenal, 160
Interspinalis, 75, 77
Interstitial, 6, 136, 161
Interthalamica, 143–45, 169, 170
Interventricular sulcus, 113
Intervertebral foramina, 28, 30, 116, 123, 145
Intestinal, 96, 120, 128, 131, 159, 168
Intestinal vein, 128
Intestine, 91, 96, 97, 120, 129, 131, 160
Iris, 151–54, 172–78
Ischiadic spine, 39
Ischial, 42, 44
Ischial tuberosity, 42
Ischiocavernosus, 135, 136
Ischiofemoral, 45
Ischium, 30, 39, 41, 42, 81, 83, 86
Isles of langerhans, 102

Jejunal, 118, 120, 129
Jejunum, 96, 102, 129
Joint
- arthroidal, 40
- articular, 20, 28, 33, 34, 39, 45, 76
- articulate, 20, 23, 24, 34, 39
- articulation, 20, 24, 28, 34, 39, 90
- ellipsoidal, 40, 45
- ginglymus, 40
- iliosacral, 39
- mandibular, 149
- menisci, 46
- meniscofemoral, 46
- meniscus, 46, 47, 88
- metacarpophalangeal, 67
- sphenopalatine, 21, 27
- symphysis, 21, 24, 39, 52, 86
- synovial, 40, 45
- temporomandibular, 47

Jugal, 24
Jugular foramen, 19–22, 28, 142
Jugular vein, 20, 27, 53, 122, 124–26, 130
Juxtaglomerular, 161

Keratin, 3
Kidney, 4, 91, 117, 120, 125, 132–34, 136, 137, 160, 161

Labia, 121
Labial, 123
Lacrimal gland, 151, 172
Lacrimals, 23
Lacteals, 97, 131
Lacunae, 6
Lambdoidal, 16, 19, 20, 53, 57, 73, 139
Lamina, 19, 24, 77, 96, 97, 170
Lamina propria, 96, 97
Larynx, 13, 24, 52, 95, 96, 106, 115, 116, 131, 142
Lateral rectus, 142, 152, 153, 174, 175
Latissimus dorsi, 49, 51, 55–58, 60, 64, 79, 145
Lens, 1, 3, 152–54, 173, 175–78
Levator, 55, 56, 58, 59, 152, 153, 173, 175
Levatores, 75
Leydig's, 161
Lienal, 116, 118, 128, 129
Ligament
- alba, 60, 78
- cruciate, 46, 47
- crural, 78
- falciform, 91, 101
- iliofemoral, 45
- ligamentum teres, 45
- linea alba, 60, 78
- meniscofemoral, 46
- nuchal, 16, 73
- pubofemoral, 45

raphe, 59
suspensory, 91, 152–54, 175
zonule fibers, 178
Ligamentum, 45, 83
Ligamentum teres, 45
Linea alba, 60, 78
Lineal, 120, 129
Lingual, 52, 114, 115
Lingualis, 52, 95, 96
Lingualis propria, 52
Linguofacial vein, 122
Liver, 91, 96, 101, 120, 125, 128, 161
Longissimus, 73–77, 79
Longissimus cervicis, 73, 75
Longus capitis, 71, 80
Longus colli, 71, 72
Lucidum, 11, 152, 153
Lumbar, 14, 28–30, 32, 33, 43, 44, 58, 76–78, 86, 121, 126, 127, 145, 147, 148
Lumborum, 86, 87, 120
Lumbosacral plexus, 147, 148
Lumbricales, 67
Lumen, 95, 97, 158
Lunate, 34
Lung, 105, 107, 108, 110, 113, 161, 162
Luteinizing hormone, 136
Lymph, 97, 116, 129, 131
Lymphatic, 129–31
Lymphatics, 92, 129
lacteals, 97, 131
lymph, 97, 116, 129, 131
peyer's, 97
spleen, 91, 96, 116, 118, 120, 129, 131, 161
thymus, 90, 95, 107, 129, 131
Lysosomes, 3, 4

Macula, 4, 173
Malleolus, 41–43
Malleus, 20, 24, 155, 156
Mammalia, xiii
Mammalian, ix, xiii, 4, 11, 12, 24, 34, 40, 91, 101, 102, 109, 132, 133, 154, 161, 172, 173
Mammillary body, 142–44, 169, 170
Mandible, 15, 16, 18, 20, 21, 24, 27, 30, 49, 51–53, 92
Mandibular foramen, 21, 24
Mandibular fossa, 19, 20
Mandibular joint, 149
Manubrium, 14, 29, 33, 35, 38, 43, 52, 53, 59, 156
Manus, 1
Masseter, 51–54, 72, 92, 115
Masseterica, 92
Mastication, 51, 53
Mastoid foramen, 19
Mastoid process, 15, 16, 19, 20, 25, 53, 76, 113
Matrix, 4, 6
Maxilla, 15, 16, 18, 19, 21–24, 27, 30, 92
Maxillae, 23
Maxillary vein, 122
Meatus, 15, 20–22, 27, 28, 106, 142, 155, 156
Mediale, 69, 88
Median eminence, 143, 144, 149
Mediastinal, 108, 111, 161
Mediastinum, 90, 91, 107, 111, 131
Medulla, 11, 12, 27, 116, 120, 132–34, 139–44, 160, 166, 168–70
Medulla oblongata, 139, 140, 142–44, 166, 168–70
Meiotic, 135
Melanophores, 158
Melanotropin, 158
Melatonin, 158
Membrane
arachnoid, 139, 167
basale, 11
basilar membrane, 155, 156
chorion, 137
choroid layer, 177
choroidea, 141, 143, 170
corneum, 11, 12
corpus luteum, 161
dorsal mesentery, 90, 91, 129
dura mater, 16, 139, 166, 167
ectoderm, 11
ectomesenchymal, 110
endothelium, 4, 110
epidermal, 10–12, 157
epidermis, 4, 11, 12
epithelial, 4, 96
epithelium, 24, 95, 101, 109, 115, 139, 142, 168, 177
granulosum, 11
intermesenteric, 149
interosseus, 67, 89
mesenteric, 90, 117, 118, 120, 128, 129, 149
mesentery, 90, 91, 96, 101, 120, 128, 129, 131, 133, 161
mesocolon, 91
mesoduodenum, 91, 101
mesogastrium, 91
mesorchium, 91
mesovarium, 91
omentum, 91, 96, 118, 128, 129, 131
parietal pericardium 91, 111, 161
parietal peritoneum, 90, 91, 133
parietal pleura, 90, 91, 105
pellucidum, 142–44, 169, 170
pericardia, 90, 91, 111, 161
pericardial, 105, 111, 112, 161
pericardium, 91, 107, 111, 112, 161
periosteal, 14
peritoneal, 90
peritoneum, 90, 91, 111, 133
pia mater, 139, 167
pleura, 90, 91, 105, 111
pleural, 90, 91, 105, 161
plica semilunaris, 151, 172
pseudostratified, 4, 5
septum, 111–13, 142–44, 161–64, 169, 170
septum pellucidum, 142–44, 169, 170
serous, *90–92, 96*
submucosa, 96, 97
synovial, 40, 45
tectorial, 155, 156
tela choroidea, 141, 143, 170
tentorium, 16, 21, 22, 139, 167
tympanica, 16, 20
tympanum, 156
Meningeal, 139, 170
Meninges, 139, 145, 166
Menisci, 46
Meniscofemoral, 46
Meniscus, 46, 47, 88
Mental foramen, 18
Mesencephali, 141
Mesencephalic, 145, 170
Mesencephalon, 138, 139
Mesenchyme, 6
Mesenteric, 90, 117, 118, 120, 128, 129, 149
Mesenteries, 90, 91
Mesentery, 90, 91, 96, 101, 120, 128, 129, 131, 133, 161
Mesocolon, 91
Mesoderm, 12
Mesoduodenum, 91, 101
Mesogastrium, 91
Mesorchium, 91
Mesovarium, 91
Metacarpal, 10, 14, 29, 30, 35, 37–39, 43, 44, 66–68
Metacarpophalangeal, 67
Metacromion, 37, 57, 58
Metanephric, 132, 133
Metatarsal, 10, 29, 30, 40–42, 88, 89
Metencephalon, 139
Miacidae, xiv
Microtubules, 4
Microvascular, 110
Microvilli, 4
Mitochondria, 3, 105
Mitosis, 3, 4, 12
Mitral, 113, 164
Molar, 24, 26, 51, 92
Mongooses, xiv
Mons, 12, 156
Morphometric, 109
Mouth, 4, 10, 24, 49, 52, 92, 95, 106
Mucinous, 101
Mucosa, 96, 156, 159
Mucosal, 159
Mucous, 91, 92, 173
Mucus, 92, 95, 96, 101, 106
Multifidus spinae, 77
Munro, 145, 170
Muscle
abdominis, 60, 70, 76, 78–80, 87
abductor cruris, 83
abductor digiti, 68, 69
acromialis, 56, 57, 63

actin, 4
adductor, 48, 58–60, 65, 67, 68, 70, 80, 84–87, 119, 121
adductor brevis, 70, 87
adductor femoris, 48, 60, 80, 85, 119
adductor longus, 70, 86, 87
adductor magnus, 70, 86, 87
anconeus, 56, 64, 65, 80
antebrachialis, 61, 64, 69
antebrachii, 64
appendicular, 14, 34
arrector pili, 11, 12
axial, 14, 58
biceps, 51, 58, 61, 64, 69, 70, 79–84, 116, 125, 145
biceps brachii, 61, 64
biceps femoris, 51, 79–84
biventer, 73, 74
brachialis, 48, 51, 56, 58, 63, 64, 145
brachiocephalic, 107, 112–16, 121–23, 126, 162, 165
brachiocephalicus, 55, 57, 58, 61, 69
brachioradialis, 63, 66, 69, 70, 80
brevis, 66–70, 84, 86, 87, 89
Bruke's muscle, 178
bulbocavernosum, 135
capitis, 57, 58, 71, 73–76, 80
cardiac, 4, 7, 48, 95, 161, 164
carnae, 112, 162
carpi radialis, 51, 61, 63, 66, 67, 69, 79
ceratohyoideus, 52
cervicalis, 55–57, 79
cervicis, 59, 73–75
cervicothoracic, 148
cleidobrachialis, 34, 54–57, 63
cleidocephalicus, 51, 54, 55, 57, 72, 79, 80
cleidomastoideus, 34, 53, 54, 72, 80
cleidotrapezius, 34
complexus, 73, 74
coracobrachialis, 58, 61, 69, 145
costarum, 58, 60, 75
costocervical, 114, 116, 122, 123
cranialis, xi, xii, 58, 59, 79, 80, 84–86
cremaster, 135
deltoid, 34, 37, 57, 125
deltoideus, 51, 55–57, 60, 63, 65, 70, 79, 80
diaphragm, 1–3, 77, 78, 90, 91, 107, 116, 125, 131, 142, 145, 147, 149
digastricus, 51, 53, 54, 72, 142
digiti, 65–69, 88
digiti minimi, 68
digitorum communis, 51, 61, 63, 66, 68, 69, 79
digitorum longus, 47, 79, 80, 84, 89
digitorum profundus, 65, 67–69, 79
effector, 138, 139, 148
extensor carpi, 51, 61, 63, 66, 67, 69, 79
extensor digitorum, 47, 51, 61, 63, 66–69, 79, 84, 89
external oblique, 51, 60, 70, 74–76, 79, 87
femoris, 39, 48, 51, 60, 70, 79–87, 119, 121, 148
flexor carpi, 61, 67, 69
flexor digitorum, 48, 61, 65, 67–69, 79, 80, 88
flexors, 64, 148
gastrocnemius, 51, 70, 79, 84, 85, 88
gemellus, 85, 86
genioglossus, 25, 52, 96
geniohyoideus, 25, 52, 96
gluteal, 121, 127
gluteus, 48, 70, 80–86
gracilis, 60, 70, 86, 87, 121
hyoglossus, 52, 96
iliacus, 70, 86, 87, 120
iliocostalis, 48, 74, 76, 80
iliopectineal, 39, 86
iliopsoas, 86, 148
inferior oblique, 152
inferior rectus, 152
infraspinatus, 34, 56, 57, 65, 79
intercostal, 77, 116, 117, 121, 123
intercostales, 76
intercostalis, 70, 75
internal oblique, 60, 70, 76
interossei, 67
interspinalis, 75, 77
ischiocavernosus, 135, 136
ischiofemoral, 45
lateral rectus, 142, 152, 153, 174, 175
latissimus dorsi, 49, 51, 55–58, 60, 64, 79, 145
levator, 55, 56, 58, 59, 152, 153, 173, 175
levatores, 75
lingualis propria, 52
longissimus, 73–77, 79
longissimus cervicis, 73, 75
longus capitis, 71, 80
longus colli, 71, 72
lumborum, 86, 87, 120
lumbricales, 67
masseter, 51–54, 72, 92, 115
multifidus spinae, 77
muscularis externa, 96
mylohyoideus, 51, 53, 54, 72
myofibrils, 4
myology, 102, 110
myosin, 4
omotransversarius, 55, 56, 58, 63, 72, 74, 79, 80
orbicularis oculi, 176
palmaris brevis, 67
palmaris longus, 68
pectineus, 39, 70, 86, 87, 121
pectoralis, 49, 51, 53, 55, 58–60, 62, 70, 74, 75, 116, 145, 148
peroneus, 70, 79, 84, 88, 89
piriform, 168
piriformis, 83
platysma, 49
popliteus, 46, 47, 88
pronator, 61, 66, 67, 69, 70, 148
psoas, 70, 86, 87, 120, 148
pterygoideus, 51, 53
quadratus, 67, 80, 85–87, 120
quadratus femoris, 80, 85
quadriceps, 83, 121
rectus, 39, 60, 70–72, 78, 79, 83–85, 87, 142, 148, 152, 153, 174, 175
rectus abdominis, 60, 70, 78, 79, 87
rectus femoris, 39, 60, 70, 83–85, 87, 148
retractor oculi, 152
rhomboideus, 48, 56–58, 62, 71, 79
sartorius, 60, 70, 79–87, 148
scalenus, 73–76
scapularis, 51, 57, 79
semimembranosus, 60, 80, 83–86, 121
semispinalis, 73, 74
semitendinosus, 60, 80, 83–86
serratus, 58–60, 62, 70, 71, 73–77, 79, 145
serratus dorsalis, 71, 74–76, 79
serratus ventralis, 62
soleus, 85, 88
sphincter, 96, 101, 178
spinalis, 74, 76, 77
spinalis thoracis, 74
splenius, 73–75
sternocleidomastoideus, 53, 70
sternohyoideus, 25, 52–56, 60
sternomastoid, 53
sternomastoideus, 53–56, 58, 60, 72
sternothyroideus, 25, 52–56
styloglossus, 25, 52
stylohyal, 24, 52
stylohyoideus, 25, 51, 52, 72, 142
subvertebral, 127
superficialis, 48, 58–61, 67, 69, 81, 84, 85, 88, 96
superior oblique, 152
superior rectus, 152
supinator, 65, 66, 148
supraspinatus, 34, 56–58, 61, 63, 65, 69, 80, 145
temporalis, 24, 51–53, 139
teres major, 48, 55–58, 63, 65, 69, 80, 145
teres minor, 56, 57, 65
thyrohyoideus, 25, 52–54
tibialis anterior, 51, 70
trabeculae carneae, 113, 164
transversus abdominis, 76, 78, 80
trapezius, 51, 55–58, 62, 63, 70, 71, 79, 80
triceps brachii, 63, 64, 80
triceps surae, 88
ulnaris, 51, 61, 63, 66, 67, 69, 79
vastus, 60, 70, 80, 81, 83–85, 87, 121, 148
Muscle action
abduct, 45, 57, 59, 66–68, 81–83, 86
adduct, 45, 57–59, 67, 68, 81, 86
circumduct, 45
extension, 48
flexion, 40, 45, 49, 88

pronation, 49
supination, 49
Muscularis externa, 96
Musculocutaneous, 145
Mustelidae, xiv
Myelencephalon, 139
Myelin, 8
Mylohyoideus, 51, 53, 54, 72
Myofibrils, 4
Myology, 102, 110
Myopia, 178
Myosin, 4

Nares, 11, 23, 24, 95, 106
Nasal, 11, 15, 17–19, 21–24, 95, 106, 115, 139, 142, 168
Nasopalatine, 27
Nasopharynx, 23, 95, 106
Natal, 137
Navicular, 39–41
Neopallium, 139, 171
Nerve, xiii, 4, 14, 20, 23, 27, 28, 55, 85, 86, 92, 119, 124, 139, 140–43, 145–49, 151–57, 166, 168–70, 173, 175, 177
abducens, 27, 28, 142, 143, 153, 169, 175
accessory, 23, 27, 34, 37, 55, 67, 68, 92, 108, 110, 140, 143, 166, 169
acoustic, 15, 19–22, 27, 28, 142, 156
autonomic, 138, 146–48, 160
axillary nerve, 146
axon, 8, 157, 177
cauda equina, 145
chiasma, 141, 143, 144, 168, 169
communicantes, 148
dendrites, 177
dorsal root, 145
facial, 20–23, 27, 28, 49, 92, 115, 123, 142, 143, 156, 169
filum, 145
ganglia, 148, 149, 155, 171
ganglion, 145–49, 154, 158, 177
ganglionated, 148
genitofemoralis, 147
glossopharyngeal, 23, 27, 143, 169
intermesenteric plexus, 149
lumbosacral plexus, 147, 148
musculocutaneous, 145
myelin, 8
nasopalatine, 27
neuron, 4, 8, 138, 139, 145, 151, 156, 158
obturator, 147
oculomotor, 27, 28, 142, 143, 152, 153, 169, 175
olfactory, 169
olfactory bulb, 24, 140–43, 167–69
olfactory tract, 140, 143
optic, 23, 27, 140, 151–54, 168, 169, 173, 175, 177
optic chiasma, 141, 143, 144, 168, 169
parasympathetic, 138, 148
paravertebral, 148
phrenic, 146
plexi, 145, 148, 149
plexus, 86, 135, 145–49
plexuses, 115, 116, 124, 148, 149
prevertebral, 148
rami communicantes, 148
saphenous, 119
sciatic, 41, 85, 147, 148
somesthetic, 151
spinal, 28, 138, 145, 148
splanchnic, 147, 149
subcommissural, 158
sympathetic, 138, 146–49, 160
synapses, 139, 145, 148, 170
synaptic, 149
thoracolumbar, 58
tracts, 135, 139, 141, 142, 145, 168, 170, 171
trigeminal, 23, 27, 28, 140, 142, 143, 156, 169
ulnar, 124, 146
vagal, 149
vagus, 23, 27, 140, 142, 143, 146, 148, 149
Nervosa, 141, 143, 144
Neural, 16, 28, 139, 142, 157, 160, 170
Neural arch, 28
Neural canal, 28, 139
Neural spine, 28
Neurocranium, 13, 14, 16, 20, 23
Neurohypophysis, 141, 144, 157
Neuron, 4, 8, 138, 139, 145, 151, 156, 158
Norepinephrine, 160, 161
Notochord, xiii
Nuchal, 16, 73
Nuclear, 3, 142, 170, 177
Nuclei, 145, 157, 170, 171
Nucleoplasm, 3
Nucleus, 3, 4, 7, 8, 145, 171

Obturator foramen, 39, 41, 42, 44
Obturator nerve, 147
Occipital, xii, 15–17, 19, 20, 22, 27, 28, 30, 44, 53, 57, 73, 76, 115, 124, 139, 167, 168
Occipital condyle, 15, 16, 19, 20, 53
Ocular lens, 1
Oculi, 152, 176
Oculomotor, 27, 28, 142, 143, 152, 153, 169, 175
Oddi, 101
Odontoid, 28
Olecranon, 34, 44, 64, 67
Olfactorius, 168
Olfactory bulb, 24, 140–43, 167–69
Olfactory nerve, 169
Olfactory tract, 140, 143
Omentum, 91, 96, 118, 128, 129, 131
Omotransversarius, 55, 56, 58, 63, 72, 74, 79, 80
Optic, 16, 18, 21–23, 27, 28, 140–44, 151–54, 158, 168–70, 173, 175, 177
Optic chiasma, 141, 143, 144, 168, 169
Optic foramen, 16, 18, 21–23, 142
Optic lobe, 158
Optic nerve, 23, 27, 140, 151–54, 168, 169, 173, 175, 177
Oral, 52, 92, 106, 110, 129, 157
Ora serrata, 173, 175, 177
Orbicularis oculi, 176
Orbital, 16, 18, 19, 22, 23, 27, 28, 142
Orbital fissure, 16, 18, 19, 22, 23, 28, 142
Orbitosphenoids, 23
Orifice, 132
Os coxae, 14, 29, 39
Ossicle, 155, 156
Ossified, 16, 28, 77, 139
Osteocyte, 6
Ostium, 137
Ovarian, 117, 120, 125, 127, 161
Ovaries, 91, 120, 137, 156
Ovary, 117, 136, 137, 158
Oviduct, 4, 137
Ovum, 135, 137
Oxytocin, 157

Palate, 24, 92, 95, 102, 106, 109, 115
Palatine, 19, 21, 24, 27, 51, 92, 95
Pallidus, 171
Pallium, 170
Palmar, 67
Palmaris brevis, 67
Palmaris longus, 68
Palpebrae, 152, 153, 173, 175
Palpebrum, 172
Pampiniform, 135
Pancreas, 91, 96, 101, 102, 116, 118, 120, 128, 129, 156, 159, 161
Pancreaticoduodenal, 120, 128, 129
Pancreaticoduodenalis, 118, 128
Pancreozymin, 160
Pandas, xiv
Panthera, xiv
Papilla, 91, 92, 95, 96, 113, 133, 164
Papillae, 95, 96, 113, 133, 164
Papillary, 95, 112, 113, 163, 164
Paraconal, 162, 164
Paraflocculus, 28
Paramastoid, 20, 28
Paramedian, 167, 168
Parasympathetic, 138, 148
Parathyroid, 159
Paraventricular, 157
Paravertebral, 148
Parfocal, 1
Parietal cell, 96
Parietal lobe, 167, 168
Parietal pericardium, 91, 111, 161
Parietal peritoneum, 90, 91, 133
Parietal pleura, 90, 91, 105
Parietals, 16
Parotid, 4, 79, 91, 92
Pars distalis, 141, 143, 144, 157
Pars nervosa, 141, 143, 144
Patella, 29, 30, 39, 41–43, 47, 81–83, 88
Pectinate, 164

Pectineus, 39, 70, 86, 87, 121
Pectoral, 10, 14, 34, 35, 37, 38, 49, 53, 55, 56, 58, 59, 61, 69, 71, 90, 116, 145, 148
Pectoral girdle, 34, 35, 37, 38, 53, 55, 71
Pectoralis, 49, 51, 53, 55, 58–60, 62, 70, 74, 75, 116, 145, 148
Peduncle, 141–43, 167, 169
Pellucidum, 142–44, 169, 170
Pelvic, 10, 14, 39, 41, 42, 45, 81, 121, 131, 136, 149
Pelvic girdle, 39, 41, 42, 81, 136
Pelvis, 121, 127, 133, 134
Penis, 11, 121, 127, 132, 135–37
Pericardia, 90, 91, 111, 161
Pericardial, 105, 111, 112, 161
Pericardial cavity, 105
Pericardium, 91, 107, 111, 112, 161
Perilymph, 156
Perilymphatic, 156
Perineum, 127
Periosteal, 14
Peritoneal, 90
Peritoneum, 90, 91, 111, 133
Peroneal, 148
Peroneus, 70, 79, 84, 88, 89
Perpendiculares, 96
Pes, 10, 40
Petromastoid, 20, 27
Petrosal, 20–23, 28, 156
Peyer's, 97
Phalanges, 10, 30, 33, 35, 37–44, 67
Phalanx, 66–68, 88, 89
Pharyngeal, xiii, 52, 115, 124, 142
Pharynx, 27, 95, 96, 102, 106, 110, 129, 141, 142, 156, 158
Philtrum, 10
Phrenic artery, 117
Phrenic nerve, 146
Phylogenetically, 95, 171
Pia mater, 139, 167
Pineal, 141, 144, 158, 170, 171
Pinna, 10, 156
Piriform, 168
Piriformis, 83
Pisiform, 34, 39, 67, 68
Pituitary, 23, 28, 136, 139, 141, 143, 144, 156, 157, 168, 169, 170
Placenta, xiii, 121, 137
Plantar, 88, 127
Plantigrade, 10
Plasma, 159, 160
Platysma, 49
Pleura, 90, 91, 105, 111
Pleural, 90, 91, 105, 161
Plexi, 145, 148, 149
Plexiform, 154, 177
Plexus, 86, 135, 145–49
Plexuses, 115, 116, 124, 148, 149
Plica semilunaris, 151, 172
Pollex, 39, 66, 67
Pons, 139–41, 143, 144, 168, 169, 171
Popliteal vein, 127
Popliteus, 46, 47, 88
Pore, 95
Portal vein, 101, 118, 120, 129
Portal vessel, 128
Posterior chamber, 154, 175, 176
Postorbital, 15–18, 24
Premolar, 16, 23, 24, 26, 51, 92, 93
Prepuce, 11, 135, 136
Preputial, 135
Presphenoid, 19, 20, 22–24, 27
Prevertebral, 148
Primates, xiii
PRL, 157
Process
  acromion, 34, 35, 37, 38, 57
  ala, 23
  aspera, 44, 86, 88
  capitulum, 34, 43
  clinoid 22
  condyle, 15, 16, 19, 20, 27, 28, 39, 44, 46, 53, 71, 88
  condyles, 20, 27, 28, 39, 46, 71
  condyloid, 15, 19, 21, 22, 24, 27, 39
  coronoid, 15, 24, 51
  cribriform, 21, 22, 24, 27, 28
  dendrites, 177
  epicondyle, 34, 35, 37, 38, 64, 66, 67, 83, 88, 89
  epiphysis, 158
  glenoid, 34, 37, 43, 57, 64
  hamulus, 51
  iliac crest, 76, 78, 81, 86, 120
  iliac spine, 39, 42, 81
  ischiadic, 39
  ischial, 42
  lambdoidal, 16, 19, 20, 53, 57, 73, 139
  malleolus, 41–43
  mastoid, 15, 16, 19, 20, 25, 53, 76, 113
  metacromion, 37, 57, 58
  neural arch, 28
  neural spine, 28
  occipital condyle, 15, 16, 19, 20, 53
  odontoid, 28
  olecranon, 34, 44, 64, 67
  paramastoid, 20, 28
  postorbital, 15–18, 24
  sella turcica, 21–23, 28, 139, 141, 170
  spine, 16, 28, 34, 38, 39, 42, 44, 56–58, 76–78, 81, 86
  spinous, 28, 37, 57, 73, 76, 77
  styloid, 15, 19–21, 24, 37, 66
  supraoccipital, 16, 19, 20
  trochanter, 39, 41–44, 81, 83, 86
  tubercle, 38, 39, 78
  tuberculum, 28
  tuberosity, 34, 35, 37, 38, 42–44, 57, 58, 64, 81, 83, 86, 88
  xiphoid, 33, 78
  zygomatic, 16–19, 22–24, 51, 53
Procyonidae, xiv
Prolactin, 157
Pronation, 49
Pronator, 61, 66, 67, 69, 70, 148
Pronephric, 132
Prosencephalon, 138, 139
Prostate, 121, 127, 135, 136
Prostatic, 127
Protoplasm, 3
Prototheria, xiii
Pseudostratified, 4, 5
Psoas, 70, 86, 87, 120, 148
Pterygoid, 16, 19, 22, 23, 51, 53, 95
Pterygoideus, 51, 53
Pubic, 39, 121, 127
Pubis, 30, 39, 41, 42, 78, 86
Pubofemoral, 45
Pudendal artery, 119, 121
Pudendal vein, 119, 127
Pudendoepigastric vein, 127
Pulmonary artery, 112, 113, 163–65
Pupil, 151, 172, 174–76
Purkinje, 164
Putamen, 170, 171
Pyloric, 95, 96, 101, 129
Pylorus, 96, 160

Quadratus, 67, 80, 85–87, 120
Quadratus femoris, 80, 85
Quadriceps, 83, 121
Quadrigemina, 141, 142, 168–70
Quadruped, xi, xii, 10, 16, 64

Raccoons, xiv
Radial, 34, 43, 44, 67, 116, 125, 146, 148
Radiale, 34, 39
Radius, 10, 14, 29, 30, 34–40, 43, 44, 64, 66, 67
Rami communicantes, 148
Ramus, 83, 86, 167
Raphe, 59
Recta, 133, 135, 136
Rectal, 129
Rectum, 87, 96, 101, 120, 121, 127, 129
Rectus, 39, 60, 70–72, 78, 79, 83–85, 87, 142, 148, 152, 153, 174, 175
Rectus abdominis, 60, 70, 78, 79, 87
Rectus femoris, 39, 60, 70, 83–85, 87, 148
Renal, 116, 117, 120, 125, 127, 132–36
Renal artery, 117, 133, 134
Renal vein, 125, 127, 133, 134, 136
Renin, 161
Reproductive, xiii, 9, 91, 132, 135–37, 156
Respiratory, 9, 13, 23, 59, 77, 95, 105, 110, 151, 168
Rete, 115, 124, 135
Retina, 142, 151–54, 173, 175–78
Retractor oculi, 152
Retroperitoneal, 133
Rhinal, 167
Rhinencephalon, 139, 171
Rhombencephalon, 138, 139
Rhomboideus, 48, 56–58, 62, 71, 79

Rostral palatine foramen, 19, 95
Rostrum, xiv
Rotunda, 155, 156
Rugae, 99–101

Sacculus, 156, 158
Sacral, 33, 77, 81, 83, 119, 121, 127, 145, 148
Sacrum, 14, 29, 30, 33, 39, 43, 44
Salivary, 49, 51, 91, 92, 102, 116, 160
Saphenous artery, 60, 119
Saphenous nerve, 119
Saphenous vein, 119, 127
Sartorius, 60, 70, 79–87, 148
Scala, 155, 156
Scalenus, 73–76
Scaphoid, 34
Scapula, 29, 30, 34–38, 44, 55, 57–59, 64
Scapulae, 34, 55, 56, 58, 59, 125
Scapular, 55–57, 123, 125, 126
Scapularis, 51, 57, 79
Schlemm, 175
Sciatic, 41, 85, 147, 148
Sclera, 151–54, 172–77
Scrotal, 135
Scrotum, 11, 121, 135
Sebaceous, 11, 12
Secretin, 160
Segmental, 110
Sellar, 40
Sella turcica, 21–23, 28, 139, 141, 170
Semilunaris, 151, 172
Semilunar notch, 34, 37, 57, 64, 66, 67
Semilunar valve, 113, 162, 164
Semimembranosus, 60, 80, 83–86, 121
Semispinalis, 73, 74
Semitendinosus, 60, 80, 83–86
Sense organ, 16, 19, 20, 24, 27, 28, 115, 155, 156
  accommodation, 177, 178
  anterior chamber, 151–54, 175, 176
  aqueous, 154, 175, 176
  auditory, 20, 27, 28, 95, 115, 141–43, 155, 156, 170
  ciliary body, 152–54, 173, 175, 177, 178
  circumvallate, 95
  cochlea, 155, 156
  conjunctiva, 151, 154, 172
  cornea, 151–54, 173–77
  cristae, 3
  endolymphatic, 155, 156
  eustachian, 20, 27, 115, 156
  exteroceptors, 139, 151
  eye, xii, 1, 10, 13, 24, 49, 115, 142, 151, 152, 153, 154, 156, 161, 168, 172–78
  foliate, 95, 96
  fovea, 45, 173
  fungiform, 95, 96
  fungiform papillae, 95, 96
  haversian, 6
  hyperopia, 178
  incus, 20, 24, 155, 156
  iris, 151–54, 172–78
  lens, 1, 3, 152–54, 173, 175–78
  macula, 4, 173
  myopia, 178
  pinna, 10, 156
  plexiform, 154, 177
  posterior chamber, 154, 175, 176
  pupil, 151, 172, 174–76
  retina, 142, 151–54, 173, 175–78
  sacculus, 156, 158
  scala, 155, 156
  sclera, 151–54, 172–77
  stapes, 20, 24, 155, 156
  tympani, 155, 156
  tympanica, 16, 20
  tympanum, 156
  uvea, 176, 177
Sensory, 8, 11, 12, 20, 138, 139, 141, 142, 145, 156, 168
Septum, 111–13, 142–44, 161–64, 169, 170
Septum pellucidum, 142–44, 169, 170
Serous, 90–92, 96
Serratus, 58–60, 62, 70, 71, 73–77, 79, 145
Serratus dorsalis, 71, 74–76, 79
Serratus ventralis, 62
Sertoli's, 161
Sesamoid, 34, 39
Sigmoid, 129
Simplex, 168
Sinus, 112, 115, 133, 134, 162, 164
Sinusoid, 101, 128
Skeleton
  acetabulum, 39, 41, 45, 86
  acromion, 34, 35, 37, 38, 57
  ala, 23
  antebrachii, 64
  antebrachium, 10, 57, 64, 66
  appendage, 10, 34, 39, 58, 81, 127, 157
  appendicular, 14, 34
  articularis, 34
  arytenoid, 106
  axial, 14, 58
  bicipital, 58, 59
  bicuspid, 112, 113, 163–65
  canaliculi, 101
  caput, 28, 34, 136
  coccyx, 30, 33, 34, 43
  coxae, 14, 29, 39
  cranium, xii, 23, 125, 139, 156, 158, 166
  dental, 26
  dentition, xiii, 47
  dermal, 10, 12, 13, 14, 23, 24
  diastema, 51
  digit, 10, 39, 40, 66–68, 88, 89
  digitigrade, 10
  endochondral, 14, 24
  facet, 28
  fenestra, 155, 156
  fenestrae, 156
  fissure, 16, 18, 19, 22, 23, 27, 28, 91, 110, 141, 142, 166, 167, 168
  foramen, 15–24, 27, 28, 39, 41, 42, 44, 95, 115, 116, 142, 145, 156, 170
  fossa, 19, 20, 23, 34, 37–39, 43–45, 51, 57, 58, 83, 86, 115
  hemifacet, 28
  intercondyloid, 39
  neurocranium, 13, 14, 16, 20, 23
  ossified, 16, 28, 77, 139
  pectoral girdle, 34, 35, 37, 38, 53, 55, 71
  pelvic girdle, 39, 41, 42, 81, 136
  pelvis, 121, 127, 133, 134
  plantigrade, 10
  pollex, 39, 66, 67
  sacral, 33, 77, 81, 83, 119, 121, 127, 145, 148
  scapular, 55–57, 123, 125, 126
  splanchnocranium, 13, 23, 24
  supraspinous fossa, 37, 38, 57
Socketed, 23, 24
Soleus, 85, 88
Somatic, 138, 139, 157, 168
Somesthetic, 151
Spermatic, 120, 121, 125–27
Sphenopalatine, 21, 27
Spheroidal, 40, 45
Sphincter, 96, 101, 178
Spigelian, 101
Spinae, 77
Spinal, 14, 20, 23, 27, 28, 55, 105, 131, 138–43, 145, 148, 166, 168, 169
Spinalis, 74, 76, 77
Spinalis thoracis, 74
Spinal nerves, 28, 138, 145, 148
Spine, 16, 28, 34, 38, 39, 42, 44, 56–58, 76–78, 81, 86
Spinosum, 11, 19
Spinous, 28, 37, 57, 73, 76, 77
Splanchnic, 147, 149
Splanchnocranium, 13, 23, 24
Spleen, 91, 96, 116, 118, 120, 129, 131, 161
Splenic vein, 128, 129
Splenium, 143
Splenius, 73–75
Squamosal, 15–18, 22, 24, 51
Squamous, 4, 5, 20
Stage, 1–3, 14
Stapedius, 156
Stapes, 20, 24, 155, 156
Sternal, 77, 90
Sternebrae, 14, 29, 33, 59
Sternebrium, 33
Sternocleidomastoideus, 53, 70
Sternohyoideus, 25, 52–56, 60
Sternomastoid, 53
Sternomastoideus, 53–56, 58, 60, 72
Sternothyroideus, 25, 52–56
Sternum, 14, 28, 29, 33, 34, 36, 43, 47, 53, 59, 76–78, 90, 91, 121
Stomach, 4, 91, 95, 96, 101, 102, 116, 118, 120, 128, 129, 131, 142, 159, 160
Strata, 170
Stratified, 4, 5, 170

Striated, 4, 7, 48, 49, 91, 92
Striatum, 170
Styloglossus, 25, 52
Stylohyal, 24, 52
Stylohyoideus, 25, 51, 52, 72, 142
Styloid, 15, 19–21, 24, 25, 27, 37, 43, 66
Styloid process, 15, 19–21, 24, 37, 66
Stylomastoid, 19, 20, 27
Subarachnoid, 139, 167
Subclavian, 90, 107, 113–17, 122–26, 130, 131, 148, 165
Subclavian artery, 90, 107, 113–17, 148
Subclavian vein, 122, 124, 126, 130, 131
Subcommissural, 158
Subcutaneous, 12, 129
Subdural, 139, 167
Sublingual, 4, 91, 92, 122
Sublingual gland, 91, 92
Sublingual vein, 122
Submaxillary, 91, 92
Submaxillary gland, 91, 92
Submental, 122, 123
Submucosa, 96, 97
Subscapularis, 58, 60, 61, 69, 145
Subscapular vein, 124
Subsinosal, 161
Subsinuosal, 164
Subvertebral, 127
Sudoriferous, 11, 12
Sulci, 141, 167, 168, 170
Sulcus, 113, 140, 144, 168
Superficialis, 48, 58–61, 67, 69, 81, 84, 85, 88, 96
Superioris, 152, 153, 173, 175
Superior oblique, 152
Superior rectus, 152
Supination, 49
Supinator, 65, 66, 148
Supracondyloid, 64, 66
Supraoccipital, 16, 19, 20
Supraoptic, 157
Suprascapular, 145
Supraspinatus, 34, 56–58, 61, 63, 65, 69, 80, 145
Supraspinous fossa, 37, 38, 57
Sus, 83, 164, 177
Suspensory ligament, 91, 152–54, 177
Sympathetic, 138, 146–49, 160
Symphysis, 21, 24, 39, 52, 86
Synapses, 139, 145, 148, 170
Synaptic, 149
Synovial, 40, 45
System, xiii, 3, 4, 12, 13, 16, 48, 49, 52, 77, 90, 95, 105, 110, 111, 128, 129, 131, 132, 135, 137–39, 148, 149, 150, 151, 156–58, 160, 175
Systemic, 113, 121
Systole, 165

Talus, 39–41
Tapetum lucidum, 152, 153
Tarsal, 10, 29, 30, 39–42, 88
Tectorial, 155, 156
Tectum, 141–43, 158
Tegmen, 170
Tegmental, 142
Tela choroidea, 141, 143, 170
Telencephalon, 139, 168, 171
Temporal, 15, 16, 18, 20, 22–24, 27, 30, 51, 76, 115, 144, 156, 167, 168
Temporalis, 24, 51–53, 139
Temporomandibular, 47
Tendinous, 113
Tendon, 58, 64, 66, 67, 78, 83, 86, 88, 89, 162, 164
Tentorium, 16, 21, 22, 139, 167
Teres major, 48, 55–58, 63, 65, 69, 80, 145
Teres minor, 56, 57, 65
Terminal, 4, 24, 39, 88, 89, 101, 110, 121, 148
Terminale, 145
Terminalis, 158, 170
Tertius, 89
Testes, 11, 120, 135–37, 156, 161
Testicles, 91
Testis, 136, 137, 158
Testosterone, 136
Tetrapods, 110, 157
Thalamic, 171
Theca, 137
Thoracic, 14, 16, 28, 30, 32, 33, 35, 37, 38, 57, 58, 71, 73, 76, 77, 78, 86, 90, 91, 105, 107, 110, 113–16, 121, 122, 123, 129–31, 145, 148, 161
Thoracica, 55–57
Thoracis, 59, 74, 76, 79
Thoracolumbar, 58
Thorax, 11, 14, 71, 131
Thymus, 90, 95, 107, 129, 131
Thyrocervical trunk, 115, 116
Thyrohyal, 24, 25
Thyrohyoideus, 25, 52–54
Thyroid artery, 114, 115
Thyroid cartilage, 24, 25, 52, 106
Thyroid gland, 115, 116, 123, 158, 159
Thyroid vein, 122, 123
Thyrotropic, 158
Tibia, 10, 29, 30, 39–42, 46, 81–83, 86, 88, 121
Tibial, 83, 127, 148
Tibialis anterior, 51, 70
Tissue, 3, 4, 6, 12–14, 40, 49–51, 81, 109, 129, 131, 135, 141, 151, 153, 154, 160, 161, 170, 172, 174–76
Tonofibrils, 4
Tonsils, 92, 129
Torus, 10
Trabeculae, 112, 113, 131, 162, 164
Trabeculae carneae, 113, 164
Trachea, 4, 51, 52, 71, 90, 106–108, 110, 113, 115, 116, 123
Tracheal cartilage, 25
Tracts, 135, 139, 141, 142, 145, 168, 170, 171
Transversus abdominis, 76, 78, 80
Trapezium, 34, 67
Trapezius, 51, 55–58, 62, 63, 70, 71, 79, 80
Trapezoid, 34, 140, 143, 169
Triassic, xiii
Triceps brachii, 63, 64, 80
Triceps surae, 88
Tricuspid, 112, 113, 162, 164, 165
Trigeminal, 23, 27, 28, 140, 142, 143, 156, 169
Triquetrum, 34
Trochanter, 39, 41–44, 81, 83, 86
Trochlear, 28, 142, 153, 169, 175
Trochoid, 40
Trunk, xi, 11, 58, 71, 75–77, 79, 80, 91, 114–17, 120, 121, 123, 126, 146–49, 162
TSH, 158
Tuber cinereum, 141, 143, 168, 169
Tubercle, 38, 39, 78
Tuberculum, 28
Tuberosity, 34, 35, 37, 38, 42–44, 57, 58, 64, 81, 83, 86, 88
Tubule, 132–34, 136
Tunic, 175, 176
Turbinate, 21, 22, 24
Tympani, 155, 156
Tympanic, 16, 19, 20, 24, 27, 28, 115, 155, 156
Tympanica, 16, 20
Tympanohyal, 20, 25, 52
Tympanohyals, 24
Tympanum, 156

Ulna, 10, 14, 29, 30, 34–40, 43–45, 57, 64, 66, 67
Ulnar, 34, 67, 116, 124, 125, 146, 148
Ulnare, 34, 37, 39, 45
Ulnaris, 51, 61, 63, 66, 67, 69, 79
Ulnar nerve, 124, 146
Ultimobranchial, 96, 159
Umbilical, 119, 121, 137
Umbilical artery, 119
Unguligrade, 10
Ureter, 132–34, 136
Urethra, 4, 11, 132, 135–37
Urinary, 11, 91, 96, 117, 121, 132, 133, 136
Urinary bladder, 91, 96, 117, 121, 132, 133
Urogenital, 11, 132
Ursidae, xiv
Uteri, 120, 137
Uterine, 137, 157
Uterus, 91, 117, 121, 127, 137, 161
Utriculus, 156
Uvea, 176, 177

Vagal, 149
Vagina, 121, 127, 136, 137
Vaginalis, 127, 135
Vagus, 23, 27, 140, 142, 143, 146, 148, 149
Vallate papillae, 95, 96
Valve, 96, 97, 111–13, 120, 162–64
Varoli, 144, 168, 169

Vasa, 133, 135, 136
Vasa afferentia, 135, 136
Vasa recta, 133, 135, 136
Vascularized, 95, 158
Vasculature, 6, 40
Vas deferens, 127, 135, 136
Vasopressin, 157
Vastus, 60, 70, 80, 81, 83–85, 87, 121, 148
Vater, 96, 101
Vein
- adrenolumbar, 117, 125, 126
- axillary, 122, 125
- azygos 127
- basilic, 125
- brachial, 122, 124
- caudal 126, 127
- cavernosum, 135
- cubital, 124, 125
- facial 123
- gastric 128, 129
- gastroduodenal 118, 120, 128
- gastroepiploic, 118, 120, 128, 129
- gonadal, 119
- hepatic portal, 101, 118, 120
- hepatoduodenal, 91
- ileocolic, 129
- iliac, 119, 126, 127
- iliolumbar, 119, 126
- interlobar, 133
- intestinal, 128
- jugular, 20, 27, 53, 122, 124–26, 130
- lienal, 116, 118, 128, 129
- linguofacial, 122
- maxillary, 122
- pancreaticoduodenal, 120, 128, 129
- pancreaticoduodenalis, 118, 128
- popliteal, 127
- portal, 101, 118, 120, 129
- pudendal, 119, 127
- pudendoepigastric, 127
- renal, 125, 127, 133, 134, 136
- saphenous, 119, 127
- splenic, 128, 129
- subclavian, 122, 124, 126, 130, 131
- sublingual, 122
- subscapular, 124
- thyroid, 122, 123
- vena cava, 90, 112, 113, 116, 117, 119, 121, 122, 125–28, 136, 162, 163

Ventral mesentery, 90, 91, 129, 161
Ventral root, 145, 148
Ventricle, 112, 113, 141–45, 149, 158, 162–64, 168–70
Ventriculi, 170
Vermis, 140, 141, 144, 145, 166–68
Vertebrae, 14, 16, 28, 29, 31–34, 45, 57–59, 71, 73, 76–78, 81–83, 86, 115, 116, 139, 148
- axis, 28, 31, 40, 45, 49, 71, 77, 118, 173, 175
- centrum, 28, 33, 71
- coccygeal, 148
- coccyx, 30, 33, 34, 43

Vertebral, 14, 16, 27–30, 57–59, 73, 75–77, 86, 90, 114–16, 122, 123, 124, 139, 148
Vertebrate, xiii, 4, 132, 156, 157, 160
Vesicle, 4, 158
Vestibular, 155, 156
Vestibule, 4, 11, 93, 132, 137, 161
Vestibuli, 155, 156
Vibrissae, 11
Villi, 96, 97, 167
Visceral, 12, 48, 90, 91, 105, 111, 112, 116, 118, 128, 138, 139, 148, 161, 168
Visceral peritoneum, 90, 91
Vitrea, 175
Vitreous, 153, 154, 173, 175–77
Vitreous body, 154, 175, 176
Vitreous chamber, 153, 154, 173, 176, 177
Vitreum, 175
Vomer, 18, 21, 24
Vomeronasal, 95

Xiphihumeralis, 56, 58
Xiphisternal, 33
Xiphoid, 33, 78
Xiphoidea, 33
Xiphoideus, 33

Zonule, 177, 178
Zonule fibers, 178
Zygoma, 15, 24, 30
Zygomatic, 16–19, 22–24, 51, 53